# MOLECULAR DISSECTION of COMPLEX TRAITS

Edited by
**Andrew H. Paterson**
*Christine Richardson Professor of Agriculture*
*Department of Soil and Crop Science*
*Texas A&M University*
*College Station, Texas*

CRC Press
Boca Raton New York

Images used in cover design courtesy of John C. Crabbe and Michael Moody of the Department of Veterans Affairs Medical Center in Portland, Oregon, and Charles W. Stuber of the U.S. Department of Agriculture, Agricultural Research Service, at North Carolina State University in Raleigh.

**Acquiring Editor:** Marsha Baker
**Project Editor:** Carol Whitehead
**Marketing Manager:** Becky McEldowney
**Cover design:** Dawn Boyd
**PrePress:** Kevin Luong
**Manufacturing:** Carol Royal

**Library of Congress Cataloging-in-Publication Data**

Molecular dissection of complex traits / edited by Andrew H. Paterson.
p. cm.
Includes bibliographical references and index.
ISBN 0-8493-7686-6 (alk. paper)
1. Molecular genetics. 2. Phenotype. 3. Gene mapping.
I. Paterson, Andrew H., 1960-
QH442.M645 1997
572.8--dc21

97-12278
CIP

This book contains information obtained from authentic and highly regarded sources. Reprinted material is quoted with permission, and sources are indicated. A wide variety of references are listed. Reasonable efforts have been made to publish reliable data and information, but the author and the publisher cannot assume responsibility for the validity of all materials or for the consequences of their use.



Direct all inquiries to CRC Press LLC, 2000 Corporate Blvd., N.W., Boca Raton, Florida 33431.


International Standard Book Number 0-8493-7686-6
Library of Congress Card Number 97-12278
Printed in the United States of America 1 2 3 4 5 6 7 8 9 0
Printed on acid-free paper

# Preface

One of the great conflicts in the history of genetics was reconciled with the realization that continuous variation in phenotype could reflect the net effects of a large number of genes. In the past ten years, contemporary geneticists using new molecular tools have been able to resolve complex traits into individual genetic components and describe each such component in much detail. This volume summarizes the state of the art in molecular analysis of complex traits (QTL mapping), placing new developments in this field within the context of their historical origins. Leading authorities address central themes in analysis of complex phenotypes, and case histories of seminal work in this burgeoning field are presented by the principal investigators.

Through this volume the authors strive to convey their excitement about a new experimental approach that has empowered us to ask precise questions about complex phenomena. We and our colleagues have, in the past few years, been fortunate to participate in a rare era in the evolution of a field of science — an era in which new technology confers the rare opportunity to reinvestigate both classical dogma and speculative proposals that are deeply entrenched in the history of a field. Results to date have begun to clarify ambiguity, in a few cases resolve dichotomy, and in many cases point to the need for still more detailed study. Now is an exciting time to be a geneticist, with unprecedented new opportunities unfolding that we could only dream of a few short years ago.

The volume is written primarily for an audience of biologists familiar with genetics and is intended to be useful both to senior scientists in a wide range of biotic disciplines and to entry-level pre- or postdoctoral scientists. Lay people with a scientific bent can readily master the topics presented herein, but it is recommended that they first become more familiar with the underlying methods and principles of genome analysis by reviewing any of several recent reviews. The efficacy with which QTL mapping is done in plants tends to bias the examples chosen; however, several chapters address issues unique to animal systems, including humans.

The specific scope and objectives of this volume are:

1. To provide a complete, well-focused volume in the field of molecular genetic analysis of complex phenotypes, including case histories written by leading practitioners in the field. Such a volume will be intended to quickly bring the scientist or student up to the state of the art in an important and rapidly-growing field and to introduce the layman with a scientific bent to a new approach for investigating fundamental questions in the life sciences.
2. To place contemporary technological developments in "molecular quantitative genetics" within the context of their historical origins. Moreover, consider how recent advances in the study of genetics are being applied to better understanding of classical questions.
3. To highlight future needs and directions in utilization of "molecular quantitative genetics" across the biological sciences. The modern capability for molecular dissection of complex phenotypes is capable of addressing a plethora of questions and issues in genetics, breeding, and other areas of the life sciences. This volume will seek to explore efficient approaches to addressing these issues which maximize return on investment of public or private resources.

The book is divided into three parts:

**Chapters 1–12**, "Fundamental Principles" are intended to quickly bring the scientist or student up to the state of the art in an important and rapidly-growing field and to introduce the layman with a scientific bent to a new approach for investigating fundamental questions in the life sciences.

**Chapters 13–19**, "Case Histories" of seminal work in QTL mapping are intended to show the utility of this new research approach in detailed investigation of a wide range of biological questions. Leading practitioners of molecular dissection reveal the thought processes which led to their seminal results.

**Chapters 20–21**, "Social Impact." The modern capability for molecular dissection of complex phenotypes is capable of addressing a plethora of questions and issues in genetics, breeding, and other areas of the life sciences. These chapters will seek to explore possible long-term impacts of these new research capabilities on agriculture and medicine, respectively.

We, the authors, hope that you, the readers, gain a sense of the excitement we enjoy about the possibility of using these new tools and techniques to better describe our own experiments and more generally to better understand the biological world around us. Further, we hope that the contents of this volume are helpful to those of you who seek to join us in this endeavor, and we welcome you into the dynamic field of "molecular quantitative genetics."

**Andrew H. Paterson**

# Editor

**Andrew H. Paterson, Ph.D.,** is the Christine Richardson Professor of Agriculture in the Department of Soil and Crop Sciences and a full member of the Graduate Fields of Genetics and Plant Physiology & Plant Biotechnology at Texas A&M University in College Station, Texas.

Dr. Paterson received his B.S. in Plant Science from the University of Delaware, Newark Delaware, in 1982, under the direction of Professors James A. Hawk and Donald L. Sparks. He obtained his M.S. in 1985 and Ph.D. in 1988 from the Department of Plant Breeding and Biometry, Cornell University, Ithaca, New York, under the direction of Professor Mark E. Sorrells. After doing postdoctoral work with Professor Steven D. Tanksley at Cornell University, he joined the Agricultural Biotechnology program of E.I. duPont de Nemours as a Research Biologist, and simultaneously held an Adjunct Assistant Professorship in Plant Molecular Biology at the University of Delaware. In 1991, he joined the Department of Soil and Crop Sciences at Texas A&M University as an Assistant Professor. He received tenure and promotion to Associate Professor in 1995 and was appointed the holder of the Christine Richardson Endowment in 1996.

Dr. Paterson is a member of the American Association for the Advancement of Science, the Genetics Society of America, the Crop Science Society of America, the Brazilian Society of Genetics, and the honorary societies Sigma Xi, Phi Kappa Phi, and Gamma Sigma Delta. In 1996, he was named the Young Crop Scientist of the Year by the Crop Science Society of America. He was recently named the Faculty Lecturer of 1997 by Texas A&M University.

Dr. Paterson has delivered many invited seminars and symposium talks. He has regularly lectured at the International Plant and Animal Genome Conference, Keystone Symposia, Gordon Conference, Beltwide Cotton Conference, and at many universities and research institutions.

Dr. Paterson is currently the recipient of research grants from the United States Department of Agriculture, International Consortium for Sugarcane Biotechnology, Pioneer Hibred International, Texas Higher Education Coordinating Board, and Texas State Support Committee of Cotton, Inc. He has published about 40 refereed papers, contributed chapters to several books, and edited one other volume. His current research interests focus on genome organization and evolution in several higher plant taxa, with particular emphasis on gene manipulation in crop improvement.

# Contributors

**Kjell Andersson**
Department of Animal Breeding and Genetics
Swedish University of Agricultural Sciences
Uppsala, Sweden

**Leif Andersson**
Department of Animal Breeding and Genetics
Swedish University of Agricultural Sciences
Uppsala, Sweden

**Lena Andersson-Eklund**
Department of Animal Breeding and Genetics
Swedish University of Agricultural Sciences
Uppsala, Sweden

**William D. Beavis**
Quantitative Genetics Group
Trait and Technology Integration
Pioneer Hi-Bred International, Inc.
Johnston, Iowa

**John K. Belknap**
Portland Alcohol Research Center
Department of Veterans Affairs Medical Center
and
Department of Behavioral Neuroscience
Oregon Heath Sciences University
Portland, Oregon

**Douglas W. Bigwood**
Department of Plant Biology
University of Maryland
College Park, Maryland

**Thomas K. Blake**
Department of Plant and Soil Sciences
Montana State University
Bozeman, Montana

**H. D. Bradshaw, Jr.**
College of Forest Resources
University of Washington
Seattle, Washington

**Mark D. Burow**
Department of Soil and Crop Sciences
Texas A&M University
College Station, Texas

**Lon R. Cardon**
Sequana Therapeutics, Inc.
La Jolla, California

**Gary A. Churchill**
Department of Plant Breeding and Biometry
Cornell University
Ithaca, New York

**John C. Crabbe**
Portland Alcohol Research Center
Department of Veterans Affairs Medical Center
and
Department of Behavioral Neuroscience
Oregon Heath Sciences University
Portland, Oregon

**Rebecca W. Doerge**
Departments of Agronomy and Statistics
Purdue University
West Lafayette, Indiana

**Inger Edfors-Lilja**
Department of Engineering and Natural Sciences
University of Växjö
Växjö, Sweden

**Hans Ellegren**
Department of Animal Breeding and Genetics
Swedish University of Agricultural Sciences
Uppsala, Sweden

**Yuval Eshed**
Faculty of Agriculture
The Hebrew University of Jerusalem
Rehovot, Israel

**Michel Georges**
Department of Genetics
Faculty of Veterinary Medicine
University of Liege
Liege, Belgium

**Chris S. Haley**
Roslin Institute
Edinburgh, UK

**Ingemar Hansson**
Department of Food Science
Swedish University of Agricultural Sciences
Uppsala, Sweden

**James E. Irvine**
Department of Soil and Crop Science
Texas A&M Research and Extension Center
Weslaco, Texas

**Sara A. Knott**
Institute of Cell, Animal and
Population Biology
University of Edinburgh
Edinburgh, UK

**Zhikang Li**
Department of Soil and Crop Sciences
Texas A&M University
College Station, Texas

**Yann-Rong Lin**
Department of Soil and Crop Sciences
Texas A&M University
College Station, Texas

**Ben-Hui Liu**
Department of Forestry
North Carolina State University
Raleigh, North Carolina

**Sin-Chieh Liu**
Department of Soil and Crop Sciences
Texas A&M University
College Station, Texas

**Kerstin Lundström**
Department of Food Science
Swedish University of Agricultural Sciences
Uppsala, Sweden

**Lena Marklund**
Department of Animal Breeding and Genetics
Swedish University of Agricultural Sciences
Uppsala, Sweden

**Maria Johansson Moller**
Department of Animal Breeding and Genetics
Swedish University of Agricultural Sciences
Uppsala, Sweden

**Susan R. McCouch**
Department of Plant Breeding and Biometry
Cornell University
Ithaca, New York

**Andrew H. Paterson**
Department of Soil and Crop Sciences
Texas A&M University
College Station, Texas

**Joao L. Rocha**
Department of Animal Science
Texas A&M University
College Station, Texas

**Hakan Sakul**
Sequana Therapeutics, Inc.
La Jolla, California

**Keith F. Schertz**
U.S. Department of Agriculture
Agricultural Research Service
College Station, Texas

**Charles W. Stuber**
U.S. Department of Agriculture
Agricultural Research Service
Department of Genetics
North Carolina State University
Raleigh, North Carolina

**Jeremy F. Taylor**
Department of Animal Science
Texas A&M University
College Station, Texas

**Claire G. Williams**
College of Agriculture and Life Sciences
Texas A&M University
College Station, Texas

**Jinhua Xiao**
Department of Plant Breeding and Biometry
Cornell University
Ithaca, New York

**Dani Zamir**
Faculty of Agriculture
The Hebrew University of Jerusalem
Rehovot, Israel

# Contents

## INTRODUCTION



## PART I. FUNDAMENTAL PRINCIPLES

## *PART II. CASE HISTORIES*

## PART III. SOCIAL IMPACT OF QTL MAPPING



## EPILOGUE

# *Dedication*

---

*To Maria*

# Introduction

# 1 Of Blending, Beans, and Bristles: The Foundations of QTL Mapping

*Andrew H. Paterson*

## CONTENTS



## 1.1 THE PREHISTORY OF QUANTITATIVE GENETICS

Quantitative genetics has a rich history of contributions to life sciences research, predating not only the identification of DNA as the hereditary molecule, but also the discovery of the Darwinian and Mendelian principles.

In agriculture, for example, as early as 10,000 years ago emerging human civilizations evolved "domesticated" plant and animal strains which differred from their wild ancestors in many important ways, maintaining and improving these strains by recurrent cultivation and selection. Modern plant breeding exemplifies a transition from this implicit understanding of genetic differentiation, to explicit and highly successful efforts to change the gene pool of domesticates.

In evolution, for example, study of relationships among organisms based on morphological features could only be meaningful with an implicit understanding that taxa faithfully reproduce these features in progeny. Extensive catalogs of biota by taxonomists such as *Linneaus* are perhaps among the earliest analyses of the relationship between phenotype and genotype. In the past few decades, evaluation of variants in proteins, DNA markers, DNA sequences, or the order of genes along the chromosomes, represented a transition from implicit awareness of a relationship between phenotype and genotype, to explicit study of genetic events responsible for taxonomic divergence.

In medicine, for example, long-standing cultural taboos in many societies express an implicit awareness of the genetic consequences of mating between close relatives. Modern quantitative genetic theory, and identification of mutant alleles which are deleterious in homozygous condition represent a transition from such implicit awareness, to an explicit molecular basis for "inbreeding depression" and "genetic load."

0-8493-7686-6/98/$0.00+$.50

Abstractions of continuous phenomena into discrete categories precipitated long-term debates regarding the genetic basis of complex traits. The first of these in the era of modern genetics was the debate regarding how "blending inheritance" of intermediate phenotypes could be reconciled with the particulate Mendelian principles. This debate was ultimately resolved by calculus-like models which integrated the strengths of each philosophical extreme. (Figure 1.1). Perhaps the most enduring debate has regarded the complexity of polygenic inheritance, with the classical "gradualist" and modern "punctuational" schools only beginning to achieve consensus — many important aspects of the dichotomy are still awaiting resolution.

### 1.1.1 A Conceptual Basis for Genetic Dissection of Complex Traits

The realization that discrete tools might be used to quantify the number, location, and individual effects of "quantitative trait loci (QTLs)", chromosomal locations of individual genes or groups of genes which influenced complex traits,[1] was clearly articulated in the pioneering work of Sax[2] on the inheritance of seed size in dry beans. While such "genetic dissection" was hindered in many biota by a paucity of discrete markers, models such as *Drosophila* enabled detailed investigations of complex traits such as "sternopleural bristle number" to be conducted, demonstrating most of the basic principles of "QTL mapping" by the 1960s.[3]

## 1.2 NEW MOLECULAR TOOLS MADE POSSIBLE COMPREHENSIVE QTL MAPPING

Technological advancement of the past decade triggered the blossoming of "molecular dissection" of complex traits. In particular, the ability to visualize specific points in the hereditary DNA molecule (see Chapter 2, this volume) has impelled development of "molecular maps" of the chromosomes of many organisms. These maps, in turn, have enabled the design, execution, and analysis of "QTL mapping" experiments to describe the inheritance of complex traits in unprecedented detail.

QTL mapping experiments follow a common basic algorithm in many taxa. First, a detailed molecular map of the chromosomes of an organism is assembled, either in the pedigree of interest or in a different pedigree of the same taxon. An example is a "complete" map of DNA markers for the ten sorghum chromosomes (Figure 1.2), assembled in a cross between a cultivated sorghum genotype and its wild relative. To derive maximal information from minimal input, QTL mapping typically uses only a subset of the available DNA markers, evenly distributed across the chromosomes at moderate intervals (see Figure 1.2), to a large population of sibling progeny derived from parents which carry different alleles both at genetic markers and at QTLs.

In experimental populations, one can virtually guarantee the detection of at least some QTLs by choosing parents which differ markedly for phenotypes of interest.[4,5] It is intuitive that parents which show large, statistically significant differences in a phenotype will be a good source of allelic variation at underlying genetic loci.

However, cryptic "transgressive" alleles can be found even in parents with similar phenotypes.[4] For example, consider a hypothetical case from Figure 1.1c. Genotypes AAbb and aaBB have the same phenotype, near the center of the phenotypic distribution. However, if these two genotypes were mated to each other, and the F1 progeny selfed, the resulting F2 progeny array would span the entire range of phenotypes shown: some individuals (AABB) with higher phenotypes than either parent and others (aabb) with lower phenotypes than either parent. It is such extreme "transgressive" individuals that are often of greatest value in plant and animal breeding.[4,6]

Associations between allelic differences at a genetic marker locus and differences among individuals in phenotype, provide the basic evidence needed to identify and describe QTLs. Many algorithms suitable for evaluating association between markers and phenotypes are available,[7] a subset of which are described in more detail herein (see Chapters 3 through 7, this volume). A key feature in the interpretation of marker-trait association is the use of statistical significance thresholds which confer acceptable "experiment-wise" error rates. The large number of individual evaluations associated with QTL analysis carries an inherent risk of producing an excessive number of "false-positive" associations, which would interfere with effective communication of results through publication. Appropriate significance thresholds to preclude false-positive results can be established either *a priori*, by simulation[5,8] or *a posteriori*, by empirical evaluation of specific data sets (see Chapter 3, this volume).

Because "QTL mapping" involves extricating a genetic signal from many sources of "noise," QTL locations and effects are typically described as "likelihood intervals," chromosomal regions in which a QTL can be asserted to map with a specified level of statistical confidence. Comparison of alternative genetic models can be used to investigate the effects of QTL allele dosage on phenotype, rejecting those models which fail to account for observed data (Figure 1.3).[9] Effects of common chromosomal regions on different traits can be evaluated at the level of resolution afforded by a particular experimental design (see Chapter 10, this volume). This level of resolution can be manipulated experimentally (see Chapter 11, this volume). The influence of "epistasis," or nonlinear interaction between different genetic loci on complex phenotypes can be readily evaluated. Although classical QTL mapping experiments generally detected little such "epistasis", recent data suggest that epistasis does account for a portion of the genetic variation which could not be explained by QTL models by the collective effects of individual QTLs (see Chapter 8, this volume).

## 1.3 POPULATION STRUCTURE IN QTL MAPPING

Many QTL mapping experiments have employed populations which can be quickly derived from two- or three-generation pedigrees, such as F2 populations derived from selfing or intercrossing heterozygous individuals. In disomic species tolerant of inbreeding, these populations trace back to homozygous grandparents. They segregate for only two alleles per informative locus and are relatively simple to analyze, contain recombinational information for two different gametes in each individual, and permit evaluation of all possible QTL allele dosages (0-2). In outcrossing species (see Chapters 5 and 7, this volume), or in polysomic polyploids (see Chapter 6, this volume), larger numbers of alleles can be evaluated, but at lower precision, and with less information about gene dosage. "Backcrossing" of the heterozygous F1 to one of its parents is sometimes dictated by reproductive barriers (e.g., sterility of F2 self or intercross progeny), and serves as an excellent means for introgression of exotic germplasm into productive domestic cultivars (for example, see Reference 6 and Chapter 15, this volume). However, backcross populations contain only a subset of the possible genotypes at a locus and so, are less informative regarding gene dosage or interactions among genetic loci.

Reproducible populations of homozygotes facilitate genetic dissection of traits which are difficult to assay due either to the nature of the measurement system or to the influence of nongenetic factors such as rainfall or temperature. Such homozygous populations are derived from selfing or sibmating of heterozygous segregants in plant and mouse populations (recombinant inbreds, RI), or chromosome doubling of recombinant gametophytes in some plant taxa (doubled haploids, DH). Because experimental designs using homozygous populations can include replication of individual genotypes, the impact of nongenetic variation can be reduced at a rate proportional to the square root of the number of replications. Specific factors which might account for genotype × environment interaction can be elucidated either by designed variation of specific parameters, or by "epidemiological" *post hoc* evaluation of associations between measured environmental parameters and experimental results. Information regarding gene dosage can be obtained from RI or DH populations, but requires testcrosses of each segregant to be made and evaluated. Evaluation of heterogeneous progeny derived from individual segregants by selfing[9] (such as F3 families derived from individual F2 individuals) or outcrossing (see Chapters 5 and 7, this volume) have been employed as an alternative means of replication. Gains in precision, however, are partly sacrificed due to genetic heterogeneity of replicates.

Finally, detection of QTL with small effects can be facilitated by reducing the contribution of other genetic loci to error variance. Marker-assisted elimination of known QTLs of large effect

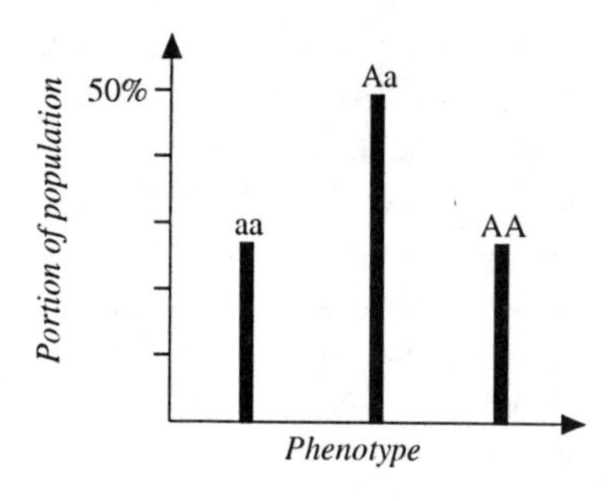


(a) Trait determined be a single genetic locus, with no influence of nongenetic factors. Graph depicts F2 progeny derived by selfing the F1 of a cross between homozygous genotypes *aa* and *AA*.

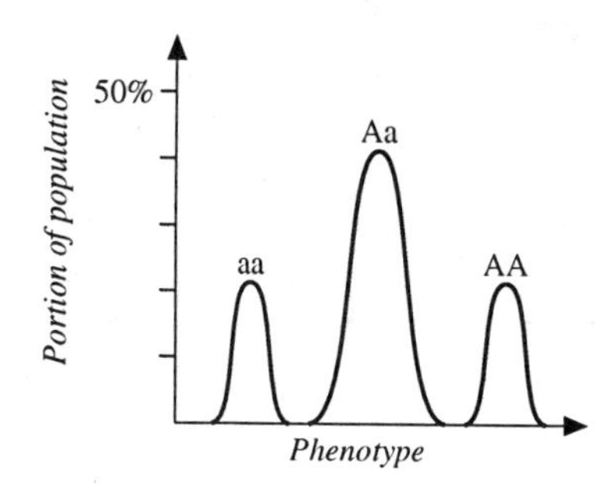


(b) Trait determined by a single genetic locus, and influenced by nongenetic factors. Graph depicts F2 progeny derived by selfing the F1 of a cross between homozygous genotypes *aa* and *AA*.

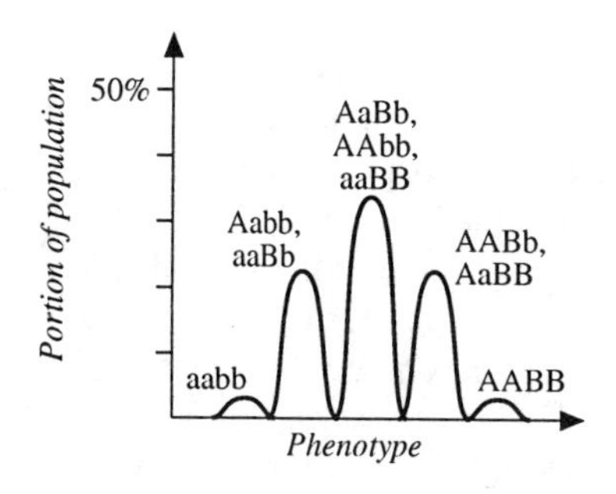


(c) Trait determined by two unlinked genetic loci, and influenced by nongenetic factors. Graph depicts F2 progeny derived by selfing the F1 of a cross between homozygous genotypes *aabb* and *AABB*.

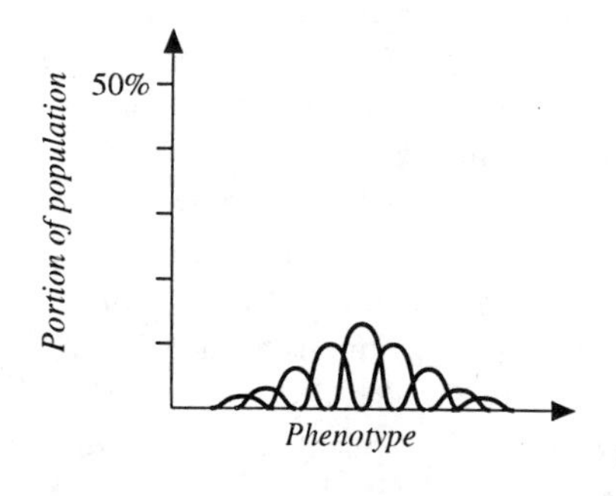


(d) Trait determined by 4 unlinked genetic loci, and influenced by nongenetic factors. Graph depicts F2 progeny derived by selfing the F1 of a cross between homozygous genotypes *aabbccdd* and *AABBCCDD*.

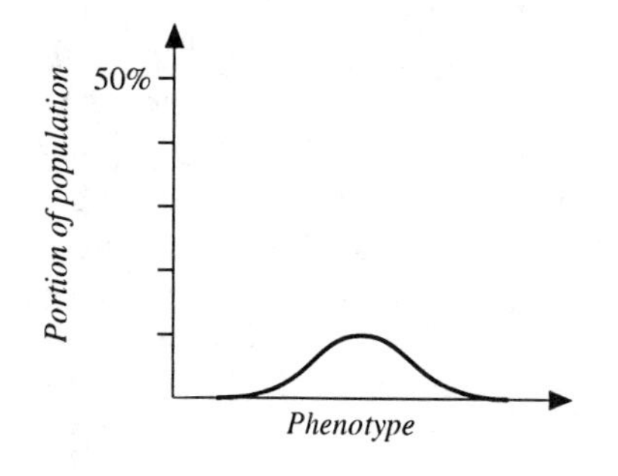


(e) Trait determined by many unlinked genetic loci, and influenced by nongenetic factors.

facilitates development of populations suitable for detecting QTLs of small effect.[10] By rendering such populations homozygous, the advantages of replication can be exploited to further improve sensitivity (see Chapter 15, this volume).

## 1.4 SHORTCUTS IN QTL MAPPING

QTL mapping experiments are notorious for their unwieldy size. Most QTL experiments have involved more than 200 individuals, and in some cases as many as 2000 or more. In Chapter 10, Beavis suggests that even 200 individuals may be too few for reliable QTL detection, and addresses issues related to choice of population size in QTL mapping experiments. Several approaches have been suggested to extract more information from analysis of smaller populations.

### 1.4.1 Selective Genotyping

An excellent approach for efficiently mapping QTLs which influence a single phenotype (only), is "selective genotyping."[5] This method is suitable if evaluation of an individual's phenotype can be done more cheaply and easily than evaluation of its DNA marker karyotype, a condition which is often true for plants and for experimental systems such as *Drosophila* or mouse, but less often applies to large animals or humans. Specifically, selective genotyping involves the identification of a subset of individuals from a genetic mapping population, which represent the most extreme phenotypes in the population. These individuals harbor more "information" than phenotypically "average" individuals, since they are more likely to contain a high proportion of the + or – alleles, respectively, at QTLs affecting the target trait. By phenotyping a large population, and then selecting only the most extreme individuals for genotyping, one can obtain equal or greater information about QTLs than from exhaustive mapping of randomly chosen individuals. Specific expectations for information gain, as a function of population size and selection intensity, have been published.[5]

A limitation of selective genotyping is the fact that it is suitable for analysis of only one phenotype at a time. This limitation often proves serious in applied experiments such as plant and animal breeding, that require evaluation for many independent characteristics.

### 1.4.2 Comparative QTL Mapping

An evolutionary approach to QTL mapping is becoming an increasingly powerful means to expedite QTL analysis. The finding that diverse taxa within common taxonomic families often share similar gene order over large chromosomal segments has been a basis for "comparative mapping", alignment of the chromosomes of different taxa based on common reference loci. The structural similarity

---

**FIGURE 1.1** Conceptual models for quantitative inheritance. The progeny of crosses between parent differing in a quantitative trait typically exhibit phenotypes ranging between those of the parents. The nature of the progeny distribution is determined jointly by the number of genes which account for the difference between the parents, together with the effects of nongenetic factors, such as micro-environmental differences and measurement error. For example, consider F2 populations derived by selfing the F1 of crosses between homozygous parents which exhibit extreme phenotypes. Alternative alleles at one or more loci will be assumed to make equal, additive contributions to phenotype. (a) A trait controlled by a single genetic locus, and which is immune to nongenetic factors, would exhibit the classical 1:2:1 distribution of phenotype. (b) If nongenetic factors partly obscure measurement of the phenotype, then even a trait under monogenic control an exhibit a more continuous distribution of progeny phenotypes. (c) Addition of a second locus to the model further obscures the relationship between phenotype and genotype. (d) Traits of intermediate complexity, influenced by four or more genes, become increasingly difficult to discern from (e) classical "polygenic traits" influenced by a virtually infinite number of genes, each with tiny effects. Differences between genotype classes are even more rapidly obscured in cases where the influence of nongenetic factors is especially large, and gene action is not equal and/or additive.

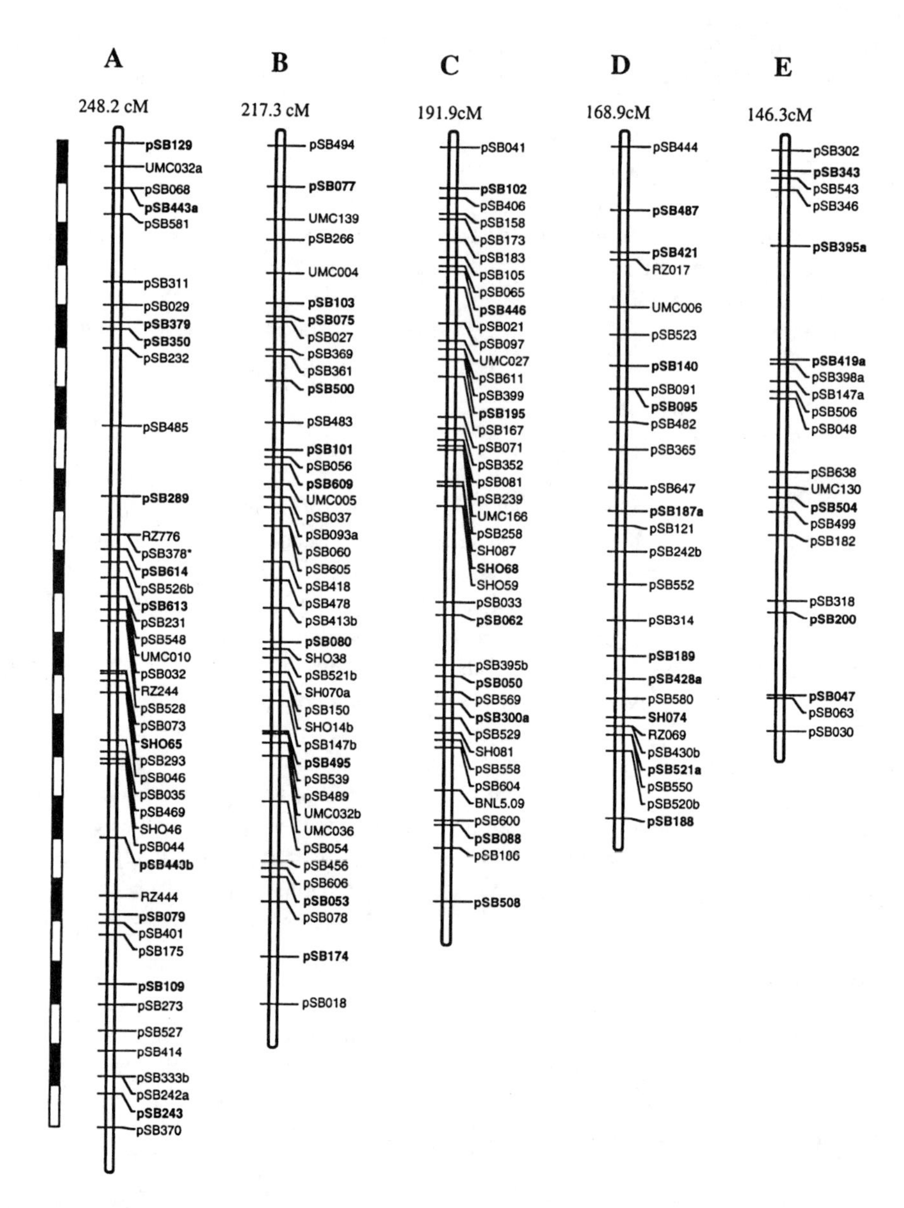


**FIGURE 1.2** Example of a complete genetic map. The sorghum genome is comprised of ten chromosomes — the ten "linkage groups" shown are drawn from the first complete molecular map of sorghum.[15] A "complete genetic map" is defined by two criteria — all available genetic markers are linked to one (and only one) of n linkage groups, and n = the number of gametic chromosomes in the organism's genome. Distances between markers along the map are measured in centiMorgans (cM) (see scale flanking map), derived from the portion of progeny of a cross which show different genotype at consecutive markers. Assembly of a complete map typically requires an average of one DNA marker per 5 cM (or about 5% recombination) — this sorghum map averaged one marker per 5.3 cM. Subsequent applications of the map to QTL analysis often use a subset of markers which are approximately equally spaced over all regions of the genome. The markers shown in **bold** were used in several QTL mapping applications of this map.[16-18]

of chromosomes in different taxa is often accompanied by functional similarity in the locations of genes influencing common phenotypes. Databases of genes previously reported to affect a phenotype are becoming increasingly powerful tools for predicting the chromosomal locations likely to account for phenotypic varation in new experimental crosses.

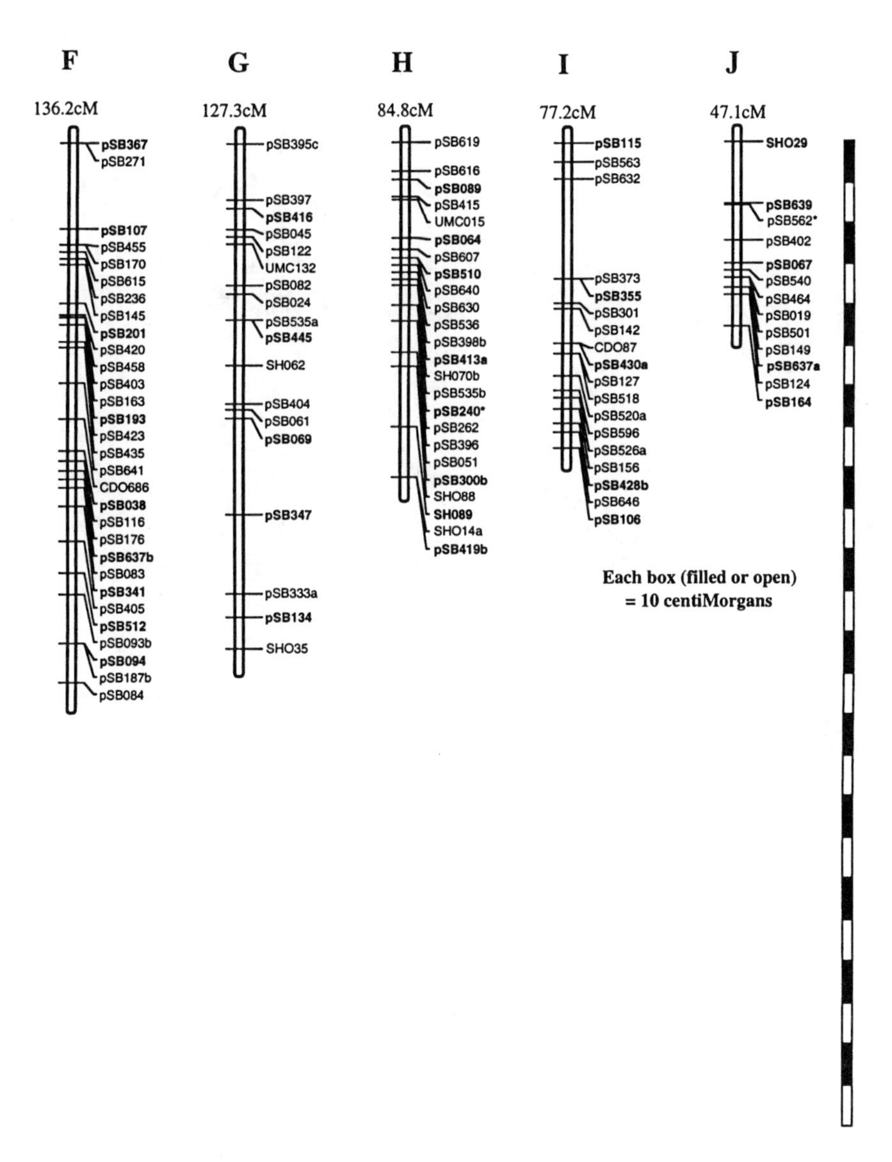


**FIGURE 1.2** (continued)

### 1.4.3 DNA Pooling

One approach which has tremendously simplified genetic mapping of simply inherited traits in plants and animals involves the *ad hoc* pooling of DNA from individuals sharing a particular characteristic. In principle, this is similar to selective genotyping, but treats the two extremes in a phenotypic distribution as a single DNA sample.

While this approach is an effective way to map genes which account for 100% of variance in a trait, both theoretical and empirical results suggest that it has limited applicability to QTL mapping.[11,12] Among the phenotypically extreme individuals for a polygenic trait, rare QTLs with unusually large effects may be fixed, and therefore may be detected as a chromosome segment which is polymorphic between the pools. However the majority of QTLs, with much smaller phenotypic effects, will remain heterogeneous in the pools and therefore will escape detection. Complicating factors such as dominance and non-Mendelian segregation reduce the feasibility of "bulked-segregant"[13] approaches to QTL mapping. While rare QTLs of very large phenotypic effect

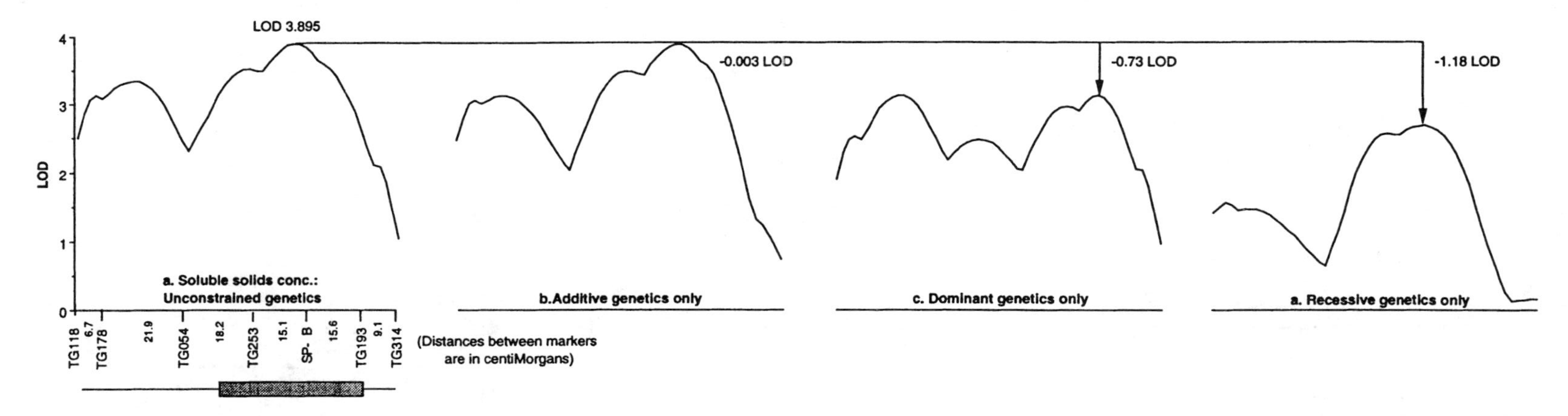


**FIGURE 1.3** QTL mapping. The figure shows four different genetic models for the effects of a tomato QTL on soluble solids concentration.[9] The "unconstrained genetics" model includes the total genetic variation which can be attributed to the maximum-likelihood location of the QTL (highest point on the curve). "Likelihood intervals" for the location of the QTL are below the graph, represented as a bar (90% likelihood) and whiskers (99% likelihood). The additive model reflects the likelihood that the observed data would occur if the locus has a strictly additive dosage effect (i.e., if the phenotype of the heterozygote is exactly intermediate between that of the alternative homozygotes). The dominant and recessive models reflect the likelihood that the observed data would occur if the phenotype of the heterozygote were equal to that of the "high" or "low" parental homozygotes, respectively.

might be reliably detected by bulked-segregant analysis, suites of many QTLs with smaller effects on a trait are best evaluated by comprehensive mapping.

## 1.5 APPLICATIONS OF QTL MAPPING INFORMATION

QTL mapping experiments are conducted toward three basic objectives. Many QTL mapping experiments are conducted for the purpose of locating genes which account for genetic variation in agriculturally important phenotypes, as a starting point for use of marker-assisted selection in plant or animal improvement. By identifying DNA markers which are diagnostic of a particular phenotype, the breeder can make selections among seedlings grown in nontarget environments and accelerate progress toward classical objectives. Significant advantages accrue for the breeder in having DNA markers for phenotypes that are difficult to measure, that can only be measured after plants have already contributed to the gene pool of the next generation, or that require unusual environments to evaluate. QTL mapping also has served as a starting point for introgression of exotic chromatin, delineating target genes and using DNA markers to reduce the portion of nontarget donor chromatin transferred (see Reference 6, and Chapter 15, this volume).

A long-term objective of some QTL mapping experiments is the molecular cloning of genes underlying specific phenotypes. Although no QTLs *per se* have been cloned yet, in principle cloning of QTLs can use logical extensions of the positional approaches used to "walk" to genes of medical importance in human, or mammalian models, or of agricultural importance in domesticated plants or animals. By making genetic stocks which are "near isogenic" for the chromosomal regions containing QTLs (see Reference 10, and Chapter 15 in this volume), individual QTLs can be rendered virtually discrete, mapped to precise locations, placed on megabase DNA contigs, associated with candidate genes, and tested for mutant complementation. Because of the small effects of individual QTLs on phenotype, single-plant tests of heterozygous primary transformants will usually be inadequate — rather, progeny testing may be necessary. QTL information may also serve as a supplement or adjunct to other molecular cloning approaches, providing a means to test whether candidate genes isolated by strategies such as tissue-specific expression, subtractive methods, insertional mutagenesis, or other methods, co-segregate with the target phenotype. In the future, the availability of detailed physical maps and transcript maps of several key taxa, together with comparative maps collating the chromosomes of a wide range of taxa, may offer a much more powerful database for identification and evaluation of candidate genes associated with QTLs.

The use of QTL mapping information in human genetic diagnosis raises a host of fascinating and controversial questions, which are addressed further by Crabbe and Belknap in Chapter 21, this volume.

Finally, QTL mapping has been used in several instances to ask basic questions about evolutionary processes. QTL analysis tends to support the generalization that many genetic phenomena are more discrete and more rapidly-evolving than classical genetic and evolutionary models would have anticipated (see Case Histories, Part II of this volume).

## REFERENCES


1. Geldermann, H., Investigations on inheritance of quantitative characters in animals by gene markers. I. Methods, *Theor. Appl. Genet.*, 46, 319, 1975.
2. Sax, K., The association of size differences with seedcoat pattern and pigmentation in Phaseolus vulgaris, *Genetics*, 8, 552, 1923.
3. Thoday, J. M., Location of polygenes, *Nature*, 191, 368, 1961.
4. Paterson, A. H., Lander, E. S., Hewitt, J. D., Peterson, S., Lincoln, S. E., and Tanksley, S. D., Resolution of quantitative traits into Mendelian factors by using a complete map of restriction fragment length polymorphisms, *Nature*, 335, 721, 1988.
5. Lander, E. S. and Botstein, D., Mapping Mendelian factors underlying quantitative traits using RFLP linkage maps, *Genetics*, 121, 185, 1989; and Corrigendum, *Genetics*, 136, 705, 1994.

6. Tanksley, S. D. and Nelson, J. C., Advanced backcross QTL analysis: a method for simultaneous discovery and transfer of valuable QTLs from unadapted germplasm into elite breeding lines, *Theor. Appl. Genet.,* 92, 191, 1996.
7. Paterson, A. H., Molecular dissection of quantitative traits: progress and prospects, *Genome Res.*, 5, 321, 1996.
8. Lander, E. S. and Kruglyak, L., Genetic dissection of complex traits: guidelines for interpreting and reporting linkage results, *Nat. Genet.,* 11, 241, 1995.
9. Paterson, A. H., Damon, S., Hewitt, J. D., Zamir, D., Rabinowitch, H. D., Lincoln, S. E., Lander, E. S., and Tanksley, S. D., Mendelian factors underlying quantitative traits in tomato: comparison across species, generations, and environments, *Genetics*, 127, 181, 1991.
10. Paterson, A. H., Deverna, J. W., Lanini, B., and Tanksley, S. D., Fine mapping of quantitative trait loci using selected overlapping recombinant chromosomes in an interspecies cross of tomato, *Genetics*, 124, 735, 1990.
11. Wang, G. and Paterson, A. H., Prospects for using DNA pooling strategies to tag QTLs with DNA markers, *Theor. Appl. Genet.*, 88, 355, 1994.
12. Darvasi, A. and Soller, M., Selective DNA pooling for determination of linkage between a molecular marker and a quantitative trait locus, *Genetics*, 138, 1365, 1994.
13. Michelmore, R. W., Paran, I., and Kesseli, R. V., Identification of markers linked to disease-resistance genes by bulked-segregant analysis: a rapid method to detect markers in specific genomic regions by using segregating populations, *Proc. Nat. Acad. Sci., U.S.A.,* 88, 9828, 1991.
14. Paterson, A. H., *Genome Mapping in Plants*, Academic Press/Landes Bioscience, Austin, TX, 1996.
15. Chittenden, L. M., Schertz, K. F., Lin, Y., Wing, R. A., and Paterson, A. H., RFLP mapping of a cross between *Sorghum bicolor* and *S. propinquum*, suitable for high-density mapping, suggests ancestral duplication of Sorghum chromosomes, *Theor. Appl. Genet.*, 87, 925, 1994.
16. Lin, Y. R., Schertz, K. F., and Paterson, A. H, Comparative mapping of QTLs affecting plant height and flowering time in the Poaceae, in reference to an interspecific *Sorghum* population, *Genetics,* 141, 391, 1995.
17. Paterson, A. H., Schertz, K. F., Lin, Y. R., Liu, S. C., and Chang, Y. L., The weediness of wild plants: molecular analysis of genes responsible for dispersal and persistence of johnsongrass (*Sorghum halepense* L. Pers.), *Proc. Nat. Acad. Sci., U.S.A.,* 92, 6127, 1995.
18. Paterson, A. H., Lin, Y. R., Li, Z., Schertz, K. F., Doebley, J. F., Pinson, S. R. M., Liu, S. C., Stansel, J. W., and Irvine, J. E., Convergent domestication of cereal crops by independent mutations at corresponding genetic loci, *Science,* 269, 1714, 1995.

# PART I

# FUNDAMENTAL PRINCIPLES

# 2 Molecular Tools for the Study of Complex Traits

*Mark D. Burow and Thomas K. Blake*

**CONTENTS**



## 2.1 INTRODUCTION

Recent advances in quantitative trait loci (QTL) analysis have been made possible by improved methods of genetic marker analysis. Prior to molecular markers, linkage of useful traits was measured frequently to morphological markers. With exceptions for maize and *Drosophila,* the paucity of such markers limited the applicability of marker analysis. Proteins became useful as markers in the 1950s for isozyme analysis, and general protein marker analysis became common later with improved methods of protein electrophoresis. However, it is the DNA technology of the past 20 years, specifically restriction fragment hybridization analysis, polymerase chain reaction (PCR), and improved cytological analysis, that have revolutionized genetic analysis and opened new possibilities in the study of complex traits. This chapter will describe briefly the most commonly used procedures for molecular marker analysis as a general background for the case studies in subsequent chapters.

0-8493-7686-6/98/$0.00+$.50

## 2.2 PROTEIN MARKERS

### 2.2.1 ISOZYME ANALYSIS

Isozyme analysis was the first type of molecular analysis widely employed. The term "isozyme", or "isoenzyme", refers to different enzymes which catalyze the same reaction. For genetic study, the term "allozyme" analysis is more proper, referring specifically to enzyme forms that are the products of different alleles, not different genes' products with similar enzymatic activities.

Isozyme analysis begins with electrophoretic separation of proteins on starch gels or, more recently, polyacrylamide gels for greater resolution.[1] The gel is then soaked in a reaction solution specific for a given enzyme, and enzymatic activity identifies enzyme location by a local change in color. Genetic analysis depends on differences in mobility among different forms of the enzyme. Depending on electrophoretic conditions, protein mobility may be influenced by protein size and/or charge. Size differences may be attributable to different subunit compositions or glycosylation; charge differences may be caused by point mutations causing amino acid substitutions, differences in glycosylation, or protein phosphorylation.[2]

Genetic analysis treats enzyme bands as representatives of alleles; contrary to early expectations, the enzyme patterns themselves generally are not the causative agents of most phenotypic differences,[3] although there are exceptions.[4] Association of potential marker alleles and the trait of interest are tested for genetic linkage in segregating progeny. Allozyme markers are typically the desirable codominant type, which allows distinguishing of heterozygotes from the homozygous dominant and the recessive allele from failure of the reaction (Figure 2.1). Occasionally, however, null alleles, caused by mutation so that no enzyme or an inactive enzyme is produced, occur.

Allozyme analysis has the advantages of being relatively inexpensive, easy to perform, and requires little preliminary work because species-specific DNA probes or PCR primers are not required. As allozyme analysis has been practiced longer than other forms of molecular analysis,

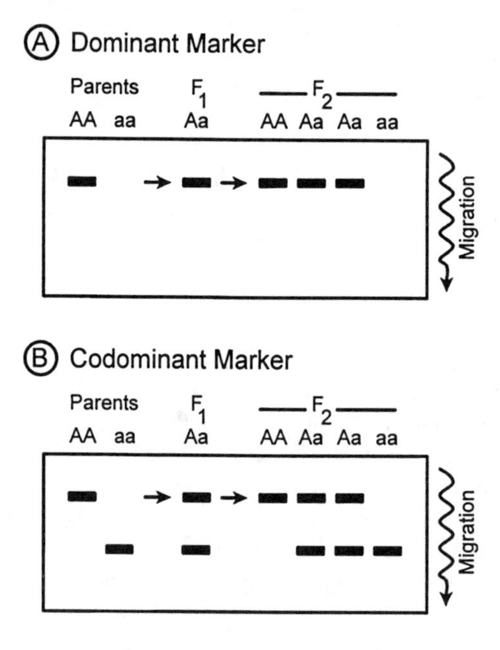


**FIGURE 2.1** Example of scoring of markers. In this example of a diploid cross, both parents are homozygous for different alleles. In the case of dominant markers (A), the recessive allele cannot be observed. The F1 progeny will be scored as identical to the dominant parent, and the F2 generation will segregate into two phenotypes. The heterozygous and homozygous dominant genotypes cannot be distinguished. In the case of codominant markers (B), the F1 will be identified as containing the contributions of both parents, and the three F2 phenotypes will correspond to the three genotypes.

there are considerable data available, and inclusion of previously identified markers allows comparison to now "classic" work.[5] For example, maize cultivars can be identified by allozyme patterns, and data from 437 elite maize varieties are available online.[6] However, the lack of potential isozyme markers (approximately 40 to 60 reactions are used commonly)[7,8] limits allozyme analysis to highly polymorphic systems, and generally precludes marker-assisted selection, most QTL analysis, fine mapping, and map-based cloning. Special attention needs to be given to the possibility of tissue-, developmentally-, and environmentally-regulated expression of isozymes which can influence results.[3,9,10] Since the 1980s, allozyme analysis has been largely supplanted by the use of DNA markers.

Many papers have been published on the applications of allozyme analysis, and many uses have been described.[11-13] The first molecular maps were performed with allozymes.[14] Likewise, in the first QTL mapping experiments with molecular markers, allozyme loci were demonstrated to be linked to yield,[15] fruit and seed weight, leaf ratio, and stigma exsertion,[16] and it was demonstrated possible to select for yield by selecting for allozyme markers.[17]

### 2.2.2 Sodium Dodecylsulfate Polyacrylamide Gel Electrophoresis (SDS-PAGE)

SDS-PAGE, a second type of protein marker analysis, detects proteins without regard to enzymatic activity. Although protein electrophoresis dates back 70 years,[18] the modern protocol dates back to 1970 with the combining of PAGE and denaturation of proteins with SDS,[19] and subsequently the substitution of multiwell glass plates for tube gels.

The detergent SDS is used to denature proteins into negatively charged polypeptide subunits. After electrophoretic separation on polyacrylamide gels, proteins are visualized with a universal protein stain such as Coomassie Brilliant Blue. Allelic differences are detected as variations in polypeptide size, and alleles are often codominant, although dominant/null allelic systems also occur.

SDS-PAGE is widely used for protein and physiological analysis because it can detect proteins whether possessing enzymatic function or not, and because the high resolution of PAGE permits discerning polypeptide size differences attributed to different lengths of coding regions, or post-translational modification of proteins, such as enzymatic cleavage and/or glycosylation.[20,21] In certain cases, proteins observed may be not only markers, but the physiologically responsible agents themselves.[22] In such instances, protein analysis yields highly useful and meaningful markers.

Notwithstanding its power, genetic mapping of proteins suffers many of the same limitations as isozyme analysis. Typically only from 20 to 50 highly abundant polypeptides soluble in the extraction buffer used can be scored per sample. Polypeptide patterns are also specific for environment, tissue, and stage of development. Cases where SDS-PAGE is used for general genetic mapping typically involve highly abundant, stable proteins. Blood-serum markers in animals are used for identification,[23] and have been included as markers for QTL analysis.[24] Plant-seed storage proteins are widely used for species and genotype identification, and can contribute to important quality characteristics such as breadmaking and digestibility. They have also been used as markers for quantitative traits such as amino acid balance, protein concentration, and yield.[25,26] Finally, protein markers have been included in comprehensive molecular marker maps of several species.[5,27]

## 2.3 DNA Restriction Fragment Mapping by Clone Hybridization

### 2.3.1 Restriction Fragment Length Polymorphism (RFLP) Analysis

Beginning in the 1980s, genetic marker analysis has shifted from use of protein markers to DNA markers, particularly as a consequence of the enhanced number of potential markers available. The first use of RFLP analysis was in construction of a human genetic map,[28] and this was suggested as a general method of genetic analysis.[29]

In this procedure, genomic DNA is digested at reproducible DNA targets using restriction endonucleases that typically recognize specific six base-pair (bp) sequences. Fragments are separated by size electrophoretically on agarose gels, and are denatured and immobilized on nylon or nitrocellulose membranes (blots). Visualization of specific DNA fragments is accomplished by use of DNA probes, synthesized enzymatically as radiolabeled complements to individual cDNA or genomic DNA clones. Blots are exposed to X-ray film, typically from several days to 2 weeks, to expose markers. Nonisotopic methods of detection are available also.[30] The basis of RFLP-detected polymorphism is the difference in the restriction enzyme cut sites. Large insertions and deletions also can be detected, but cannot be distinguished from changes in restriction sites. Point mutations in the fragment will not be detected unless in the restriction sites themselves. A variant on the standard method of electrophoresis, denaturing gradient gel electrophoresis (DGGE), makes possible detection of point mutations between restriction sites by separating samples on polyacrylamide gels containing a gradient of urea.[31,32] Although potentially useful for critical experiments, gel preparation is time consuming, fewer samples can be run simultaneously, and differences in migration are small and can be difficult to score.

RFLP analysis is currently the most widely used form of molecular marker analysis. An almost unlimited number of nonoverlapping single-copy probes are possible for most organisms, and additional polymorphisms can be identified by the use of different restriction enzymes to cleave genomic DNA. Many samples (typically up to 120) can be loaded on a single gel, and nylon blotting membranes can be reused for hybridization 10 or 20 times, gaining more information from the large quantity of DNA required (typically 5 μg DNA per lane.) Maps of approximately 1 cM resolution have been constructed from highly polymorphic parents in several taxa.[33] RFLP markers are codominant, and scoring is simple because of the low copy number of probes. However, this mandates the use of many probes, a particular problem when searching for polymorphisms among closely related genotypes such as self-pollinating cultivars or near-isogenic lines (NILs). RFLP analysis requires a cDNA or genomic library of the appropriate species. Libraries of many species currently exist; alternatively, they can be synthesized using commercially available kits. If necessary, probes from related species may be used to detect polymorphisms. In this vein, ESTs (expressed sequence tags), cDNA clones whose sequences have been published, can be used effectively if computer searches are performed to find the most conserved of these across species. This ability to identify nearly homologous markers across species boundaries is a particular strength of RFLP analysis and has been the basis for comparative mapping of genome structure across species, genus, and even family boundaries.[34,35] Identification of genes by homology has also provided evidence for ancestral duplication of chromosome number in several species.[36,37]

As with isozyme analysis, RFLP analysis has been used for many purposes including mapping QTLs affecting many traits.[38,39] Examples include soluble solids in tomato,[40] yield in maize,[41] photoperiod sensitivity in *Arabidopsis*,[42] growth, and flowering in pine and *Populus*,[43] malaria parasite susceptibility in mosquito,[44] blood pressure in rat,[45] and dyslexia in humans.[24] The availability of closely spaced markers has also allowed narrowing of QTL loci to short, specific, chromosomal segments by interval mapping.[24,46] Genetic maps, generated in large part by RFLP analysis, for many species can be found online.[38]

### 2.3.2 DNA Hybridization Using Repetitive Elements

RFLP analysis typically uses single- or low-copy number cDNA of genomic sequences for probes, and generally has a low polymorphism rate per probe. Use of multiple-copy sequences as probes offers the possibility to map multiple, highly polymorphic markers. Several classes of repetitive DNA are known. Satellite DNA consists of highly abundant DNA sequences, generally with repeat units of 100 to 300 bp and repeated up to $10^6$ copies/genome.[47,48] The most common of these are SINEs (short interspersed repetitive DNA sequences), such as the human *Alu* family. SINEs are not used commonly for genome analysis because of the difficulty in separating the large number of DNA fragments present. Two other, more useful, types of VNTR (variable number of tandem repeats) sequences are minisatellite and microsatellite DNA. Minisatellites,[49] tandem repeats from 10 to 100 bp long, have been characterized now in mammals, birds, insects, fungi, and plants, and many discovered to date share one of several common core motifs.[49-52] Microsatellites typically have repeat units of one to six nucleotides, and also occur in tandem arrays.[53] Repetitive elements may be used for mapping as either direct hybridization probes or by PCR amplification (see below).

The protocol for analysis of repetitive elements is similar to the RFLP protocol. For minisatellites, minisatellite monomers are used as as probes. For hybridization to microsatellites, the protocol may be modified by digestion of genomic DNA with a restriction endonuclease with a 4-bp recognition sequence, to obtain smaller fragments for which differences of several nucleotides in length can be distinguished.[50,54] Oligomers of the microsatellite repeats are used as probes because the monomers are too short to hybridize specifically.

VNTR probes may detect markers mapping to multiple sites throughout the genome, and typically have a high rate of polymorphism per locus than RFLPs. Polymorphism is based on the number of repeat units between restriction sites. The human minisatellite pλg3 has a 37-bp repeat unit, and was detected in multimers of from 15 to 500 tandem copies, making restriction fragments from 1.8 to 29 kb in size.[55] The distribution of such loci may vary among species; minisatellite sequences were reported to be highly clustered in proterminal regions of human chromosomes,[56] but are dispersed throughout the bovine genome.[57] Microsatellites are dispersed more widely throughout the genome, with the most abundant elements spaced an average of 10 to 20 kbp apart.[58] For pλg3, polymorphism among individuals was 97%, much higher than for most RFLP analyses. The high polymorphism among repetitive sequences is believed to be the consequence of reduced selection pressure for noncoding sequences, meiotic unequal exchange, sister chromatid exchange, or DNA slippage during replication.[49]

The primary disadvantages of repetitive elements are difficulty in scoring large numbers of bands, prompting some researchers to look for repetitive elements that hybridize to fewer bands,[57] and the lack of suitable probes in many species. The latter particularly applies to minisatellite sequences. Isolation of these sequences requires screening of cDNA libraries with potential probes or by PCR.[59] Some minisatellite sequences are similar to the human or m13 core sequences, hence these may be used as a starting point.[50,51,59] Relative to RFLP analysis, few minisatellite probes have been identified, with the exception of the bovine genome for which 36 were identified in one report.[57] Consequently, minisatellite analysis frequently must be used in combination with other types of analysis. Microsatellite sequences are easier to identify. They can be obtained by searching EMBL or GenBank databases for microsatellite sequences in clones of the target species, such as *Arabidopsis*, where many clones sequences are available, or by using synthetic oligonucleotides of common sequences to screen libraries.[60]

Mini- and microsatellite hybridization has been used extensively in human DNA fingerprinting and in genetic mapping in humans and cattle.[50,57,61] Use is less common in plants.[62-65] Repetitive DNA probes may be of special benefit where genetic diversity of coding genes is limited and it is difficult to find significant (in some cases, any) polymorphism, as in some self-pollinated crops.

## 2.4 RESTRICTION FRAGMENT MAPPING WITHOUT HYBRIDIZATION

### 2.4.1 Restriction Landmark Genome Scanning (RLGS)

The RLGS technique is based on earlier attempts to simultaneously map multiple protein or DNA fragments by two-dimensional electrophoresis.[2,66-70] Rather than relying on probes to detect specific sequences from the large number of restriction fragments available, the RLGS method labels all fragments of one restriction digestion, and subsequently employs a series of digests and electrophoretic separations to resolve the potential markers.

The RLGS procedure begins by filling in enzymatically the sheared ends of large, high-quality DNA to prevent high background. Typically, the DNA is then digested with the 8-bp recognition-sequence restriction endonuclease *NotI* to produce a relatively-small number ($10^2$ to $10^4$) of large DNA fragments, which are then end-labeled. These are digested with a 6-bp cutting enzyme, then separated electrophoretically in agarose tube gels. DNA is digested in-gel with a third nuclease, followed by PAGE in the second dimension, and the gel is exposed to X-ray film for visualization of spots. Polymorphism is based primarily on mutations in any of three potential restriction recognition sequences of fragments containing the labeled *NotI* site.

The resolved marker pattern typically consists of from 1000 to 2000 spots, which are derived from all chromosomes.[66] Using studies of mouse and human DNA, 1100 (3 enzyme combinations) and 352 (1 enzyme combination) polymorphic spots were observed between parents using one set of gels.[71,72] RLGS may be especially useful for identifying polymorphisms distinguishing parents, near-isogenic lines, or pooled DNA samples of contrasting phenotypes (bulked segregant analysis). As for RFLP analysis, it is expected that markers are codominant; in practice, only half were, with the second allele presumed to have fallen outside the resolving range of the gel.[72] Unlike RFLP or mini- or microsatellite analysis, no DNA libraries or sequence information is required. Little DNA is required, approximately 1 μg for analysis of most genomes, thus considerably less DNA is needed than for Southern blots or even for a large number of PCR reactions. Large genomes ($>3 \times 10^9$ bp), however, require additional steps (and substantially more DNA) to reduce background signal caused by shearing of genomic DNA during extraction.[73] RLGS markers have a special property that may be used to facilitate map-based cloning. *NotI* sites occur in proximity to CpG islands, which are frequently located upstream of functional genes. As *NotI* sites are rare, occurring approximately once every Mbp, a closely-linked marker, perhaps mapping 1 cM from a trait, may be very close upstream physically of the target gene and relatively easy to clone.

Perhaps the greatest obstacle to the greater use of RLGS is the technical expertise required. High-molecular weight DNA, three complete digests, and two electrophoreses are required. Comparative analysis of the larger number of spots is potentially difficult (as with 2-D PAGE of proteins), but can be aided by preparing and running gels simultaneously in the same custom-made apparatus, and analyzing with high-quality scanning equipment and software. An alternative to running large numbers of RLGS gels is to screen progeny to identify polymorphic markers, clone,[74,75] and use them as RFLP or STS (sequence tagged site) markers to screen segregating populations.

RLGS has been used to construct genome maps for mouse and Syrian hamster.[71,77] It has been used also to identify genome differences between normal and cancerous cells, identifying both amplified and methylated sequences using *PstI* as second enzyme.[76,78] RLGS has proven useful for fine mapping of markers, detecting 31 loci mapping within an 11.7-cM region around the mouse *reeler* locus.[79] Mapping of QTL remains to be attempted.

## 2.5 DIRECT HYBRIDIZATION OF CLONES TO CHROMOSOMES

### 2.5.1 Fluorescent *In Situ* Hybridization (FISH)

FISH is an improvement of older techniques for radioactive labeling of chromosomes,[80] which involved incorporation of radiolabeled nucleotides[81] or hybridization of radiolabeled probes to fixed chromosomes, followed by autoradiography.[82,83]

FISH requires slides of chromosome preparations made from enzymatically digested cell preparations or from cell squashes of either mitotic or meiotic cells.[30,80] Probes are synthesized enzymatically, incorporating either biotinylated deoxyuridine triphosphate (dUTP) or nucleotides bound to fluorescent tags,[30] and are hybridized to denatured DNA fixed on the slides. Samples are either irradiated directly with ultraviolet light or after addition of fluorochrome-conjugated avidin to bind biotinylated probes. Greater sensitivity is possible by addition of biotinylated goat anti-avidin and fluorescein-avidin or by detection using charge-coupled electronic cameras.[80] Fluorescent labeling allows much quicker detection than isotopic labeling, which requires weeks or months for image development.[81,83]

FISH has the unique property of mapping markers physically to specific chromosomes, providing a means to develop marker-chromosome associations in taxa that lack cytological chromosome markers. It is also not necessary to score segregating populations to map markers by *in situ* hybridization. Physical markers are identified with specific chromosomes by their cytological appearance, or by mapping to R-banded regions of chromosomes. The latter has been used to map markers to regions of approximately 1 to 3 cM or 10 Mbp in size.[84-86] For greater resolution, use of the longer interphase or pachytene chromosomes,[87,88] or use of different fluorochromes to simultaneously distinguish and order three closely-linked markers has been proposed.[89,90] More commonly, YAC (yeast artificial chromosomes) or BAC (bacterial artificial chromosomes) markers, determined by FISH to map to the same R-band, are ordered by cross-hybridizing. FISH can also test YACs for chimerism, as chimeric YACs may map to more than one chromosomal locus.

Use of FISH is hindered by several factors. It requires more skill than many other methods of marker analysis, and setup costs are high for microscope, detection, and image analysis systems. Probe selection is problematic, as the single-copy probes used in RFLP analysis are often too short to use directly, due to low signal intensity.[88] Detection frequently requires the use of larger DNA fragments as probes, such as cosmids, YACs, BACs, VNTRs, or multigene families mapping to a single locus.[84,91-94] For physical markers to be useful for genetic analysis, each must contain a mappable genetic marker (typically an RFLP or STS marker). Scoring physical markers themselves in a large segregating population is not routine due to the effort required.

Markers mapped by FISH are candidates for use in localizing QTL loci, and for map-based cloning thereof. The greatest effort has been applied to mapping mammalian genomes, especially the human genome.[86,91,93,95] Chromosomes are large and YAC libraries are available. Although of great potential benefit in many plant species, analysis of plant chromosomes is most common in grasses.[96] For many plant species, chromosomes are small, satisfactory chromosome preparations can be difficult to obtain,[88,97,98] and YAC or BAC libraries are not yet available.

## 2.6 DEFINED-SEQUENCE PCR AMPLIFICATION SYSTEMS

### 2.6.1 MICROSATELLITES

"Satellite" or "repetitive element" DNA was first observed when eukaryotic genomic DNA was subjected to isopycnic cesium chloride density gradient centrifugation. Distinctive "bands" of DNA of lesser or greater density than the bulk of the genomic DNA are frequently observed.[99] The sequences comprising these most commonly lower-than-average density (AT-rich) genomic elements tend to be repetitive and often derived from centromeric heterochromatin. Their repetitive nature has limited their utility as intraspecific genetic markers, although carefully selected cloned satellite sequences have proven informative and useful.[100]

Most well-developed crop plant maps have been developed through the brute-force application of RFLP analysis.[101-104] Significant gaps exist in many of these maps, and extensive efforts to fill these marker-poor regions by randomly adding additional RFLP markers has often proven ineffective (for example, see the >30cM gap on barley chromosome 7L). Ellegren and Basu[105] faced a similar problem with marker-poor porcine chromosome 18. Using a primer sharing homology with a known short-interspersed-nuclear element (SINE), an array of elements were amplified from a flow-sorted chromosome 18 sample. These were then cloned and screened with a $(CA)_{15}$ probe. They identified 11 markers, 8 of which proved polymorphic, and 2 of which were found to actually reside on chromosome 18. While obviously a laborious process, this targeted approach to gap filling proved partially successful, while untargeted approaches have yet to fill many significant (and some practically important) gaps.

Microsatellites are composed of tandem repeats of one to six nucleotides.[53] These short tandem repeat elements have been found to be both abundant and widely distributed throughout the genomes of many higher plants and animals.[106-118] Microsatellite analysis is performed by amplification of genomic DNA using pairs of specific primers flanking tandem arrays of microsatellite repeats.

Products are typically separated on polyacrylamide gels to obtain the resolution needed to separate DNA bands differing by a few nucleotides, although specialized agaroses may suffice. Polymorphism is based on differences in the number of tandem repeats in the amplified regions.

Markers of this class have proven their worth for making a high-density map of the human genome,[27] genomic maps of rat and mosquito,[119,120] and as a primary QTL mapping tool in livestock species, including cattle and swine.[121,122] Microsatellites tend to be remarkably informative, apparently due to the ability of tandem repeat sequences to expand or contract during DNA replication. Microsatellite instability has been associated with several cancers including human ovarian cancer and primary bladder cancer.[123,124] While Jeffreys et al.[49] reported a 2% per generation mutation rate for human minisatellite sequences, this author is unaware of similar estimates of mutation rates gathered for microsatellite sequences from mapping projects.

Microsatellite polymorphisms are generally relatively small in size.[117,125] In order to obtain maximum informativeness, high resolution analysis is generally required. While initially appearing to be a disadvantage, this limitation has been utilized to advantage by livestock genome analysis. Multiplexed microsatellite PCR reaction systems have been developed which can be analyzed by automated DNA sequencing systems. These carefully designed primer sets direct amplification of microsatellite-containing sequences of distinct size ranges. Informativeness, genome-wide distribution, and the ability to evaluate multiple markers of known location make these markers extraordinarily useful in map construction.

A fundamental problem inherent in the development of microsatellite markers is the need for characterized sequences containing microsatellite sequences. While many have been directly identified in sequence databases,[112] plant sequence databases are generally too small to provide many useful starting points. Several techniques have been developed to permit enrichment of genomic libraries for microsatellite-containing sequences. The most commonly utilized approach was developed by Ostrander et al.[126] This 'second-strand protection' approach results in libraries with a frequency of microsatellite-containing clones near 50% (about a 100-fold enrichment). Another novel approach is the triplex affinity capture method developed by Nishikawa et al.[127] In this approach, triplexes formed by the intercalation of a biotinylated $GA_{17}$ probe were captured using streptavidin-coated magnetic beads prior to cloning. This enrichment procedure was reported to result in libraries in which the frequency of microsatellite-containing clones approached 80%.

### 2.6.2 Sequence-Tagged Sites

The term STS was originally coined by Olsen et al.[128] The term refers to the use of PCR primer sets which direct amplification of a sequence from a specific locus. Many synonyms exist for this term, including amplicon length polymorphism (ALP) and specifically amplified polymorphism (SAP). Since these products may be restricted and electrophoretically evaluated, they are conceptually similar to RFLPs.

STS analysis consists of PCR amplification of a specific marker, using a pair of specific (18 to 22 bp in length) PCR primers. Suitable primers are designed on the basis of the sequence of cloned DNA fragments, usually cloned RFLP fragments. Amplification products are then separated by agarose gel electrophoresis and stained with ethidium bromide for visualization. Polymorphism in amplification is generally the result of mutations in the primer binding site, resulting in one allele failing to amplify.

Often, a pair of primers amplifies a fragment which does not differ among two alleles distinguished by RFLP analysis. These apparently monomorphic products may actually be polymorphic in sequence between the primer binding sites. Several methods are available to detect this polymorphism. The most common is to perform test restriction digests on the amplified fragment with different enzymes until an enzyme is found that will produce different-sized fragments that distinguish alleles. If sufficiently important, single base sequence differences may be evaluated using single-strand conformation polymorphism analysis (SSCP),[129] denaturing gradient gel electrophoresis (DGGE),[31] or RNAse protection. Each of these latter techniques is cumbersome.

Many workers have developed PCR primer sets which direct amplification of alternative alleles which differ sufficiently in sequence to make restriction-site analysis relatively simple. While relatively cumbersome to develop, markers of this type efficiently utilize prior RFLP information. Like other PCR methods which are difficult to multiplex, this technique is relatively inefficient in genome mapping. However, these do provide easy-to-discriminate markers which are relatively straightforward to use in marker-assisted selection applications. Fidelity appears to be similar to that of RFLP analysis.

The most comprehensive application of this class of markers has been to help orient YAC and P1 contigs onto human chromosomes. Recently, a 15,086 STS map of the human genome was published.[130] Several groups have utilized markers of this type to characterize germplasm and to perform marker-assisted selection.[131-134] The advantages of this approach is that the robust RFLP databases already developed for many species may be utilized to select markers for development, and that restriction-site polymorphisms are relatively easy to detect using moderate-resolution analytical techniques. The primary disadvantages include the effort required to comprehensively sequence large numbers of RFLP markers, and the time and effort required to survey amplified fragments for restriction-site polymorphisms. While several hundred primer sets of this sort have been developed and published during the past 4 years (see *primers* in the Graingenes database http://wheat.pw.usda.gov/graingenes.html), no comprehensive survey of germplasm using these markers has yet been published. Utilization of these markers remains an *ad hoc* process.

## 2.7 AMPLIFICATION OF UNDEFINED ELEMENTS

### 2.7.1 Random Amplification of Polymorphic DNA (RAPD)

The theoretical underpinnings of RAPD analysis were thoroughly developed by Williams et al.[135] Low-stringency annealing of short (10 base) primers of arbitrary sequence is utilized to direct amplification of products of unknown sequence. Unlike other PCR-based systems, only one primer is used. The short sequence of the primers makes possible a multitude of potential primer binding sites throughout the genome, and efficient amplification of DNA fragments may occur when two primer binding sites occur in close proximity.

In RAPD analysis, numerous short (approximately 300 to 2000 bp) DNA sequences are amplified from small (approximately 10 ng) genomic DNA samples using one primer. Amplified DNA is typically separated electrophoretically on agarose gels, stained with ethidium bromide, and DNA visualized under ultraviolet light, although radiolabeling of bands, separation on polyacrylamide gels, and detection by autoradiography may detect more, weaker amplification products. Variations on this technique include DAF (DNA amplification fingerprinting) (using primers 5 to 8 bases long) and AP-PCR.[136,137]

The advantages of this approach include the limited investment of time and training required to get the technique running. Sets of several hundred primers are available commercially, and no clones or sequence information from the target species is required. A typical amplification may produce from 5 to 15 bands detectable using ethidium bromide, and more using polyacrylamide gels and radioactive detection. Only small amounts of DNA are required, eliminating the need for large numbers of extractions and allowing use of samples of limited availability. This has made simpler the construction of genetic maps of species for which tissue availability is a limiting factor, such as honey bee and mosquito.[138-139] Many thermocyclers accept 96-well sample plates, allowing running of large numbers of samples simultaneously. Polymorphisms are typically detected as the failure of one allele to amplify due to mutations in the primer binding site. RAPD markers are therefore generally dominant in nature, and only approximately 5% are estimated to be codominant.[135]

Disadvantages include the inherently low reliability of low-stringency annealing, inability to discern differences in sequence homology among similarly sized fragments and therefore limited utility across germplasm resources, inefficiency of utilizing unmapped markers for genetic analysis, and possible clustering of markers in some instances. Variations in DNA quality, concentration, and optimal primer concentrations also may contribute to lack of reproducibility in marker patterns.[140] As the drawbacks of RAPD analysis have become clear, this procedure, although initially promising, has declined in usage.

Products of RAPD amplification have been cloned and sequenced, and the sequence information used to develop STS primers.[138,141-143] While having the advantage of ease of use, these markers show limitations in transferrability across populations within species, show high endogenous error rates, especially when used with organisms with large genomes,[101] and (like other low-stringency techniques) are sensitive to variation in DNA contaminants and DNA concentration.

### 2.7.2 Amplified Fragment-Length Polymorphism (AFLP)

DNA analysis without prior sequence information can be potentially extremely useful. AFLP analysis provides an extraordinarily efficient approach to map construction and provides the raw materials needed for STS derivation.

To perform AFLP analysis, total plant DNA is restricted with two enzymes, then the cut ends are ligated to synthetic linkers with overhangs complementary to the cut ends.[144,145] Following ligation, the DNA is 'preamplified' using primer sets 16 to 17 bases in length and homologous to the linkers, but carrying at the 3′ end of each primer one additional arbitrary base. This step reduces later background and increases the amount of DNA available for the succeeding step. Amplification using one $^{32}$P-labeled primer and one nonlabeled primer, each carrying two additional 'selective bases' at the 3′ end, is then performed. The products of amplification are then evaluated on a DNA sequencing gel. Silver-staining of gels is an alternative to radiolabeling.[146]

The advantages conferred by AFLP analysis are many. The technique permits a flexible method to survey a large number of restriction sites for polymorphisms without requiring cloned probes or sequence information. Use of longer primers promotes more reliable amplification than RAPD analysis, selective nucleotides allow examination of a subset of potential markers, and 256 possible combinations of selective nucleotides allows examination of a large number of potential markers. In a typical lane, from 50 to 100 bands may be observed.[144,147] Amplification is less susceptible to artifacts from DNA concentration than is RAPD analysis.[144] Markers found to be linked to genes of interest may be converted to STSs by band recovery, amplification, cloning, and sequencing. Disadvantages include a higher error rate, and consequent map expansion, than RFLP, microsatellite, or STS analysis. In addition, the methodology is more involved than for some types of marker analysis, requiring a double digest, ligation, and two amplifications. Incomplete digests may be an important cause of false markers. Finally, markers are typically dominant, as is the case with RAPD analysis.

As a comparatively recent innovation, AFLP-based work is just beginning to reach the published literature. AFLP has been used to identify markers tightly linked to *Cladosporium* resistance in tomato,[148] mapping the barley genome,[149] and enrichment for markers closely linked to nematode resistance in potato.[150,151]

## 2.8 SUMMARY

The past decade has proven to be a golden era for eukaryotic genetics. Technical advances have provided geneticists with the ability to infer the locations of many important genes, and to manipulate genes and genotypes with a level of precision previously impossible. As maps and markers progress from laboratory curiosities to practically important tools, the technologies utilized for genotype analysis will become progressively more robust and reliable. Currently, in addition to RFLP analysis, the most promising new methods for map construction appears to be a combination of AFLP and multiplexed microsatellite analysis. The tools of preference for marker-assisted selection appear to be RFLP, microsatellite, and STS analysis. As these and other tools are developed, we can look forward to a remarkable future in manipulative and descriptive genetics.

## REFERENCES


1. Hunter, R. L. and Markert, C. L., Histochemical demonstration of isozymes separated by zone electrophoresis in starch gels, *Science,* 125, 1294, 1957.
2. Anderson, N. L. and Anderson, N. G., Microheterogeneity of serum transferrin, haptoglobin, and $\alpha_2$HS glycoprotein examined by high resolution two-dimensional electrophoresis, *Biochem. Biophys. Res. Commun.*, 88, 258, 1979.
3. Millikin, D. E., Plant isozymes: a historical perspective, in *Isozymes in Plant Genetics and Breeding,* Tanksley, S. D. and Orton, T. J., Eds., Elsevier, Amsterdam, 1983, pp. 3–13.
4. Goodman, M. M., Newton, K. J., and Stuber, C. W., Malate dehydrogenase: viability of cytosolic nulls and lethality of mitochondrial nulls in maize, *Proc. Natl. Acad. Sci. U.S.A.*, 78, 1783, 1981.
5. Kleinhofs, A., Kilian, A., Saghai-Maroof, M. A., Biyashov, R. M., Hayes, P., Chen, F. Q., Lapitan, N., Fenwick, A., Blake, T. K., and Kanazin, V., A molecular, isozyme, and morphological map of the barley (*Hordeum vulgare*) genome, *Theor. Appl. Genet.*, 86, 705, 1993.
6. Polacco, M. L., Yerk-Davis, G., Byrne, P., Hancock, D., Coe, E. H., Berlyn, M., and Letovsky, S., The maize genome database, MaizeDB: internet gateway to maize genetics and biology, *Agron. Abstr.*, p. 65, 1995.
7. Gabriel, O., Locating enzymes on gels, *Meth. Enz.*, 22, 578, 1971.
8. Gottlieb, L. D., Conservation and duplication of isozymes in plants, *Science*, 216, 373, 1982.
9. Markert, C. L. and Moller, F., Multiple forms of enzymes: tissue, ontogenetic, and species-specific patterns, *Proc. Natl. Acad. Sci. U.S.A.*, 45, 753, 1959.
10. Scandalios, J. G., Isozymes in development and differentiation, *Annu. Rev. Plant Physiol.*, 25, 225, 1974.
11. Tanksley, S. D. and Orton, T. J., Eds., *Isozymes in Plant Genetics and Breeding,* Elsevier, Amsterdam, 1983.
12. Nielsen, G., The use of isozymes as probes to identify and label plant varieties and cultivars, in *Isozymes: Current Topics in Biological and Medical Research,* vol. 12, Rattazzi, M. M., Scandalios, J. G., and Whitt, G. S., Eds., Alan R. Liss, New York, 1985, pp. 1–32.
13. Weeden, N. F., Applications of isozymes in plant breeding, *Plant. Breed. Rev.*, 6, 11, 1989.
14. Tanksley, S. D. and Rick, C. M., Isozymic gene linkage map of the tomato: applications in genetics and breeding, *Theor. Appl. Genet.*, 57, 161, 1980.
15. Stuber, C. W. and Moll, R. H., Frequency changes of isozyme alleles in a selection experiment for grain yield in maize (*Zea mays* L.), *Crop. Sci.*, 12, 337, 1972.
16. Tanksley, S. D., Medina-Filho, H., and Rick, C. M., The effect of isozyme selection of metric characters in an interspecific backcross of tomato — basis of an early screening procedure, *Theor. Appl. Genet.*, 60, 291, 1981.
17. Stuber, C. W., Goodman, M. M., and Moll, R. H., Improvement of yield and ear number resulting from selection at allozyme loci in a maize population, *Crop Sci.*, 22, 737, 1982.
18. Kendall, J., Separations by the ionic migration method, *Science,* 67, 163, 1928.
19. Laemmli, U. K., Cleavage of structural proteins during the assembly of the head of bacteriophage T4, *Nature,* 227, 680, 1970.
20. Brown, J. W. S., Ersland, D. R., and Hall, T. C., Molecular aspects of storage protein synthesis during seed development, in *The Physiology and Biochemistry of Seed Development, Dormancy, and Germination*, Khan, A. A., Ed., Elsevier, Amsterdam, 1982, pp. 3–42.
21. Marks, M. D. and Larkins, B. A., Analysis of sequence microheterogeneity among zein messsenger RNAs, *J. Biol. Chem.*, 257, 9976, 1982.
22. Osborn, T. C., Alexander, D. C., Sun, S. M., Cardona, C., and Bliss, F. A., Insecticidal activity and lectin homology of arcelin seed protein, *Science,* 240, 207, 1988.
23. Davis, B. J., Disc electrophoresis. II. Method and application to human serum proteins, *Ann. N.Y. Acad. Sci.,* 121, 404, 1964.
24. Cardon, L. R., Smith, S. D., Fulker, D. W., Kimberling, W. J., Pennington, B. F., and DeFries, J. C., Quantitative trait locus for reading disability on chromosome 6, *Science*, 266, 276, 1994.
25. Bliss, F. A. and Brown, J. W. S., Breeding common bean for improved quantity and quality of seed protein, *Plant Breed. Rev.,* 1, 59, 1983.
26. Burow, M. D., Ludden, P. W., and Bliss, F. A., Suppression of phaseolin and lectin in seeds of common bean, *Phaseolus vulgaris* L.: increased accumulation of 54kD polypeptides is not associated with higher seed methionine concentrations, *Mol. Gen. Genet.*, 241, 431, 1993.

27. Murray, J. C., Buetow, K. H., Weber, J. L., Ludwigsen, S., Scherpbier-Heddema, T., Manion, F., Quillen, J., Scheffield, V. C., Sunden, S., Duyk, G. M., Weissenbach, J., Gyapay, G., Dib, C., Morrissette, J., Lathrop, G. M., Vignal, A., White, R., Matsunami, N., Gerken, S., Melis, R., Albertsen, H., Plaetke, R., Odelberg, S., Ward, D., Dausset, J., Cohen, D., and Cann, H., A comprehensive human linkage map with centimorgan density, *Science*, 265, 2049, 1994.
28. Botstein, D., White, R. L., Skolnick, M., and Davis, R. W., Construction of a genetic linkage map in man using restriction fragment length polymorphisms, *Am. J. Hum. Genet.*, 32, 314, 1980.
29. Soller, M. and Beckmann., J. S., Genetic polymorphism in varietal identification and genetic improvement, *Theor. Appl. Genet.*, 67, 25, 1983.
30. Isaac, P. G., *Protocols for Nucleic Acid Analysis by Nonradioactive Probes,* Humana Press, Totowa, NJ, 1994.
31. Myers, R. M., Maniatis, T., and Lerman, L. S., Detection and localization of single base changes by denaturing gradient gel electrophoresis, *Meth. Enz.*, 155, 501, 1987.
32. Gray, M., Charpentier, A., Walsh, K., Wu, P., and Bender, W., Mapping point mutations in the Drosophila rosy locus using denaturing gradient gel blots, *Genetics*, 127, 139, 1991.
33. Tanksley, S. D., Ganal, M. W., Prince, J. P., de Vicente, M. C., Bonierbale, M. W., Broun, P., Fulton, T. M., Giovannoni, J. J., Grandillo, S., Martin, G. B., Messeguer, R., Miller, J. C., Miller, L., Paterson, A. H., Pineda, O., Röder, M. S., Wing, R. A., Wu, W., and Young, N. D., High-density molecular linkage maps of the tomato and potato genomes, *Genetics*, 132, 1141, 1992.
34. Paterson, A. H., Lin, Y.-R., Li, Z., Schertz, K. F., Doebley, J. F., Pinson, S. R. M., Liu, S. C., Stansel, J. W., and Irvine, J. E., Convergent domestication of cereal crops by independent mutations at corresponding genetic loci, *Science,* 269, 1714, 1995.
35. Johansson, M., Ellegren, H., and Andersson, L., Comparative mapping reveals extensive linkage conservation — but with gene order rearrangements — between the pig and the human genomes, *Genomics*, 25, 682, 1995.
36. Whitkus, R., Doebley, J., and Lee, M., Comparative genetic mapping of sorghum and maize, *Genetics*, 132, 1119, 1992.
37. Kowalski, S. P., Lan, T.-H., Feldmann, K. A., and Paterson, A. H., Comparative mapping of *Arabidopsis thaliana* and *Brassica oleracea* chromosomes reveals islands of conserved organization, *Genetics*, 138, 1, 1994.
38. U.S.D.A., Plant Research Genome Participants, USDA plant genome research program, *Adv. Agron.*, 55, 113, 1995.
39. Lee, M., DNA markers and plant breeding programs, *Adv. Agron.*, 55, 265, 1995.
40. Tanksley, S. D. and Hewitt, J., Use of molecular markers in breeding for soluble solids content in tomato — a re-examination, *Theor. Appl. Genet.*, 75, 811, 1988.
41. Stuber, C. W., Lincoln, S. E., Wolff, S. W., Helentjaris, T., and Lander, E. S., Identification of genetic factors contributing to heterosis in a hybrid from two elite maize inbred lines using molecular markers, *Genetics*, 132, 823, 1992.
42. Kowalski, S. P., Lan, T.-H., Feldmann, K. A., and Paterson, A. H., QTL mapping of naturally occurring variation in flowering time of *Arabidopsis thaliana, Mol. Gen. Genet.*, 245, 548, 1994.
43. Bradshaw, H. D., Jr. and Stettler, R. F., Molecular genetics of growth and development in Populus. IV. Mapping QTLs with large effects on growth, form, and phenology traits in a forest tree, *Genetics*, 139, 963, 1995.
44. Severson, D. W., Thathy, V., Mori, A., Zhang, Y., and Christensen, B. M., Restriction fragment length polymorphism mapping of quantitative trait loci for malaria parasite susceptibility in the mosquito *Aedes aegypti, Genetics*, 139, 1711, 1995.
45. Rapp, J. P., Wang, S.-M., and Dene, H., A genetic polymorphism in the renin gene of Dahl rats cosegregates with blood pressure, *Science*, 243, 542, 1989.
46. Paterson, A. H., DeVerna, J. W., Lanini, B., and Tanksley, S. D., Fine mapping of quantitative trait loci using selected overlapping recombinant chromosomes in an interspecies cross of tomato, *Genetics*, 124, 735, 1990.
47. Rinehart, F. P., Ritch, T. G., Deininger, P. L., and Schmid, C. W., Renaturation rate studies of a single family of interspersed repeated sequences in human deoxyribonucleic acid, *Biochem.,* 20, 3003, 1981.
48. Cox, R. D., Copeland, N. G., Jenkins, N. A., and Lehrach, H., Interspersed repetitive element polymerase chain reaction product mapping using a mouse interspecific backcross, *Genomics*, 10, 375, 1991.
49. Jeffreys, A. J., Wilson, V., and Thein, S. L., Hypervariable 'minisatellite' regions in human DNA, *Nature,* 317, 67, 1985.

50. Tokarskaya, O. N., Kalnin, V. K., Panchenko, V. G., and Ryskov, A. P., Genetic differentiation in a captive population of the endangered Siberian crane (*Grus leucogeranus Pall.*), *Mol. Gen. Genet.*, 245, 658, 1994.
51. Tourmente, S., Deragon, J. M., Lafleuriel, J., Tutois, S., Pélissier, T., Cuvillier, C., Espagnol, M. C., and Picard, G., Characterization of minisatellites in *Arabidopsis thaliana* with sequence similarity to the human minisatellite core sequence, *Nucl. Acids Res.*, 22, 3317, 1994.
52. Harris, A. S. and Wright, J. M., Nucleotide sequence and genomic organization of cichlid fish minisatellites, *Genome*, 38, 177, 1995.
53. Litt, M. and Luty, J. A., A hypervariable microsatellite revealed by *in vitro* amplification of a dinucleotide repeat within the cardiac muscle actin gene, *Am. J. Hum. Genet.*, 44, 397, 1989.
54. Vergnaud, G., Mariat, D., Apiou, F., Aurias, A., Lathrop, M., and Lauthier, V., The use of synthetic tandem repeats to isolate new VNTR loci: cloning of a human hypermutable sequence, *Genomics*, 11, 135, 1991.
55. Wong, Z., Wilson, V., Jeffreys, A. J., and Thien, S. L., Cloning of a selected fragment from a human DNA "fingerprint": isolation of an extremely polymorphic minisatellite, *Nucl. Acids Res.*, 14, 4605, 1986.
56. Royle, N. J., Clarkson, R. E., Wong, Z., and Jeffreys, A. J., Clustering of hypervariable minisatellites in the proterminal regions of human autosomes, *Genomics*, 3, 352, 1988.
57. Georges, M., Gunawardana, A., Threadgill, D. W., Lathrop, M., Olsaker, I., Mishra, A., Sargeant, L. L., Schoeberlein, A. A., Steele, M. R., Terry, C., Threadgill, D. S., Zhao, X., Holm, T., Fries, R., and Womack, J. E., Characterization of a set of variable number of tandem repeat markers conserved in Bovidae, *Genomics*, 11, 24, 1991.
58. Stallings, R. A., Ford, A. F., Nelson, D., Torney, D. C., Hildebrand, C. E., and Moyzis, R. K., Evolution and distribution of $(GT)_n$ repetitive sequences in mammalian genomes, *Genomics*, 10, 807, 1991.
59. Rogstad, S. H., Surveying plant genomes for variable number of tandem repeat loci, *Meth. Enz.*, 224, 278, 1993.
60. Depeiges, A., Goubely, C., Lenoir, A., Cocherel, S., Picard, G., Raynal, M., Grellet, F., and Delseny, M., Identification of the most represented repeat motifs in *Arabidopsis thaliana* microsatellite loci, *Theor. Appl. Genet.*, 91, 160, 1995.
61. Jeffreys, A. C., Brookfield, J. F. Y., and Semeonoff, R., Positive identification of an immigration test-case using human DNA fingerprints, *Nature*, 317, 818, 1985.
62. Hamann, A., Zink, D., and Nagl, W., Microsatellite fingerprinting in the genus *Phaseolus*, *Genome*, 38, 507, 1995.
63. Broun, P. and Tanksley, S. D., Characterization of tomato DNA clones with sequence similarity to human minisatellites 33.6 and 33.15, *Plant Mol. Biol.*, 23, 231, 1993.
64. Zhou, A. and Gustafson, J. P., Genetic variation detected by DNA fingerprinting with a rice minisatellite probe in *Oryza sativa*, *Theor. Appl. Genet.*, 91, 481, 1995.
65. Winberg, B. C., Zhou, Z., Dallas, J. F., McIntyre, C.L., and Gustafson, J. P., Characterization of minisatellite sequences from *Oryza sativa*, *Genome*, 36, 978, 1993.
66. Hatada, I., Hayashizaki, Y., Hirotsune, S., Komatsubara, H., and Mukai, T., A genomic scanning method for higher organisms using restriction sites as landmarks, *Proc. Natl. Acad. Sci. U.S.A.*, 88, 9523, 1991.
67. O'Farrell, P. H., High-resolution two-dimensional electrophoresis of proteins, *J. Biol. Chem.*, 250, 4007, 1975.
68. Colas des Francs, C. and Thiellement, H., Chromosomal location of structural genes and regulators in wheat by 2D electrophoresis of ditelosomic lines, *Theor. Appl. Genet.*, 71, 31, 1985.
69. Fisher, S. and Lerman, L., Length-independent separation of DNA restriction fragments in two-dimensional gel electrophoresis, *Cell*, 16, 191, 1979.
70. Uitterlinden, A. G., Slagboom, P. E., Knook, D. L., and Vijg, J., Two-dimensional DNA fingerprinting of human individuals, *Proc. Natl. Acad. Sci. U.S.A.*, 86, 2742, 1989.
71. Hayashizaki, Y., Hirotsune, S., Okazaki, Y., Shibata, H., Akasako, A., Muramatsu, M., Kawai, J., Hirasawa, T., Watanabe, S., Shiroishi, T., Moriwaka, K., Taylor, B. A., Matsuda, Y., Elliott, R. W., Manly, K. F., and Chapman, V. M., A genetic linkage map of the mouse using restriction landmark genome scanning (RLGS), *Genetics*, 138, 1207, 1994.
72. Kuick, R., Asakawa, J.-I., Neel, J. V., Satoh, C., and Hanash, S., High yield of restriction fragment length polymorphisms in two-dimensional separations of human genomic DNA, *Genomics*, 25, 345, 1995.

73. Okuizumi, H., Okazaki, Y., Sasaki, N., Muramatsu, M., Nakashima, K., Fan, K., Ohba, K., and Hayashizaki, Y., Application of the RLGS method to large-size genomes using a restriction trapper, *DNA Res.*, 1, 99, 1994.
74. Hirotsune, S., Shibata, H., Okazaki, Y., Sugino, H., Imoto, H., Sasaki, N., Hirose, K., Okuizumi, H., Muramatsu, M., Plass, C., Chapman, V. M., Tamatsukuri, S., Miyamoto, C., Furuichi, Y., and Hiyashizaki, Y., Molecular cloning of polymorphic markers on RLGS gel using the spot target cloning method, *Biochem. Biophys. Res. Commun.*, 194, 1406, 1993.
75. Ohsumi, T., Okazaki, Y., Hirotsune, S., Shibata, H., Muramatsu, M., Suzuki, H., Taga, C., Watanabe, S., and Hayashizaki, Y., A spot cloning method for restriction landmark genome scanning, *Electrophoresis*, 16, 203, 1995.
76. Hayashizaki, Y., Shibata, H., Hirotsune, S., Sugino, H., Okazaki, Y., Sasaki, N., Hirose, K., Imoto, H., Okuizumi, H., Muramatsu, M., Komatsubara, H., Shiroishi, T., Moriwaka, K., Katsuki, M., Hatano, N., Sasaki, H., Ueda, T., Mise, N., Takagi, N., Plass, C., and Chapman, V. M., Identification of an imprinted U2af binding protein related sequence in mouse chromosome 11 using the RLGS method, *Nat. Genet.*, 6, 33, 1994.
77. Okazaki, Y., Okuizumi, H., Ohsumi, T., Nomura, O., Takada, S., Kamiya, M., Sasaki, N., Matsuda, Y., Nishimura, M., Tagaya, O., Muramatsu, M., and Hayashizaki, Y., A genetic linkage map of the Syrian hamster and localization of cardiomyopathy locus on chromosome 9qa2.1-b1 using RLGS spot-mapping, *Nat. Genet.*, 13, 87, 1996.
78. Hirotsune, S., Hatada, I., Komatsubara, H., Nagai, H., Kuma, K., Kobayakawa, K., Kawara, T., Nakagawara, A., Fujii, K., Mukai, T., and Hayashizaki, Y., New approach for detection of amplification of cancer DNA using restriction landmark genome scanning, *Cancer Res.*, 52, 3642, 1992.
79. Okazaki, Y., Hirose, K., Hirotsune, S., Okuizumi, H., Sasaki, N., Ohsumi, T., Yoshiki, A., Kusakabe, M., Muramatsu, M., Kawai, J., Katsuki, M., and Hayashizaki, Y., Direct detection and isolation of restriction landmark genomic scanning (RLGS) spot DNA markers tightly linked to a specific trait by using the RLGS spot-bombing method, *Proc. Natl. Acad. Sci. U.S.A.*, 92, 5610, 1995.
80. Pinkel, D., Straume, T., and Gray, G. W., Cytogenetic analysis using quantitative, high-sensitivity fluorescence hybridization, *Proc. Natl. Acad. Sci. U.S.A.*, 83, 2934, 1986.
81. Cairns, J., The chromosome of *E. coli*, *Cold Spring Harbor Symp. Quant. Biol.*, 28, 43, 1963.
82. Henderson, A. S., Warburton, D., and Atwood, K. C., Location of ribosomal DNA in human chromosome complement, *Proc. Natl. Acad. Sci. U.S.A.*, 69, 3394, 1972.
83. Prescott, D. M., Bostock, C. J., Hatch, F. T., and Mazrimas, J. A., Localization of satellite DNAs in the chromosomes of the kangaroo rat (*Dipodomys ordii*), *Chromosoma*, 42, 205, 1973.
84. Lichter, P., Tang, C.-J. C., Call, K., Hermanson, G., Evans, G. A., Housman, D., and Ward, D. C., High-resolution mapping of human chromosome 11 by *in situ* hybridization with cosmid clones, *Science*, 247, 64, 1990.
85. Fan, Y.-S., Davis, L. M., and Shows, T. B., Mapping of small DNA sequences by fluorescence *in situ* hybridization directly on banded chromosomes, *Proc. Natl. Acad. Sci. U.S.A.*, 87, 6223, 1990.
86. Moir, D. T., Dorman, T. E., Day, J. C., Ma, N. S.-F., Wang, M.-T., and Mao, J.-I., Toward a physical map of human chromosome 10: isolation of 183 YACs representing 80 loci and regional assignment of 94 YACs by fluorescence *in situ* hybridization, *Genomics*, 22, 1, 1994.
87. Trask, B., Pinkel, D., and van den Engh, G., The proximity of DNA sequences in interphase cell nuclei is correlated to genomic distance and permits ordering of cosmids spanning 250 kilobase pairs, *Genomics*, 5, 710, 1989.
88. Shen, D.-L., Wang, Z.-F., and Wu, M., Gene mapping on maize pachytene chromosomes by *in situ* hybridization, *Chromosoma*, 95, 311, 1987.
89. Ferguson-Smith, M. A., From chromosome number to chromosome map: the contribution of human cytogenetics to genome mapping, in *Chromosomes Today*, Vol. 11, Sumner, A. T. and Chandley, A. C., Eds., Chapman and Hall, London, 1993.
90. Speicher, M. R., Ballard, S. B., and Ward, D. C., Karyotyping human chromosomes by combinatorial multifluor FISH, *Nat. Genet.*, 12, 368, 1996.
91. Driesen, M. S., Dauwerse, J. G., Wapenaar, M. C., Meershoek, E. J., Mollevenger, P., Chen, K. L., Fischbeck, K. H., and van Ommen, G. J. B., Generation and fluorescent *in situ* hybridization mapping of yeast artificial chromosomes of 1p, 17p, 17q, and 19q from a hybrid cell line by high-density screening of an amplified library, *Genomics*, 11, 1079, 1991.

92. Hanson, R. E., Zwick, M. S., Choi, S., Islam-Faridi, M., McKnight, T. D., Wing, R. A., Price, H. J., and Stelly, D. M., Fluorescent *in situ* hybridization of a bacterial artificial chromosome, *Genome*, 38, 646, 1995.
93. Moyzis, R. K., Albright, K. L., Bartholdi, M. F., Cram, L. S., Deaven, L. L., Hildebrand, C. E., Joste, N. E., Longmire, J. L., Meyne, J., and Schwarzacher-Robinson, T., Human chromosome-specific repetitive DNA sequences: novel markers for genetic analysis, *Chromosoma*, 95, 375, 1987.
94. Fuchs, J., Joos, S., Lichter, P., and Schubert, I., Localization of vicilin genes on field bean chromosome II by fluorescent *in situ* hybridization, *J. Hered.*, 85, 487, 1994.
95. Iannuzzi, L., Di Meo, G. P., Gallagher, D. S., Ryan, A. M., Ferrara, L., and Womack, J. E., Chromosomal localization of omega and trophoblast interferon genes in goat and sheep by fluorescent *in situ* hybridization, *J. Hered.*, 84, 301, 1993.
96. Heslop-Harrison, J. S. and Schwarzacher, T., Molecular cytogenetics — biology and applications in plant breeding, in *Chromosomes Today*, Vol. 11, Sumner, A. T. and Chandley, A. C., Eds., Chapman and Hall, London, 1993.
97. Crane, C. F., Price, J. H., Stelly, D. M., and Czeschin, D. C., Jr., Identification of a homeologous chromosome pair by *in situ* DNA hybridization to ribosomal RNA loci in meiotic chromosomes of cotton *(Gossypium hirsutum), Genome* 36, 1015, 1993.
98. Schubert, I., Dolezel, J., Houben, A., Scherthan, H., and Wanner, G., Refined examination of plant metaphase chromosome structure at different levels made feasible by new isolation methods, *Chromosoma,* 102, 96, 1993.
99. Beridze, T., *Satellite DNA* (transl.), Springer-Verlag, Berlin, 1986.
100. Santos-Rosa, H. and Aguliera, A., Isolation and genetic analysis of extragenic suppressors of the hyper-deletion phenotype of the *Saccharomyces cerevisiae hpr1 delta* mutation, *Genetics*, 139, 57, 1995.
101. Kleinhofs, A., Kilian, A., Saghai-Maroof, M. A., Biuashev, R. M., Hayes, P. M., Chen, F. Q., Lapitan, N., Fenwick, A., Blake, T. K., Kanazin, A., Ananiev, E., Dahleen, L., Kudrna, D., Bollinger, J., and Knapp, S. J., A saturated medium density map of the barley genome, *Theor. Appl. Genet.,* 86, 705, 1993.
102. McCouch, S. R., Kochert, G., Yu, Z. H., Wang, Z. Y., Khush, G. S., Coffman, W. R., and Tanksley, S. D., Molecular mapping of rice chromosomes, *Theor. Appl. Genet.,* 76, 815, 1988.
103. Tanksley, D. D. and Bernatzky, R., Molecular markers for the nuclear genome of tomato, *Plant Biol.*, 4, 37, 1987.
104. Helentjaris, T., Slocum, M., Wright, S., Schaefer, A., and Nienhuis, J., Construction of genetic linkage maps in maize and tomato using restriction fragment length polymorphisms, *Theor. Appl. Genet.*, 72, 761, 1986.
105. Ellegren, H. and Basu, T., Filling the gaps in the porcine linkage map: isolation of microsatellites from chromosome 18 using flow sorting and SINE-PCR, *Cytogenet. Cell Genet.,* 71, 370, 1995.
106. Holmes, N. G., Microsatellite markers and the analysis of genetic disease, *Br. Vet. J.,* 150, 411, 1994.
107. Wu, K. S. and Tanksley, S. D., Abundance, polymorphism and genetic mapping of microsatellites in rice, *Mol. Gen. Genet.*, 241, 225, 1003, 1993.
108. Devos, K. M., Bryan, G. J., Collins, A. J., Stephenson, P., and Gale, M. D., Application of two microsatellite seqences in wheat storage proteins as molecular markers, *Theor. Appl. Genet.*, 90, 247, 1995.
109. Broun, P. and Tanksley, S. D., Characterization and genetic mapping of simple repeat sequences in the tomato genome, *Mol. Gen. Genet.,* 250, 39, 1996.
110. Decker, R. A., Moore, J., Ponder, B., and Weber, J. L., Linkage mapping of human chromosome 10 microsatellite polymorphisms, *Genome*, 12, 604, 1992.
111. Wang, Z., Weber, J. L., Zhong, Z., and Tanksley, S. D., Survey of plant short tandem DNA repeats, *Theor. Appl. Genet.,* 88, 1, 1994.
112. Yagil, G., The frequency of two base tracts in eukaryotic genomes, *J. Mol. Evol.,* 37, 123, 1993.
113. Akkaya, M. S., Shoemaker, R. C., Specht, J. E., Bhagwat, A. A., and Cregan, P. B., Integration of simple sequence repeat DNA markers into a soybean linkage map, *Crop Sci.,* 35, 1439, 1995.
114. Smith, D. N. and Devey, M. E., Occurrence and inheritance of microsatellites in *Pinus radiata, Genome,* 37, 977, 1994.
115. Roder, M. S., Plaschke, J., Konig, S. U., Borner, A., Sorrells, M. E., Tanksley, S. D., and Ganal, M. W., Abundance, variability and chromosomal location of microsatellites in wheat, *Mol. Gen. Genet.,* 246, 327, 1995.

116. Morgante, M., Rafalski, A., Biddle, P., Tingey, S., and Olivieri, A. M., Genetic mapping and variability of seven soybean simple sequence repeat loci, *Genome,* 37, 763, 1994.
117. Saghai-Maroof, M. A., Biyashev, R. M., Yang, G. P., Zhang, Q., and Allard, R. W., Extraordinarily polymorphic microsatellite DNA in barley: species diversity, chromosomal locations and population dynamics, *Proc. Nat. Acad. Sci. U.S.A.,* 91, 5466, 1994.
118. Durward, E., Shiu, O. Y., Luczak, B., and Mitchelson, K. R., Identification of clones carrying minisatellite-like loci in an *Arabidopsis thaliana* YAC library, *J. Exp. Bot.,* 46, 271, 1995.
119. Jacob, H. J., Brown, D. M., Bunker, R. K., Daly, M. J., Dzau, V. J., Goodman, A., Koike, G., Kren, V., Kurtz, T., Lernmark, È., Levan, G., Mao, Y-P., Pettersson, A., Pravenec, M., Simon, J. S., Szpirer, C., Szpirer, J., Trolliet, M. R., Winer, E. S., and Lander, E. S., A genetic linkage map of the laboratory rat, *Rattus norvegicus, Nat. Genet.,* 9, 63, 1995.
120. Zheng, L., Benedict, M. Q., Cornel, A. J., Collins, F. H., and Kafatos, F. C., An integrated genetic map of the African human malaria vector mosquito, *Anopheles gambiae, Genetics,* 143, 941, 1996.
121. Georges, M., Nielsen, D., Mackinnon, M., Mishra, A., Okimoto, R., Pasquino, A. T., Sargeant, L. S., Sorenson, A., Steele, M. R., Zhao, X., Womack, J. E., and Hoeschele, I., Mapping quantitative trait loci controlling milk production in dairy cattle by exploiting progeny testing, *Genetics,* 139, 907, 1995.
122. Andersson, L., Haley, C. S., Ellegren, H., Knott, S. A., Johansson, M., Andersson, K., Andersson-Eklund, L., Edfors-Lilja, I., Fredholm, M., Hansson, I., Håkansson, J., and Lundström, K., Genetic mapping of quantitative trait loci for growth and fatness in pigs, *Science,* 263, 1771, 1994.
123. Fujita, M., Enomoto, T., Yoshino, K., Nomura, T., Buzard, G. S., Inoue, M., and Okudaira, Y., Microsatellite instability and alterations in the *hMSH2* gene in human ovarian cancer, *Int. J. Cancer,* 64, 361, 1995.
124. Mao, L., Shoenberg, M. P., Scicchitano, M., Erozan, Y. S., Merlo, A., Schwab, D., and Sidransky, D., Molecular detection of primary bladder cancer by microsatellite analysis, *Science,* 271, 659, 1996.
125. Akkaya, M. S., Bhagwat, A. A., and Cregan, P. B., Length polymorphisms of simple sequence repeat DNA in soybean, *Genetics,* 132, 1131, 1992.
126. Ostrander, E. A., Jong, P. M., Rine, J., and Duyk, G., Construction of small-insert genomic DNA libraries highly enriched for microsatellite repeat sequences, *Proc. Nat. Acad. Sci. U.S.A.,* 89, 3419, 1992.
127. Nishikawa, N., Oishi, M., and Kiyama, R., Construction of a human genomic library of clones containing poly(dG-dA) poly(dT-dC) tracts by $Mg^{++}$-dependent triplex affinity capture, *J. Biol. Chem.,* 270, 9258, 1995.
128. Olson, M., Hood, L., Cantor, C., and Botstein, D., A common language for physical mapping of the human genome, *Science,* 245, 1434, 1989.
129. Iizuka, M., Mashiyama, S., Oshimura, M., Sekiya, T., and Hiyashi, K., Cloning and polymerase chain reaction–single-strand conformation analysis of anonymous *Alu* repeats on chromosome 11, *Genomics,* 12, 139, 1992.
130. Hudson, T. J., Stein, L. D., Gerety, S. G., Ma, J., Castle, A. B., Silva, J., Slonim, D. K., Baptista, R., Kruglyak, L., Xu, S.-H., Hu, X., Colbert, A. M. E., Rosenberg, C., Reeve-Daly, M. P., Rozen, S., Hui, L., Wu, X., Vestergaard, C., Wilson, K. M., Bae, J. S., Maitra, S., Ganiatsas, S., Evans, C. A., DeAndelis, M. M., Ingalls, K. A., Nahf, R. W., Horton, L. T., Jr., Anderson, M. O., Collymore, A. J., Ye, W., Kouyoumjian, V., Zemsteva, I. S., Tam, J., Devine, R., Courtney, D. F., Reynaud, M. T., Nguyen, H., O'Connor, T. J., Fizames, C., Fauré, S., Gyapay, G., Dib, C., Morissette, J., Orlin, J. B., Birren, B. W., Goodman, N., Weissenbach, J., Hawkins, T. L., Foote, S., Page, D. C., and Lander, E. S., An STS-based map of the human genome, *Science,* 270, 1945, 1995.
131. Ghareyazie, B., Huang, N., Second, G., Bennett, J., and Khush, G. S., Classification of rice germplasm. I. Analysis using ALP and PCR-based RFLP, *Theor. Appl. Genet.,* 91, 218, 1995.
132. Chee, P. W., Lavin, M., and Talbert, L. E., Molecular analysis of evolutionary patterns in U genome wild wheats, *Genome,* 38, 290, 1995.
133. Tragoonrung, S., Kanazin, V., Hayes, P. M., and Blake, T. K., Sequence-tagged-site facilitated PCR for barley genome mapping, *Theor. Appl. Genet.,* 84, 1002, 1992.
134. Van Campenhout, S., Vander Stappen, J., Sagi, L., and Volckaert, G., Locus-specific primers for LMW glutenin genes on each of the group 1 chromosomes of hexaploid wheat, *Theor. Appl. Genet.,* 91, 313, 1995.
135. Williams, J. G. K., Kubelik, A. R., Livak, K. J., Rafalski, A., and Tingey, S. V., DNA polymorphisms amplified by arbitrary primers are useful as genetic markers, *Nucl. Acids Res.,* 18, 5631, 1990.

136. Caetano-Anollés, G., Bassam, B. J., and Gresshoff, P. M., DNA amplification fingerprinting using very short arbitrary oligonucleotide primers, *Bio/Technology*, 9, 553, 1991.
137. Welsh, J. and McClelland, M., Fingerprinting genomes using PCR with arbitrary primers, *Nucl. Acids Res.*, 18, 7213, 1990.
138. Dimopoulos, G., Zheng, L., Kumar, V., della Torre, A., Kafatos, F. C., and Louis, C., Integrated genetic map of *Anopheles gambiae*: use of RAPD polymorphisms for genetic, cytogenetic, and STS landmarks, *Genetics*, 143, 953, 1996.
139. Hunt, G. J. and Page, R. E., Jr., Linkage map of the honey bee, *Apis mellifera*, based on RAPD markers, *Genetics*, 139, 1371, 1995.
140. Ehrlich, H. A., Gibbs, R., and Kazazian, H. H., Jr., *Polymerase Chain Reaction*, Cold Spring Harbor Press, Cold Spring Harbor, 1989.
141. Williams, J. G. K., Reiter, R. S., Young, R. M., and Scolnik, P. A., Genetic mapping of mutations using phenotypic pools and mapped RAPD markers, *Nucl. Acids Res.*, 21, 2697, 1993.
142. Burow, M. D., Simpson, C. E., Paterson, A. H., and Starr, J. L., Identification of peanut (*Arachis hypogaea* L.) RAPD markers diagnostic of root-knot nematode (*Meloidogyne arenaria* (Neal) Chitwood) resistance, *Mol. Breeding*, 2, 369, 1996.
143. Talbert, L. E., Blake, N. K., Storlie, E. W., and Lavin, M., Variability in wheat based on low-copy DNA sequence comparisons, *Genome*, 38, 951, 1995.
144. Vos, P., Hogers, R., Bleeker, M., Reijans, M., van de Lee, T., Hornes, M., Frijters, A., Pot, J., Peleman, J., Kuiper, M., and Zabeau, M., AFLP: a new technique for DNA fingerprinting, *Nucl. Acids. Res.*, 23, 4407, 1995.
145. Liscum, M. and Oeller, P., AFLP: not only for fingerprinting, but for positional cloning, http://carnegiedpb.stanford.edu/methods/aflp.html, 1996.
146. Falcone, E., Spadafora, P., deLuca, M., Ruffolo, R., Brancati, C., and de Benedictus, G. DYS19, D12S67, and D1S80 polymorphisms in population samples from southern Italy and Greece, *Human Biol.*, 67, 689, 1995.
147. Lin, J. J., Kuo, J., Ma, J., Saunders, J. A., Beard, H. S., MacDonald, M. K., Kenworth, W., Ude, G. N., and Matthews, B. F., Identification of molecular markers in soybean comparing RFLP, RAPD, and AFLP DNA mapping techniques, *Plant Mol. Biol. Rep.*, 14, 156, 1996.
148. Thomas, C., Vos, P., Zabeau, M., Jones, D. A., Norcott, K. A., Chadwick, P. J., and Jones, J. D. G., Identification of amplified restriction fragment polymorphism (AFLP) markers tightly linked to the tomato *Cf-9* gene for resistance to *Cladosporium fulvum*, *Plant J.*, 8, 785, 1995.
149. Becker, J., Vos, P., Kuiper, M., Salamini, F., and Heun, M., Combined mapping of AFLP and RFLP markers in barley, *Mol. Gen. Genet.*, 249, 65, 1995.
150. Ballvora, A., Hesselbach, J., Niewohner, J., Leister, D., Salamini, F., and Gebhardt, C., Marker enrichment and high-resolution map of the segment of potato chromosome VII harbouring the nematode resistance gene, *Gro1*. *Mol. Gen. Genet.*, 249, 82, 1995.
151. Folkertsma, R. T., vander Voort, J. N. R., de Groot, K. E., van Zandvoort, P. M., Schots, A., Gommers, F. J., Elder, J., and Bakker, J., Gene pool similarities of potato cyst nematode populations assessed by AFLP analysis, *Mol. Plant-Microb. Interact.*, 9, 47, 1996.

# 3 Mapping Quantitative Trait Loci in Experimental Populations

*Gary A. Churchill and Rebecca W. Doerge*

**CONTENTS**



## 3.1 INTRODUCTION

Mapping genes that control quantitative traits is an important problem in modern plant breeding. In this chapter, the authors examine a statistical framework for making inferences about the effects of quantitative trait loci (QTL) in experimental populations. An experimental population is obtained by crossing inbred parental lines to obtain a population of statistically independent individuals. Examples include, but are not limited to, backcross, F2, and recombinant inbred (RI) populations. The treatment presented here does not apply, without significant modification, to populations with an extended pedigree structure or to samples from natural populations.

The statistical problem of QTL mapping can be viewed as having three components. First is the *detection* of genetic factors that have effects on a trait and are segregating in population. Second is the *location* of QTL relative to marker loci. Third is the *estimation* of the QTL effects and their interactions. These problems are interdependent, but the distinction is useful in clarifying the inferential procedures used in QTL mapping.

The observable data in a typical QTL mapping experiment are trait values and marker phenotypes on each of $n$ plants. In some experiments, additional covariates such as location or time may be available. These can generally be accommodated in a linear model of QTL effects by including additional terms. A general inferential approach that has proven to be useful for QTL mapping is based on the missing data principle. In this setting, the observed data are augmented with additional information that, if available, would simplify the statistical problem. For experimental crosses designed to map QTL, the missing data are the QTL and marker genotypes for each individual. It

0-8493-7686-6/98/$0.00+$.50

will be illustrated how the missing data principle can be applied to QTL mapping problems in a general setting and in two specific instances. This general paradigm can be applied to arbitrarily complex mapping problems to yield either classical or Bayesian inference procedures.[1-3] Once an appropriate model is formulated, any of a number of computational tools[4] can be implemented to obtain solutions to inference problems.

## 3.2 MODELING QTL EFFECTS

### 3.2.1 QTL and Mixtures

Consider a population of plants indexed by $i = 1, \ldots, n$ with quantitative trait values $Y_i$ and marker phenotypes $M_i$. The missing data are $Q_i$, the QTL genotype and $G_i$, the marker genotype. Both the QTL and marker genotypes may consist of one or more loci and may include relative linkage phases as needed. The distinction between marker phenotypes and marker genotypes is necessary in cases where (1) there are dominant markers, (2) the relative phases of multiple loci cannot be determined, or (3) there are marker typing errors. The model will be defined in terms of parameters that represent linkage fractions and QTL effects. We will assume that the relative ordering of marker loci is known. The problem of simultaneously inferring marker order and QTL locations is of some interest but is beyond the scope of the present treatment.

In the simplest setting, we observe a trait value $Y_i = y$ and single diallelic marker

$$M_i = \begin{cases} 0 & \text{absent} \\ 1 & \text{present} \end{cases}$$

for each plant. We assume that the marker phenotype and marker genotype are identical so that $G_i = M_i$. A single QTL with two alleles

$$Q_i = \begin{cases} 0 & \text{low} \\ 1 & \text{high} \end{cases}$$

is assumed to be segregating in the population.

The distribution of the trait within a given QTL genotype class is typically modeled as a normal random variable. It is also possible to consider other distributions such as Poisson for count data or exponential for lifetime data. We will use the notation $\Pr(Y_i = 1 \mid Q_i = q)$ to denote the appropriate density or probability mass function evaluated at y for the known QTL genotype class q. The effect of the QTL is typically modeled as a shift in the mean trait value. If the mean trait value for individuals with $Q_i = 0$ is $\mu_0$ and the mean for individuals with $Q_i = 1$ is $\mu_1$, the effect of the QTL is a shift of magnitude $\Delta = \mu_1 - \mu_0$. Thus $\Pr(Y_i = y \mid Q_i = 0) = f(y)$ and $\Pr(Y_i = y \mid Q_i = 1) = f(y - \Delta)$ for a density function f(). It is also possible to model the effect of the QTL as linear in any monotone function g() of the mean,

$$g(EY_i) = \beta_0 + \beta_1 Q_i. \tag{3.1}$$

Additional terms may be included for dosage and/or dominance effects. Under the normal linear model, g() is typically the identity function. It is suggested the interested reader see Reference 5 for a discussion of the generalized linear model.

The allelic state of the QTL cannot be directly observed. However, we can observe the marker class $M_i$ of each plant. If the QTL and the marker are linked and we let r denote the recombination fraction, i.e., $r = \Pr(Q_i \neq M_i)$, the conditional densities of the trait value within a marker class are also mixtures

$$\Pr(Y_i = y \mid M_i = m) = r^m (1-r)^{1-m} f(y) + r^{1-m}(1-r)^m f(y-\Delta). \tag{3.2}$$

Note that the means of the conditional densities Pr $(Y_i = y \mid M_i = 0)$ and Pr $(Y_i = y \mid M_i = 1)$ will differ by $(1 - 2r)\Delta$. This location change is the key to QTL detection.

We can extend this framework to include multiple QTL. For example define the indicators $Q_{1i}$ and $Q_{2i}$ as above. The joint effects of two QTL on the mean of the trait distribution can be expressed as a linear combination

$$g(EY_i) = \beta_0 + \beta_1 Q_{1i} + \beta_2 Q_{2i} + \beta_3 Q_{1i} Q_{2i}. \tag{3.3}$$

If the QTL genotypes were directly observable, the theory of generalized linear models[5] could be applied directly to make inferences about QTL effects. In practice, we observe markers $M_{1i}$ and $M_{2i}$ that are linked to the QTL and the resulting likelihood is a mixture with four terms. In general, the likelihood will have one mixture component for each distinct QTL genotype class. If we relax the assumption that $G_i = M_i$, the situation becomes even more complex.

### 3.2.2 Augmented Data Likelihood

We can take advantage of the missing data structure of the problem by writing the likelihood of the observed data as a mixture over the missing QTL and marker genotype classes as follows

$$\prod_{i=1}^{n} \Pr(Y_i, M_i) = \prod_{i=1}^{n} \sum_{G_i} \sum_{Q_i} \Pr(Y_i, M_i, G_i, Q_i) \tag{3.4}$$

$$= \prod_{i=1}^{n} \sum_{G_i} \sum_{Q_i} \Pr(Y_i, M_i \mid G_i, Q_i) \Pr(G_i, Q_i) \tag{3.5}$$

$$= \prod_{i=1}^{n} \sum_{G_i} \sum_{Q_i} \Pr(Y_i \mid G_i, Q_i) \Pr(M_i \mid G_i, Q_i) \Pr(G_i, Q_i) \tag{3.6}$$

$$= \prod_{i=1}^{n} \sum_{G_i} \sum_{Q_i} \Pr(Y_i \mid Q_i) \Pr(M_i \mid G_i) \Pr(G_i, Q_i). \tag{3.7}$$

Dependence on a vector $\theta$ of model parameters is implicit throughout. The likelihood is expressed as a mixture in Equation 3.4 and factored using the definition of conditional probability in Equation 3.5. Conditional independence of the trait value and the marker phenotype is assumed in Equation 3.6 which seems reasonable in most cases. Conditional independence of $Y_i$ and $G_i$ given $Q_i$ is used to derive Equation 3.7. This assumption may be questionable in some cases. For example, if there are additional loci in the genome that affect the trait value distribution and are linked to the marker(s), this conditional independence will not hold. We will proceed by making this assumption but note that it may be worthwhile to pursue models which do not.

Using this factorization of the likelihood, there are three components of the model that must be specified.

1. The conditional distribution of the trait value given the QTL genotype Pr $(Y_i \mid Q_i)$ may be taken to be any distribution. Restricting attention to the (very broad) class of exponential family distributions, however, will allow us to take full advantage of the theory of

generalized linear models.[5] In practice, most QTL analyses assume a normal distribution. Other distributions may be more appropriate in some cases. An example with a multivariate normal trait distribution is considered below.

2. The conditional distribution of marker phenotypes given marker genotypes Pr $(M_i \mid G_i)$ will be multinomial on the marker phenotype classes. In some cases, it will be degenerate with all of the class probabilities equal to 0 except one that is equal to 1. Nondegenerate cases arise when there are untyped markers or multiple markers of unknown relative phase. Lander and Green[6] discuss the problem of restoring the missing genotype data. Another interesting nondegenerate case arises when marker typing errors are introduced into the model.[7]
3. The joint distribution of the QTL and marker genotypes Pr $(G_i, Q_i)$ reflects the segregation process that gave rise to the experimental population. The classes are discrete and individuals are assumed to be independent, thus the distribution will be multinomial with class probabilities defined as functions of linkage fractions. For simple designs, e.g., a backcross, these are readily computed. More complex designs can also be accommodated. See Fisch[3] for an example. In principle, segregation distortion and/or crossover interference could be introduced into the model by modifying this term.

The augmented data likelihood

$$\prod_{i=1}^{n} \Pr(Y_i, M_i, G_i, Q_i) = \prod_{i=1}^{n} \Pr(Y_i \mid Q_i) \Pr(M_i \mid G_i) \Pr(G_i, Q_i) \tag{3.8}$$

is the product of these three terms and will, in general, take the form of a generalized linear model. Once these components of the model are defined, augmented data methods such as EM[8] or Markov chain Monte Carlo[9] can be applied. An example using an EM algorithm is provided.

## 3.3 INFERENCE PROBLEMS

### 3.3.1 Detecting QTL Effects

We first consider the problem of detecting QTL effects at a fixed point in the genome. We refer to a location in the genome at which the test statistic is calculated as an analysis point. In a single marker analysis, all of the analysis points are markers. If analysis points between markers are used, the analysis is an interval analysis. The maximal value (over all analysis points in the genome) of the test statistics can be used as an overall test for QTL effects.

There are three hypotheses relevant to the QTL detection problem[10] these being

$H_0^1 : \ \Delta = 0$; no QTL is present

$H_0^2 : \ r = 1/2, \Delta > 0$; a QTL is present but is not linked to the marker

$H_A : \ r < 1/2, \Delta > 0$; a QTL is present and is linked to the marker.

There are two types of errors that can occur in the QTL detection problem. A type I error occurs when no linked QTL exists but we (incorrectly) declare that QTL are present. A type II error occurs when there are linked QTL but we fail to detect them. The type I error rate is set by the experimenter. The type II error rate is then a function of sample size and the magnitude of QTL effects. Criterion for setting the type I and type II rates will vary depending on the application. Further discussion of this issue is presented by Lander and Schork.[11] An alternative Bayesian approach is discussed by Hoeschele and van Raden.[10]

Procedures for detecting a QTL are typically based on a statistic that has some power to detect a shift in the trait means between classes of individuals as defined by a marker or marker interval. For a given density f(), the likelihood ratio test can be computed to compare $H_A$ to either of the two null hypotheses. However, this test can present some computational and analytic difficulties (e.g., see Reference 12). A number of approaches to testing for QTL effects have been presented in the literature.[10,13-17] Most of these approaches are based on regression or likelihood methods. While many discussions have arisen as to which test statistic is "best", in the end the key issues are power to detect QTL and robustness of the procedures to model assumptions. In practice, a t-test or an ANOVA F-test is often used.

The problem that presents itself is that of obtaining an appropriate critical value for the test statistic. This choice determines the type I error rate of the test. The defining feature of a critical value is that, under the assumptions of no QTL effects ($H_0^1$) or no linked QTL ($H_0^2$), the value of the test statistic should exceed the critical value with probability not to exceed some nominal level $\alpha$ (e.g., $\alpha = 0.05$). A permutation based method for determining an appropriate critical value has been described.[18] Individuals in the experiment are indexed from 1 to n. The data are shuffled by computing a random permutation of the indices 1,…,n and assigning the ith trait value to an individual whose index is given by the *i*th element of the permutation. The shuffled data are then analyzed for QTL effects. The resulting test statistics are stored and the entire procedure (shuffling and analysis) is repeated N times. At the end of this process we will have stored the results of QTL analyses on N shuffled data sets. Two types of threshold values can be estimated from these results. The first is a comparisonwise threshold that can be estimated separately for each analysis point and provides a $100( - \alpha)\%$ critical value for the test at that point. The second is an experimentwise threshold that provides an overall $100(1 - \alpha)\%$ critical value that is valid simultaneously for all analysis points. Results of the QTL analysis on the original data can be compared to these critical values to determine statistical significance. Alternative approaches exist to compute critical values.[14,19,20]

### 3.3.2 Locating QTL

The detection and location problems are closely connected. In a typical mapping experiment, hundreds of markers may be available and tests will be carried out at each marker. If the markers are organized into a map, tests may also be carried out at analysis points in the intervals between markers. The location of the QTL will be inferred by identifying the marker(s) that is (are) most strongly associated with the trait. For a single QTL model, the analysis point that achieves the maximal value of the test statistic is a reasonable estimate of QTL location. When a trait is controlled by multiple QTL, as is typically the case, the problem becomes more complex. This is an active area of research. See Doerge and Churchill[21] and Satagopan et al.,[9] for current approaches to this problem.

### 3.3.3 Estimating QTL Effects

Maximum likelihood parameter estimates can be obtained by the following algorithm, a special case of the EM algorithm.[8] Starting with an initial estimate of the parameter $\theta^{(0)}$, iterate the following two steps.

*E-step.* Compute E ($Q_i$, $G_i$ | $Y_i$, $M_i$) using the current estimate $\theta^{(p)}$. The genotypes $G_i$ and $Q_i$ can be represented as indicator vectors, thus the desired expectations follow directly from the conditional probability density

$$\begin{aligned} \Pr(Q_i, G_i \mid Y_i, M_i) &\propto \Pr(Y_i, M_i \mid Q_i, G_i)\Pr(Q_i, G_i) \\ &= \Pr(Y_i \mid Q_i)\Pr(M_i \mid G_i)\Pr(Q_i, G_i) \end{aligned} \tag{3.9}$$

The constant of proportionality $\Sigma_{G_i} \Sigma_{Q_i} \Pr(Y_i, M_i \mid Q_i, G_i)\Pr(Q_i, G_i)$, will be tractable for single markers or small sets of markers. For large sets of markers more elaborate algorithms may be required.[6] For an alternative approach to the estimation problem see Hoeschele and van Raden.[2]

*M-step.* Obtain new parameter estimates $\theta^{(p+1)}$ replacing $Q_i$ and $G_i$ by their conditional expectations in the augmented data likelihood. For exponential family distributions, the estimation becomes a standard problem in generalized linear models.[5]

The E-step and the M-step are iterated until convergence is obtained in the parameter estimates. A number of well-placed starting values should be tested to ensure that convergence to a global maximum has been obtained.

## 3.4 EXAMPLES

### 3.4.1 SINGLE QTL IN A BACKCROSS POPULATION

We first consider the problem of estimating the recombination fraction between a single marker locus A and a quantitative trait locus Q in a backcross population. The genotypic state of a backcross individual i is specified by two indicator functions for the presence/absence of the nonrecurrent parental allele,

$$Q_i = \begin{cases} 0 & \text{absent} \\ 1 & \text{present} \end{cases} \quad \text{and} \quad A_i = \begin{cases} 0 & \text{absent} \\ 1 & \text{present.} \end{cases}$$

The marker phenotype and marker genotype are identical in this design so the component $\Pr(M_i \mid G_i)$ can be dropped from the model. Let r denote the probability of a recombination between Q and A per chromosome per generation and assume regular Mendelian segregation. The linkage and segregation component of the model is specified by enumerating the four possible genotype configurations and counting recombination events. Thus

$$\begin{aligned} \Pr(A=1,Q=1) &= \Pr(A=0,Q=0) = (1-r)/2 \\ \Pr(A=1,Q=0) &= \Pr(A=0,Q=1) = r/2. \end{aligned} \tag{3.10}$$

We will assume that the trait distributions $\Pr(Y_i \mid Q_i)$ are normal within each QTL genotype class and that the classes have a common variance. Thus

$$Y_i \sim N(\mu_i, \sigma^2) \tag{3.11}$$

where

$$\mu_i = \begin{cases} \nu_0 & \text{if } Q_i = 0 \\ \nu_1 & \text{if } Q_i = 1. \end{cases} \tag{3.12}$$

For identifiability, we assume $\nu_0 \neq \nu_1$.

This is the standard QTL model for a backcross population. Some generalizations are immediately available to us in the present framework. First, the assumption of common variance $\sigma^2$ can be relaxed with only minor changes to the analysis below. This is important as in practice both the mean and the variance of a trait may be affected by the QTL. Second, the assumption of a normal distribution within genotype classes can be replaced with any distribution. Modifications to the analysis below will be relatively minor provided we stay within the class of exponential family distributions. A number of other generalizations are possible. For example non-Mendelian segregation could be introduced as an additional parameter in the genotype class distribution, replacing the factor 1/2 in Equation 3.10.

*M-step.* If the QTL states $Q_i = q_i$ were known for each plant, we could obtain simple direct estimates of all model parameters by maximizing the augmented data likelihood,

$$\Pr(Y,M,Q) = \theta^x(1-\theta)^{n-x}\prod_{i=1}^{n}\left[\frac{q_i}{\sigma}\phi\left(\frac{y_i - v_1}{\sigma}\right) + \frac{1-q_i}{\sigma}\phi\left(\frac{y_i - v_0}{\sigma}\right)\right], \tag{3.13}$$

where

$$x = \sum_{i=1}^{n} q_i(1-a_i) + (1-q_i)a_i, \tag{3.14}$$

$a_i$ is the observed marker state, $y_i$ is the observed trait value, and $\phi()$ is the standard normal density function.

The augmented data maximum likelihood estimators are

$$\begin{aligned}
\hat{r} &= x/n \\
\hat{v}_0 &= \sum_{i=1}^{n}(1-q_i)y_i \Big/ \sum_{i=1}^{n}(1-q_i) \\
\hat{v}_1 &= \sum_{i=1}^{n} q_i y_i \Big/ \sum_{i=1}^{n} q_i \\
\hat{\sigma}^2 &= \sum_{i=1}^{n}(y_i - \hat{\mu}_i)^2 / n
\end{aligned} \tag{3.15}$$

where $\hat{\mu}_i = (1 - q_i)\,\hat{v}_0 + q_i\,\hat{v}_1$.

*E-step.* In this problem because the marker phenotype and genotype are identical, we compute the conditional expectation of the QTL genotype state given the observed phenotype, marker genotype, and the current estimate of the model parameters,

$$E(Q_i \mid Y_i = y, A_i = a) = \frac{r^{1-a}(1-r)^a\phi\left(\frac{y - v_1}{\sigma}\right)}{r^{1-a}(1-r)^a\phi\left(\frac{y - v_1}{\sigma}\right) + r^a(1-r)^{1-a}\phi\left(\frac{y - v_0}{\sigma}\right)}. \tag{3.16}$$

*Two markers.* We can extend this model to the case of two markers

$$A_i = \begin{cases} 0 & \text{absent} \\ 1 & \text{present} \end{cases} \quad \text{and} \quad B_i = \begin{cases} 0 & \text{absent} \\ 1 & \text{present.} \end{cases}$$

There are three possible arrangements of two markers and one QTL when all three are linked. However, it is only necessary to consider the case A-Q-B where the QTL is located in the interval between the two markers. This is because, with the assumption of independence between recombination events in different intervals, the cases Q-A-B and A-B-Q reduce to the single marker problems Q-A and B-Q, respectively. Let $r_A$ be the recombination fraction between the QTL and marker A and let $r_B$ be the recombination fraction between the QTL and marker B. The joint distribution of genotypes at A, Q, and B is

$$\Pr(Q_i = 0,\ A_i = 0,\ B_i = 0) = \Pr(Q_i = 1,\ A_i = 1,\ B_i = 1) = \frac{1}{2}(1 - r_A)(1 - r_B)$$

$$\Pr(Q_i = 0,\ A_i = 0,\ B_i = 1) = \Pr(Q_i = 1,\ A_i = 1,\ B_i = 0) = \frac{1}{2}(1 - r_A)r_B$$

$$\Pr(Q_i = 0,\ A_i = 1,\ B_i = 0) = \Pr(Q_i = 1,\ A_i = 0,\ B_i = 1) = \frac{1}{2}r_A(1 - r_B)$$

$$\Pr(Q_i = 0,\ A_i = 1,\ B_i = 1) = \Pr(Q_i = 1,\ A_i = 0,\ B_i = 0) = \frac{1}{2}r_A r_B.$$

*M-step.* If the QTL genotypes are known we can write the augmented data likelihood in a form similar to Equation 3.13 and obtain maximum likelihood estimates from the augmented data.

$$\hat{r}_A = \frac{x_A + x_Q}{n} \quad \text{and} \quad \hat{r}_B = \frac{x_B + x_Q}{n} \tag{3.17}$$

where

$$x_A = \sum_{i=1}^{n} a_i(1 - b_i)(1 - q_i) + (1 - a_i)b_i q_i$$

$$x_B = \sum_{i=1}^{n} (1 - a_i)b_i(1 - q_i) + a_i(1 - b_i)q_i \tag{3.18}$$

$$x_Q = \sum_{i=1}^{n} (1 - a_i)(1 - b_i)q_i + a_i b_i(1 - q_i).$$

maximum likelihood estimates of $v_0$, $v_1$ and $\sigma^2$ are obtained as in Equation 3.15.

*E-step.* The conditional expectations follow from

$$\begin{aligned}\Pr(Q_i \mid Y_i, A_i, B_i) &\propto \Pr(Q_i)\Pr(Y_i, A_i, B_i \mid Q_i) \\ &= \Pr(Q_i)\Pr(Y_i \mid Q_i)\Pr(A_i \mid Q_i)\Pr(B_i \mid Q_i)\end{aligned} \tag{3.19}$$

where we have assumed conditional independence of A, B, and Y given Q. These are

$$E(Q_i \mid Y_i = y,\ A_i = a,\ B_i = b) =$$

$$\frac{r_A^{1-a}(1 - r_A)^a r_B^{1-b}(1 - r_B)^b \phi\left(\frac{y - v_1}{\sigma}\right)}{r_A^{1-a}(1 - r_A)^a r_B^{1-b}(1 - r_B)^b \phi\left(\frac{y - v_1}{\sigma}\right) + r_A^a(1 - r_A)^{1-a} r_B^b(1 - r_B)^{1-b} \phi\left(\frac{y - v_0}{\sigma}\right)} \tag{3.20}$$

*Multiple markers.* Consider a map with m markers in a known order. For each interval in the map we can compute the maximized log-likelihood assuming a QTL is located in that interval using the EM algorithm. This approach is valid if we assume one QTL and independence of recombination events within each plant. Thus, for the kth interval we obtain a maximized likelihood $L_k(\hat{\theta})$. Likelihoods for distinct intervals can be compared and the interval with the highest likelihood is a maximum likelihood estimate of the QTL location.

### 3.4.2 Two QTL in an Intercross Population

We consider a trait Y with distribution determined by two QTL. In the augmented data setting where the QTL genotype is known and the conditional trait distributions are normal with common variance, the estimation problem is equivalent to the standard two-way ANOVA. In an intercross population there are three possible genotypes at each locus. For linked loci, we must also consider the relative phases of these loci. For unlinked QTL, the possible genotypes can be represented as a pair of indicator vectors $Q_i = (Q_{i1}, Q_{i2})$,

$$Q_{ij} = \begin{cases} (1,0,0)^T & \text{homozygous 11} \\ (0,1,0)^T & \text{heterozygous 12} \\ (0,0,1)^T & \text{homozygous 22} \end{cases}$$

for $i = 1,\ldots,n$, $j = 1,2$. The two QTL in this system are assumed to be unlinked, thus there are nine possible values for $Q_i$.

The model presented here was motivated by work on the expression of acylsugars in tomatoes derived from an intercross between a wild species and a cultivar.[22] The observed phenotype for each plant consists of a bivariate observation $Y_i = (Y_{i1}, Y_{i2})$, where $Y_{i1}$ is the total acylsugar detected in a standard assay and $Y_{i2}$ is the proportion of glucose acylsugar among the total acylsugar. Data from a population of 196 plants are shown in Figure 3.1.

The following genetic model is proposed as a working hypothesis. There are two major QTL in this system plus other modifiers that may be genetic or environmental. We assume that the two QTL are unlinked and that any additional genetic modifiers are unlinked to the two QTL. The first

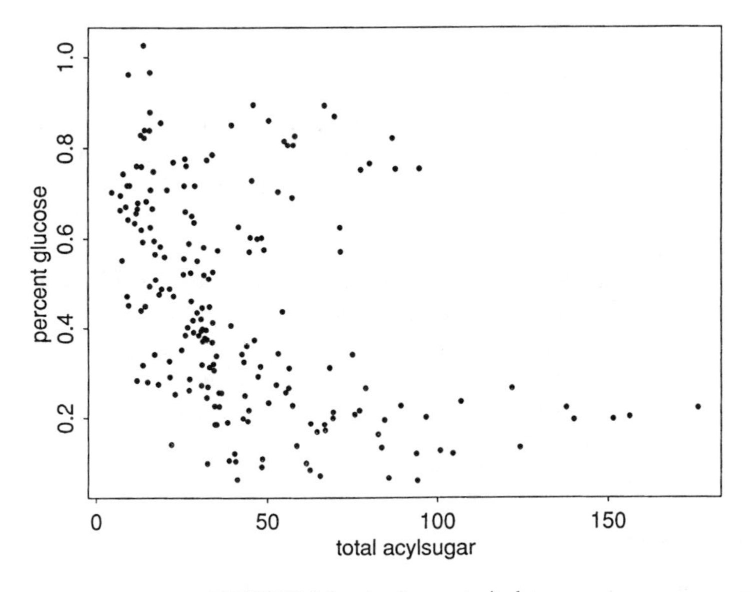


**FIGURE 3.1** Acylsugar trait data.

QTL affects the level of acylsugar production. The high production allele is dominant to low production allele. The second QTL affects the proportion of glucose among the high level producers. It has no effect on the low level producers. The low glucose allele is dominant to the high glucose allele. The genetic model is summarized in the following table.

| | | $Q_1$ 11 | $Q_1$ 12 | $Q_1$ 22 |
|---|---|---|---|---|
| $Q_2$ | 11 | I | II | II |
| | 12 | I | II | II |
| | 22 | I | III | III |

The conditional trait distributions Pr $(Y_i \mid Q_i)$ are

$$\begin{aligned} Y \mid \text{group I} &\sim N_2(\mu_1, \Sigma_1) \\ Y \mid \text{group II} &\sim N_2(\mu_2, \Sigma_2) \\ Y \mid \text{group III} &\sim N_2(\mu_3, \Sigma_3) \end{aligned} \tag{3.21}$$

where

$$\Sigma_i = \begin{bmatrix} \sigma_{i1} & 0 \\ 0 & \sigma_{i2} \end{bmatrix}.$$

We assume that we have located two markers, $M_i = (M_{1i}, M_{2i})$, such that the first is linked to $Q_{i1}$ and the second is linked to $Q_{i2}$. The distributions Pr $(M_i \mid G_i)$ and Pr $(G_i, Q_i)$ follow from standard intercross genetics. There are 100 (phase-known) genotypes to enumerate, 10 at each marker-QTL pair in all combinations. The enumeration is straightforward, and is not shown here.

## 3.5 CONCLUSION

We have described a general approach to modeling the effects of QTL and developed inference procedures based on this model. Model based methods are always subject to the criticism that the models are not correct. We acknowledge that some of the assumptions required here oversimplify the reality of quantitative genetics. However, with this formulation it is clear where the various assumptions enter into the analysis and which components of the model should be modified. The modeling approach requires the researcher to specify the number of QTL involved in a system and the nature of any interactions between multiple QTL. We do not consider this to be a disadvantage. The model, and hence the inference procedures, are specifically directed to the problem at hand. It is possible with this approach to compare various alternative models to test hypotheses about the genetic system and also to make checks on the adequacy of the model.

The QTL mapping problem presents new and interesting statistical challenges, many of which remain unsolved. Some open problems of practical importance include estimation of the number of (major) QTL in a genetic system, development of efficient algorithms for localizing multiple QTL, and modeling genetic interactions.

## ACKNOWLEDGMENTS

The authors are grateful to Martha Mutschler for providing the data used in the second example.

## REFERENCES


1. Hoeschele, I. and van Raden, P. M., Bayesian analysis of linkage between genetic markers and quantitative loci I: prior knowledge, *Theor. Appl. Genet.,* 85, 953–960, 1993.
2. Hoeschele, I. and van Raden, P. M., Bayesian analysis of linkage between genetic markers and quantitative trait loci II: combining prior knowledge with experimental evidence, *Theor. Appl. Genet.,* 85, 946–952, 1993.
3. Fisch, R. D., Ragot, M., Gay, G., A generalization of the mixture model in the mapping of quantitative trait loci for progeny from a biparental cross of inbred lines, *Genetics,* 143, 571–577, 1996.
4. Tanner, M. A., *Tools for Statistical Inference,* Springer, New York, 1991.
5. McCullagh, P. and Nelder, J. A., *Generalized Linear Models,* 2nd ed., Chapman and Hall, London, 1989.
6. Lander, E. S. and Green, Construction of multilocus genetic maps in humans, *Proc. Natl. Acad. Sci. U.S.A.,* 84, 2363–2367, 1987.
7. Lincoln, S. and Lander, E., Systematic detection of errors in genetic linkage data, *Genomics,* 14, 604–610, 1992.
8. Dempster, A. P., Laird, N. M., and Rubin, D. B., Maximum likelihood from incomplete data via the EM algorithm, *J. R. Stat. Soc. Ser. B,* 39, 1–22, 1977.
9. Satagopan, J. M., Yandell, B. S., Newton, M. S., and Osborn, T. C., A Bayesian approach to detect quantitative trait loci using Markov chain Monte Carlo, *Genetics,* in press.
10. Knott, S. A. and Haley, C. S., Aspects of maximum likelihood methods for the mapping of quantitative trait loci in line crosses, *Genet. Res.,* 60, 139–151, 1992.
11. Lander, E. S. and Schork, N. J., Genetic dissection of complex traits, *Science,* 265, 2037, 1994.
12. Hartigan, J. A., A failure of likelihood asymptotics for normal distributions, *Proc. Berkeley Conf.,* Vol. II, 1985.
13. Weller, J. I., Maximum likelihood techniques for the mapping and analysis of quantitative trait loci with the aid of genetic markers, *Biometrics,* 42,627–640, 1986.
14. Lander, E. S. and Botstein, D., Mapping Mendelian factors underlying quantitative traits using RFLP linkage maps, *Genetics,* 121, 185–199, 1989.
15. Jansen, R. C., Interval mapping of multiple quantitative trait loci, *Theor. Appl. Genet.,* 79, 583–592, 1993.
16. Zeng, Z.-B., Theoretical basis for separation of multiple linked gene effects in mapping quantitative trait loci, *Proc. Natl. Acad. Sci. U.S.A.,* 90, 10,972–10,976, 1993.
17. Zeng, Z.-B., Precision mapping of quantitative trait loci, *Genetics,* 136, 1457–1468, 1994.
18. Churchill, G. A. and Doerge, R. W., Empirical thresholds values for quantitative trait mapping, *Genetics,* 138, 963–971, 1994.
19. Lander, E. S. and Botstein, D., Corrigendum, *Genetics,* 36, 705, 1994.
20. Rebai, A., Goffinet, B., and Mangin, B., Approximate thresholds of interval mapping tests for QTL detection, *Genetics,* 138, 235–240, 1994.
21. Doerge, R. W. and Churchill, G. A., Permutation tests for multiple loci affecting a quantitative character, *Genetics,* 142, 285–294, 1996.
22. Mutschler, M. A. and Shapiro, Y., *Biochemical Systematics and Ecology,* 1994.
23. Jansen, R. C., A general mixture model for mapping quantitative trait loci by using molecular markers, *Theor. Appl. Genet.,* 85, 252–260, 1993.
24. Jansen, R. C. and Stam, P., High resolution of quantitative traits into multiple loci via interval mapping, *Genetics,* 136, 1447–1455, 1994.

# 4 Computational Tools for Study of Complex Traits

*Ben-Hui Liu*

**CONTENTS**



0-8493-7686-6/98/$0.00+$.50

## 4.1 INTRODUCTION

Quantitative or complex traits are defined traditionally as traits having continuous distribution in contrast with discrete distribution. The trait values are usually obtained by measuring instead of counting. The trait is considered controlled by many genes and each of the genes has a small effect on the trait by the traditional wisdom. However, recent findings using the combination of genomic mapping and traditional quantitative genetics show that a small number of genes can produce a trait with continuous distribution. Searching for genes controlling complex or quantitative traits plays an important role in applying the genomic information to clinical diagnosis, agriculture, and forestry because a large portion of the traits related to human diseases and agronomic importance are quantitative traits. The loci controlling quantitative traits are commonly referred to as QTL (quantitative trait locus). The procedures for finding QTL are called QTL mapping.

The genetics of quantitative traits are more complex than single factor Mendelian traits. These traits are usually controlled by more than a single gene and influenced by environmental effects. Traits controlled by a single gene with incomplete penetrance can be treated as quantitative traits in finding the gene. QTL mapping involves construction of genomic maps and searching relationships between traits and polymorphic markers. A significant association between the traits and the markers may be the evidence of a QTL near the region of the markers. A simple t-test, simple linear regression model, multiple linear regression model, log-linear model, mixture distribution model, nonlinear regression model, and interval test approach using partial regression have been proposed and used to map QTLs.[1-7] For QTL mapping using human populations a sib-pair approach has been used.[8-12] To solve the models, least square, maximum likelihood, and EM algorithms have been used. To carry out the data analysis, computer programs such as MAPMAKER/QTL,[13] QTL-STAT,[14] QTL Cartographer,[15] PGRI,[16] MAPQTL,[17] Map Manager QTL,[18] and QGENE[19] are available.

Commonly used approaches for QTL mapping, such as single-marker t-test, simple linear regression, interval mapping, and composite interval mapping, etc. are all single-QTL models. If there is no QTL interaction in the model, then the model is considered as a single-QTL model. The number of markers in the single-QTL models can vary from one to a large number. However, only one or two markers are directly related to the putative QTL and the other markers are used in the models to control genetic background effects and sampling errors. QTL mapping has been recognized as a multiple test problem. The tests are not independent among marker loci because of the linkage relationship and possible gene interactions. Traditional adjustment on test statistic cannot be applied to QTL mapping. Permutation approaches can be used to determine the empirical distributions of the statistics.[20]

There is a rich literature on QTL mapping. Table 4.1 provides a partial list. However, almost all the literature related to QTL mapping can be traced from the references. Papers by Tanksley[21] and Lander and Schork[22] and an issue of *Trends in Genetics* (December 1995, Vol. 11, No. 12, pp. 463-524)[23-27] are good sources for information on QTL mapping.

**TABLE 4.1**
**A List of the Key References for QTL Mapping**

| Methodology | Author | Ref. |
|---|---|---|
| Historical | Sturtevant 1913; Sax 1923; Penrose 1938 | 29, 30, 31 |
| Single-marker: linear model | Soller et al. 1976; Edwards et al. 1987; Stuber et al. 1987 | 32, 33, 34 |
| Single-marker: likelihood | Weller 1986 | 35 |
| Interval mapping: regression | Knapp et al. 1990; Knott & Haley 1992; Martinez & Curnow 1992; Jansen 1992 & 1993 | 36, 37, 38, 39, 40 |
| Interval mapping: likelihood | Lander & Botstein 1989; Jensen 1993; Lou & Kearsey 1989; Knott & Haley 1992 | 3, 41, 42, 43 |
| Interval mapping: composite | Jansen 1993; Rodolphe & Lefort 1993; Zeng 1993 & 1994 | 41, 44, 6, 7 |
| Experimental design | Knapp & Bridges 1990; Knapp 1994; | 36, 45 |
| Multi-QTL | Moreno-Gonzalez 1992; Jansen 1993; Rodolphe & Lefort 1993; Zeng 1993 & 1994 | 46, 41, 44, 6, 7 |
| Sib-pair: single marker | Haseman & Elston 1972; Cockerham & Weir 1983; Lange 1986; Weeks & Lange 1988 | 8, 47, 9, 10 |
| Sib-pair: interval mapping | Fulker & Cardon 1994; Cardon & Fulker 1994 | 11, 12 |
| Sib-pair: multi-locus | Weeks & Lange 1992; Fulker et al. 1995 | 48, 49 |
| Resampling | Churchill & Doerge 1994 | 20 |
| QTL-environment interactions | Hayes et al. 1993; Knapp 1994; Jiang & Zeng 1995 | 28, 45, 50 |
| Statistical power and resolution | Soller et al. 1976; Rebai et al. 1994 & 1995; Lander & Botstein 1989; Zeng 1993 & 1994; Boehnke 1994; Jansen & Stam 1994; Kruglyak & Lander 1995 | 32, 51, 52, 3, 6, 7, 53, 54, 55 |
| | **Computer Software** | |
| MAPMAKER/QTL | Lander et al. 1987; Lander & Botstein 1989 | 13, 3 |
| QTLSTAT | Knapp et al. 1992 | 4 |
| LINKAGE | Terwilliger & Ott 1994 | 56 |
| PGRI | Liu 1995 | 16 |
| QTL Cartographer | Basten 1996 | 15 |
| MAPQTL | Van Ooijen & Maliepaard 1996 | 17 |
| Map Manager QT | Manly & Cudmore 1996 | 18 |
| QGENE | Tanksley & Nelson 1996 | 19 |
| | **Experiments** | |
| Drosophila | Mackay 1995 | 23 |
| Mice | Frankel 1995; Schork et al. 1995 | 24 |
| Cattle | Haley 1995 | 25 |
| Human | Lander & Schork 1994 | 22 |
| Maize | Stuber 1995 | 26 |
| Tomato | Nienhuis et al. 1987; Paterson et al. 1988 & 1991 | 57, 58, 59 |
| Rice | McCouch & Doerge 1995 | 27 |
| Barley | Hayes et al. 1993 | 28 |
| Trees | Groover et al. 1994; Bradshaw & Stettler 1995; Grattapaglia et al. 1995 | 60–61 |

## 4.2 GENETIC MODELS FOR COMPLEX TRAITS

Procedures of QTL mapping have been derived from some specific hypothetical models. These models include the genetic models of the traits, and models for the relationship between the hypothetical QTL and genetic markers. QTL mapping is a process to implement statistical hypothesis tests and parameter estimations for the models using observations on the traits and genetic markers in certain genetic designs. Usually, the mapping populations determine the genetic designs.

**TABLE 4.2**
**Notations for Single-QTL Models in Backcross (Qq × QQ) and F2 (Qq × Qq) Populations**

| Model | | Value |
|---|---|---|
| Backcross Model | QQ | $\mu_1$ |
| | Qq | $\mu_2$ |
| | Genetic effect | $g = 0.5 (\mu_1 - \mu_2)$ |
| F2 Model | QQ | $\mu_1$ |
| | Qq | $\mu_2$ |
| | qq | $\mu_3$ |
| | Additive effect | $a = 0.5 (\mu_1 - \mu_3)$ |
| | Dominance effect | $d = 0.5 (\mu_1 + \mu_3 - 2\mu_2)$ |

### 4.2.1 Single-QTL Model

One QTL mapping strategy is to search the whole genome by hypothesis test on single markers or single genome positions and then to build multiple-QTL model based on the results from single QTL analysis. Certainly, searching the whole genome simultaneously is better than scanning individual points if information content is adequate to do so. However, this has happened rarely. Here, let us focus on a single-QTL model first. The definitions of the gene effects for single-QTL models are same as traditional quantitative genetic definitions.[63,64] The genotypic values for the three genotypes (QQ, Qq, and qq) in an F2 population, which is selfed progeny of a heterozygous Qq, are $\mu_1$, $\mu_2$, and $\mu_3$, respectively, as shown in Table 4.2. The additive and dominance effects are defined as

$$\begin{cases} a = 0.5(\mu_1 - \mu_3) \\ d = 0.5(\mu_1 + \mu_3 - 2\mu_2) \end{cases} \tag{4.1}$$

The additive effect is same as the average effect of the gene-substitution because the expected allelic frequencies for the two alleles are same (0.5) in F2 population. For backcross progeny produced by cross between a heterozygous parent Qq and a homozygous parent QQ, the additive and dominance effects are confounded. The mixed effect is defined as

$$g = 0.5(\mu_1 - \mu_2) \tag{4.2}$$

From Equation 4.1, we have $\mu_2 = 0.5\ (\mu_1 + \mu_3 - 2d)$ and

$$g = 0.5(\mu_1 - \mu_2) = 0.5\mu_1 - 0.25(\mu_1 + \mu_3 - 2d) = 0.5(a + d) \tag{4.3}$$

The genetic effect defined in backcross progeny is a combination of additive and dominance effects.

### 4.2.2 Multiple-Locus Model (A Perfect Model)

Genetic model for a quantitative trait (QT) is usually defined in terms of the number of genes, gene actions, relationship among the genes, and relationship between environments and the gene actions. For gene actions and relationship among the genes, there are additive, dominance, and epistatic genetic effects by classical quantitative genetic definitions. Classical quantitative genetics has been focusing on additive and dominance genetic variation. Epistatic interactions, which is the relationship among the genes, at least relatively high level of epistatic interactions, has been very difficult to be estimated and detected (however, see Chapter 8 for new development). Let us assume that $l$ genes control a QT. For a conventional F2 population, there are four main effects, two additive effects, and dominance effects for each of the locus and there are four epistatic interactions, one additive by additive, one dominance by dominance, and two additive by dominance interactions for a two-locus model. In general, there are

$$2^i\binom{l}{i} = \frac{2^i l!}{i!\,(l-i)!} \tag{4.4}$$

possible $i$-way effects for a $l$-locus model among a total number of genetic effects, which is

$$\sum_{i=1}^{l} 2^i\binom{l}{i} = 3^l - l \tag{4.5}$$

Using matrix notation, the multiple-locus model for a QT can be written as

$$Y = A + D + I + E \tag{4.6}$$

where Y is trait value, A, D, I, and E are additive genetic, dominance genetic, epistatic genetic, and error effects for the trait, respectively. Using notations of Cockerham,[63] components of the matrices are

$$\begin{cases} A = \sum_{i=1}^{n} f_i c_{1i} a_i, \\ D = \sum_{i=1}^{n} f_i c_{2i} d_i \\ I = \sum_{i=2}^{n}\sum_{j=1}^{i-1} f_j c_{1i} c_{1j} a_{ij} + \sum_{i=2}^{n}\sum_{j=1}^{i-1} c_{2i} c_{2j} d_{ij} + \sum_{i=1, i\neq j}^{n}\ \sum_{j=1, j\neq l}^{n} c_{1i} c_{2j} e_{ij} \end{cases} \tag{4.7}$$

and $\varepsilon$ for A, D, I and E, respectively. Where, $a_i$ and $d_i$ are additive and dominance main effects for locus i, $a_{ij}$, $d_{ij}$, and $e_{ij}$ are additive by additive, dominance by dominance, and additive by dominance interactions between loci i and j, respectively. Definitions of the coefficients (dummy variables) are listed in Table 4.3. Equation 4.7 is a complete model for a QT. However, this model cannot be obtained using traditional quantitative genetical approaches. Even for the recent QTL mapping approaches, the model is difficult to obtain when the number of genes is more than two.

## 4.3 STATISTICAL MODELS FOR QTL MAPPING

### 4.3.1 RATIONALE

The relationship between quantitative trait variation and qualitative traits can be modeled using linear models. Qualitative traits usually can be observed more easily and more accurately than quantitative traits, because quantitative traits are usually controlled by a number of genes and the gene effects are usually interactive with the environment. The genetics of qualitative traits usually can be inferred at an individual level. However, the genetics for quantitative traits can be inferred

**TABLE 4.3**
**Definitions of the Coefficients (Dummy Variables) and Gene Effects**

| QTL Genotype | Frequency | $c_1$ | $c_2$ | Effects |
|---|---|---|---|---|
| QQ | $f_1$ | 1 | 1 | a |
| Qq | $f_2$ | 0 | −1 | d |
| qq | $f_3$ | −1 | 1 | −a |

at a population level. Scientists have tried gaining more understanding on the inheritance of complex traits through searching the relationships between complex traits and simple traits with known genetics, for example relating complex human diseases to blood types or relating economically important traits of field crops with simple morphological traits. The rationale behind these simple experiments is the fundamental basis of QTL mapping. Genetic markers can be considered as traits with simple inheritance.

Through linkage analysis we know the mode of marker inheritance and their genomic locations. Now the question becomes: can we use this large amount of linkage information to infer genetics of quantitative traits? The underlying genetics of finding the relationship between quantitative trait inheritance and the genetic markers are: (1) Genes controlling the quantitative traits are located on the genome, just like simple genetic markers. (2) If the markers cover a large portion of the genome then there is a large chance that some of the genes controlling the quantitative traits are linked with some of the genetic markers. (3) If the genes and the markers are segregating in a genetically defined population, then the linkage relationship among them may be resolved by studying the association between trait variation and marker segregation pattern. Certainly if the genotypes of the genes controlling the traits can be observed through experiments, then the question becomes simple linkage analysis. In practice, the genotypes of the genes cannot be observed — instead what we observe are the continuous trait values.

### 4.3.2 Single Marker Linear Model (Backcross Model)

#### 4.3.2.1 Model

Early work on finding the association between the trait value and marker segregation patterns has been based on linear models, such as

$$y_j = \mu + f(\text{marker}_j) + \varepsilon_j \tag{4.8}$$

where $y_j$ is the trait value for the jth individual in the population, $\mu$ is the population mean, f(marker$_j$) is a function of marker genotype, and $\varepsilon_j$ is the residual associated with the jth individual. The marker genotypes can be treated as classification variables for a t-test or analysis of variance (ANOVA). The marker genotypes can also be coded as dummy variables for regression analysis. Model 4.8 also can be resolved using the likelihood approach by finding the joint distribution of marker genotypes and the putative QTL genotypes.

Single-marker analysis for QTL mapping is a set of procedures to solve Equation 4.8 when the marker function term only involves one segregating marker. The single-marker analysis can be implemented as a simple *t*-test, ANOVA, linear regression, and likelihood ratio test and maximum likelihood estimation. The single-marker analysis has the full feature of QTL mapping. It is a good start not only for learning QTL mapping but also for most of the practical data analysis. Single-marker analysis can be performed using commonly used statistical software, such as SAS. Gene orders and a complete linkage map are not required in terms of the methodology for single-marker

**TABLE 4.4**
**Expected QTL Gentoypic Frequency Conditional on Genotypes of a Nearby Marker in Backcross Populations with no Double Crossover**

| Marker genotype | QTL genotype QQ | QTL genotype Qq | Expected trait value |
|---|---|---|---|
| AA | 1 – r | r | $(1 - r)\mu_1 + r\mu_2$ |
| Aa | r | 1 – r | $r\mu_1 + (1 - r)\mu_2$ |

*Note:* r is recombination frequency between the marker and the QTL. See Table 4.2 for the other notations.

analysis. However, gene orders and linkage maps will help to present the results. The limitations of the single-marker analysis are: (1) The putative QTL genotypic means and QTL positions are confounded. This causes a biased estimator of QTL effects and a low statistical power when linkage map density is low. (2) QTL positions cannot be precisely determined due to the nonindependence among the hypothesis tests for linked markers that confounds with QTL effect and position.

#### 4.3.2.2 Analysis of Variance and *t*-Test

For a classical backcross design, which is the population generated by a heterozygous F1 back-crossed to one of its homozygous parents (for example a cross of AaQq × AAQQ), the rationale behind the single marker analysis can be explained using the cosegregation listed in Table 4.4. Marker A and the QTL are assumed to be linked at a distance of r recombination units. The expected frequencies for the four marker-QTL genotypes (AAQQ, AAQq, AaQQ, and AAQq) are listed in Table 4.4. The conditional frequencies of the QTL genotypes (QQ and Qq) on the marker genotypes (AA and Aa) can be obtained by dividing the joint marker-QTL genotypic frequencies by the marginal marker genotypic frequencies (they are 0.5 for this case). The expected phenotypic values for the observable marker genotypes can be obtained by modifying the conditional frequencies by the expected trait values (see definitions in Table 4.2). For example, the expected trait value for marker genotype AA is

$$y_{AA} = p(QQ|AA)\mu_1 + p(Qq|AA)\mu_2 = (1-r)\mu_1 + r\mu_2 \tag{4.9}$$

where p(QQ|AA) and p(Qq|AA) are probabilities that an individual with marker genotype AA is QQ or Qq genotype, respectively; and $\mu_1$ and $\mu_2$ are expected genotypic value for the two QTL genotypes, respectively.

The expectation of difference between the two marker classes is

$$E[\mu_{AA} - \mu_{Aa}] = (1-2r)(\mu_1 - \mu_2) = 2g(1-2r) = (a+d)(1-2r) \tag{4.10}$$

There are two possible interpretations for the null hypothesis $H_0$: $[\mu_{AA} - \mu_{Aa}] = 0$. One is that $(a + d) = 0$ and the other is that $(1 - 2r) = 0$ or $r = 0.5$. The biological meaning for the first one is that there is no genetic effect at the marker position, and for the second is that the QTL and the marker are independent (no linkage). These are what we want to test. So the single-marker analysis for the backcross progeny is valid. However, the power of the test is low when the marker is loosely linked with the QTL and an unbiased estimate of genetic effect cannot be obtained.

#### 4.3.2.3 Linear Regression

The model can also be tested using simple linear regression by regressing the trait values on a dummy variable for the marker genotypes. The regression model is

$$y_j = \beta_0 + \beta_1 x_j + \varepsilon_j \tag{4.11}$$

where $y_j$ is the trait value for the jth individual in the population, $x_j$ is the dummy variable taking 1 if the individual is AA and –1 for Aa, $\beta_0$ is the intercept for the regression which is the overall mean for the trait, $\beta_1$ is the slope for the regression line, and $\varepsilon_j$ is random error for the jth individual. The expected means, variances, and covariance needed for estimating regression coefficients for the two variables are

$$\begin{cases} E(\bar{x}) = 0,\ E(\hat{s}_x^2) = 1, E(\bar{y}) = 0.5(\mu_1 + \mu_2) \\ E(\hat{s}_{xy}) = 0.5(1-2r)(\mu_1 - \mu_2) \end{cases} \tag{4.12}$$

The estimated intercept and slope for model 4.11 are

$$\begin{cases} \beta_0 = 1 \\ \beta_1 = 0.5(1-2r)(\mu_1 - \mu_2) \end{cases} \tag{4.13}$$

It is not difficult to see that the expectation of the slope is the expectation for the difference between the two marker classes

$$E(\beta_1) = (1-2r)g = 0.5(a+d)\,(1-2r) \tag{4.14}$$

The hypothesis test $H_0$:$\beta_1$ = 0 is equivalent to testing that the marker and the QTL are unlinked (r = 0.5) or that the genetic effects equal zero (g = 0.5 (a + d) = 0).

### 4.3.3 Single Marker Linear Model (F2 Model)

#### 4.3.3.1 Model

Table 4.5 shows the expected frequencies of the three possible QTL genotypes conditional on the marker genotypes in an F2 progeny assuming the QTL Q and the marker A are linked at a distance of r recombination units. The conditional frequencies are the joint frequencies divided by the marginal frequencies corresponding to the marker genotypes,

$$p(Q_j|M_i) = p_{ij}/p_{i\circ} \tag{4.15}$$

where p ($Q_j$ | $M_i$) is the frequency of the jth putative QTL genotypic class conditional on the ith genotypic class of marker A, $p_{ij}$ is the joint frequency of the jth putative QTL genotypic class and the ith genotypic class of marker A, and $p_{i\circ}$ is the marginal frequency of the ith genotypic class of marker A. For the F2 progeny, the marginal frequencies for the marker genotypes AA, Aa, and aa are 0.25, 0.5, and 0.25, respectively.

The trait values for the putative QTL genotypes were defined in Table 4.2. For example, the expected trait values for the putative QTL genotypes QQ, Qq, and qq are $\mu_1$, $\mu_2$, and $\mu_3$, respectively,

**TABLE 4.5**
**Expected QTL Genotypic Frequency Conditional on Genotypes of a Nearby Marker in F2 Populations**

| | $n_i$ | $p_{io}$ | QQ | Qq | qq |
|---|---|---|---|---|---|
| AA | $n_1$ | 0.25 | $(1-r)^2$ | $2r(1-r)$ | $r^2$ |
| Aa | $n_2$ | 0.5 | $r(1-r)$ | $(1-r)^2+r^2$ | $r(1-r)$ |
| aa | $n_3$ | 0.25 | $r^2$ | $2r(1-r)$ | $(1-r)^2$ |

and $a = 0.5(\mu_1 - \mu_3)$ and $d = 0.5(\mu_1 + \mu_3 - 2\mu_2)$ were defined as additive and dominance genetic effects of the QTL. The expected trait values for the marker genotypic class is the summation of the product of the putative QTL genotypic value and the frequency conditional on the marker genotypes, for example the expected trait value for marker class AA is

$$\mu_{AA} = \sum_{j=1}^{3} \mu_j p(Q_j|M_1) = \mu_1(1-r)^2 + 2\mu_2 r(1-r) + \mu_3 r^2 \quad (4.16)$$

If we use $\mu$ to denote the population mean then we have

$$\mu_1 = \mu + a,\ \mu_2 = \mu + d,\ \mu_3 = \mu - a \quad (4.17)$$

The expected trait values in terms of the putative QTL genetic effects can be derived as

$$\begin{cases} \mu_{AA} = \mu + (1-2r)a + 2r(1-r)d \\ \mu_{Aa} = \mu + \left[(1-r)^2 + r^2\right]d \\ \mu_{aa} = \mu - (1-2r)a + 2r(1-r)d \end{cases} \quad (4.18)$$

The expectation for the two contrasts are

$$\begin{cases} E[\text{"additive"}] = E[\mu_{AA} - \mu_{aa}] = 2(1-2r)a \\ E[\text{"dominant"}] = E[\mu_{AA} + \mu_{aa} - 2\mu_{Aa}] = -2(1-2r+\mu)^2 d \end{cases} \quad (4.19)$$

The hypothesis test based on the contrasts for the marker genotypes corresponds to the marker and the QTL are independent or the no genetic effects can be detected for the putative QTL.

#### 4.3.3.2 Linear Regression

A linear regression model also can be used for single marker analysis using F2 progeny. The model is

$$y_j = \beta_0 + \beta_1 x_{1j} + \beta_2 x_{2j} + \varepsilon_j \quad (4.20)$$

where $y_j$ is a trait value for the jth individual in the population, $x_{1j}$ is the dummy variable for the marker "additive" effect taking 1, 0, and –1 for marker genotypes AA, Aa, and aa, respectively, $x_{2j}$ is a dummy variable for the marker "dominance" effect taking 1, –2, and 1 for marker genotypes AA, Aa, and aa, $\beta_0$ is the intercept for the regression which is the overall mean for the trait, $\beta_1$ is the slope for the additive regression line, $\beta_2$ is the slope for the dominance regression line, and $\varepsilon_j$ is random error for the jth individual. The expectation of the intercept and the slopes are

$$E\begin{bmatrix}\hat{\beta}_0\\ \hat{\beta}_1\\ \hat{\beta}_2\end{bmatrix}=\begin{bmatrix}1 & 0 & 0\\ 0 & 2 & 0\\ 0 & 0 & 1\end{bmatrix}\begin{bmatrix}0.25(\mu_1+2\mu_2+\mu_3)\\ 0.5(1-2r)(\mu_1-\mu_3)\\ 0.5(1-2r)^2(\mu_1+\mu_3-2\mu_2)\end{bmatrix}=\begin{bmatrix}0.25(\mu_1+2\mu_2+\mu_3)\\ (1-2r)a\\ (1-2r)^2 d\end{bmatrix} \tag{4.21}$$

Hypothesis tests can be performed using an F-statistic which is the ratio between the residual mean squares for the reduced-model and the full-model.

### 4.3.4 Single Marker Likelihood Function

Likelihood approach is also used for the single-marker analysis.[35] For the backcross model, the likelihood is

$$L=\frac{1}{\{\sqrt{2\pi}\sigma\}^N}\prod_{i=1}^{N}\sum_{j=1}^{2}p(Q_j|M_i)\exp\left[-\frac{(y_i-\mu_j)^2}{2\sigma^2}\right] \tag{4.22}$$

if we assume that each of the four marker-QTL classes has equal variance, $\sigma^2$, and the trait values are normally distributed, where $y_i$ is the observed trait phenotypic value for the ith individual, $p(Q_j|M_i)$ is the conditional probability listed in Table 4.4, and $\mu_j$ is the trait value for the jth QTL genotype. The likelihood for the single-marker analysis using F2 progeny is

$$L=\frac{1}{\{\sqrt{2\pi}\sigma\}^N}\prod_{i=1}^{N}\sum_{j=1}^{3}p(Q_j|M_i)\exp\left[-\frac{(y_i-\mu_j)^2}{2\sigma^2}\right] \tag{4.23}$$

if we assume that each of the nine marker-QTL classes has equal variance, $\sigma^2$, where $y_i$ is the observed trait phenotypic value for ith individual, $p(Q_j|M_i)$ is the conditional probability listed in Table 4.5, and $\mu_j$ is the trait value for the jth QTL genotype.

### 4.3.5 Interval Mapping Model (Backcross Model)

#### 4.3.5.1 Likelihood Function

Table 4.6 shows the conditional frequencies of QTL genotype on the marker genotypes and the expected values for the marker genotypes. $\rho$ in the table is defined as the relative position of the putative QTL in the genome segment flanked by the two markers. For example, if $\rho = 0$ the putative QTL is located on marker A, if $\rho = 0.5$ the QTL is located in the middle of the segment, and if $\rho = 1$ the QTL is located on marker B.

The likelihood function for interval mapping is constructed based on cosegregation among the putative QTL and the two flanking markers. The likelihood function is

$$L=\frac{1}{\{\sqrt{2\pi}\sigma\}^N}\prod_{i=1}^{N}\sum_{j=1}^{2}p(Q_j|M_i)\exp\left[-\frac{(y_i-\mu_j)^2}{2\sigma^2}\right] \tag{4.24}$$

**TABLE 4.6**
**Expected QTL Genotypic Frequency Conditional on Genotypes of the Flanking Markers in Backcross Populations with no Double Crossover**

| Marker genotype | Frequency $p_{i\cdot}$ | QQ | Qq | Expected value ($g_i$) |
|---|---|---|---|---|
| AABB | 0.5 (1 − r) | 1 | 0 | $\mu_1$ |
| AABb | 0.5r | $r_2/r = 1 - \rho$ | $r_1/r = \rho$ | $(1 - r)\ \mu_1 + \rho\mu_2$ |
| AaBB | 0.5r | $r_1/r = \rho$ | $r_2/r = 1 - \rho$ | $\rho\mu_1 + (1 - \rho)\ \mu_2$ |
| AaBb | 0.5 (1 − r) | 0 | 1 | $\mu_2$ |

*Note:* See Table 4.2 for some notations.

if we assume that each of the eight marker-QTL classes has equal variance $\sigma^2$ and the trait values are normally distributed, where $y_i$ is the observed trait phenotypic value for the ith individual, $p(Q_j|M_i)$ is the conditional probability listed in Table 4.6, and $\mu_j$ is the trait value for the jth QTL genotype.

#### 4.3.5.2 Nonlinear Regression

A nonlinear regression model for the trait value can be written as

$$y_j = X_1\mu_1 + X_2\frac{1}{r}(r_2\mu_1 + r_1\mu_2) + X_3\frac{1}{r}(r_1\mu_1 + r_2\mu_2) + X_4\mu_2 + \varepsilon_j \tag{4.25}$$

where $X_1$, $X_2$, $X_3$, and $X_4$ are the coefficients for the four marker genotypes as listed in Table 4.7; $g_i$ is the expected trait value for marker genotypic class i as shown in Table 4.6; r, $r_1$, and $r_2$ were defined as the recombination frequencies between the two markers, the putative QTL and marker A, and the putative QTL and marker B, respectively; and $\varepsilon_j$ is experimental error associated with the individual j. If we reparameterize the recombination frequency between marker A and the putative QTL as $\rho$ then we have

$$y_j = X_1\mu_1 + X_2[(1-\rho)\mu_1 + \rho\mu_2] + X_3[\rho\mu_1 + (1-\rho)\mu_2] + X_4\mu_2 + \varepsilon_j \tag{4.26}$$

where $\rho = r_1/r$ and $0 \le \rho \le 1$. Three unknown parameters are involved in the model, the two QTL genotypic means and the parameter for relative QTL position.

**TABLE 4.7**
**Coefficients for Interval Mapping Using Regression Analysis for Backcross Progeny**

| Maker genotype | $X_1$ | $X_2$ | $X_3$ | $X_4$ |
|---|---|---|---|---|
| AABB | 1 | 0 | 0 | 0 |
| AABb | 0 | 1 | 0 | 0 |
| AaBB | 0 | 0 | 1 | 0 |
| AaBb | 0 | 0 | 0 | 1 |

#### 4.3.5.3 Linear Regression

Knapp et al.[36] suggested using linear models for interval mapping using regression analysis. If we define

$$\begin{cases} \theta_1 = \mu_1 \\ \theta_2 = (1-\rho)\mu_1 + \rho\mu_2 \\ \theta_3 = \rho\mu_1 + (1-\rho)\mu_2 \\ \theta_4 = \mu_2 \end{cases} \tag{4.27}$$

Equation 26 can be rearranged as

$$y_j = X_1\theta_1 + X_2\theta_2 + X_3\theta_3 + X_4\theta_4 + \varepsilon_j \tag{4.28}$$

where the θs are trait means for the marker genotypes. If no constrains are imposed the estimators of $\hat{\theta}_1$ and $\hat{\theta}_4$ are the estimates for the two QTL genotypic means ($\hat{\mu}_1$ and $\hat{\mu}_2$). If constrain $\theta_1 + \theta_4 = \theta_2 + \theta_3$ is imposed, we have $\theta_3 = \theta_1 + \theta_4 - \theta_2$ and linear model

$$y_j = \theta_1(X_1 + X_3) + \theta_2(X_2 - X_3) + \theta_4(X_3 + X_4) + \varepsilon_j \tag{4.29}$$

Solving Equation 4.29 with minimizing the residual sum of squares leads us to the estimates $\hat{\theta}_1$, $\hat{\theta}_2$ and $\hat{\theta}_4$. In practical interval mapping, ρ has been considered as a known parameter. Using relation $\theta_2 = (1 - \rho)\mu_1 + \rho\mu_2$, Equation 4.29 becomes

$$y_j = \theta_1[X_1 + (1-\rho)X_2 + \rho X_3] + \theta_4[\rho X_2 + (1-\rho)X_3 + X_4] + \varepsilon_j \tag{4.30}$$

The independent variables in the linear regression model 4.30 are the probabilities that the individual is a QQ or Qq for the putative QTL conditioning on the flanking marker genotypes. The solutions for the two parameters are $\theta_1 = \mu_1$ and $\theta_4 = \mu_2$. So the estimates are unbiased estimators for the QTL means.

### 4.3.6 Interval Mapping Model (F2 Model)

#### 4.3.6.1 Likelihood Function

The likelihood function for the data is

$$L = \frac{1}{\{\sqrt{2\pi}\sigma\}^N} \prod_{i=1}^{N} \left\{ \sum_{k=1}^{3} p(Q_j|M_i) \exp\left[ -\frac{(y_i - \mu_j)^2}{2\sigma^2} \right] \right\} \tag{4.31}$$

if we assume that each of the 27 marker-QTL classes has equal variance $\sigma^2$ and the trait values are normally distributed, where $y_i$ is the observed trait phenotypic value for the ith individual, $p(Q_j|M_i)$ is the conditional probability listed in Table 4.8, and $\mu_j$ is the trait value for the jth QTL genotype.

#### 4.3.6.2 Nonlinear Regression

Following the rationale for the regression models for the backcross progeny, the models for F2 progeny can be constructed. The marker genotypes in F2 progeny can be coded as shown in Table 4.9. A nonlinear regression model for the trait value can be written as

**TABLE 4.8**
**Expected QTL Genotypic Frequency Conditioning on Genotypes of the Flanking Markers in F2 Populations with no Double Crossover**

| Marker genotype | Frequency | $p(Q_j \mid M_i)$ | | |
|---|---|---|---|---|
| | | QQ | Qq | qq |
| AABB | $0.25(1-r)^2$ | 1 | 0 | 0 |
| AABb | $0.5r(1-r)$ | $r_2/r$ | $r_1/r$ | 0 |
| AAbb | $0.25r^2$ | $(r_2/r)^2$ | $2r_1r_2/r^2$ | $(r_1/r)^2$ |
| AaBB | $0.5r(1-r)$ | $r_1/r$ | $r_2/r$ | 0 |
| AaBb | $0.5[(1-r)^2+r^2]$ | $\frac{r_1r_2}{(1-r)^2+r^2}$ | $\frac{(1-r)^2+r_1^2+r_2^2}{(1-r)^2+r^2}$ | $\frac{r_1r_2}{(1-r)^2+r^2}$ |
| Aabb | $0.5r(1-r)$ | 0 | $r_2/r$ | $r_1/r$ |
| aaBB | $0.25r^2$ | $(r_1/r)^2$ | $2r_1r_2/r^2$ | $(r_2/r)^2$ |
| aaBb | $0.5r(1-r)$ | 0 | $r_1/r$ | $r_2/r$ |
| aabb | $0.25(1-r)^2$ | 0 | 0 | 1 |

**TABLE 4.9**
**Coefficients for the Regression Approaches for QTL Interval Mapping Using F2 Progeny**

| Marker genotype | $X_1$ | $X_2$ | $X_3$ | $X_4$ | $X_5$ | $X_6$ | $X_7$ | $X_8$ | $X_9$ |
|---|---|---|---|---|---|---|---|---|---|
| AABB | 1 | 0 | 0 | 0 | 0 | 0 | 0 | 0 | 0 |
| AABb | 0 | 1 | 0 | 0 | 0 | 0 | 0 | 0 | 0 |
| AAbb | 0 | 0 | 1 | 0 | 0 | 0 | 0 | 0 | 0 |
| AaBB | 0 | 0 | 0 | 1 | 0 | 0 | 0 | 0 | 0 |
| AaBb | 0 | 0 | 0 | 0 | 1 | 0 | 0 | 0 | 0 |
| Aabb | 0 | 0 | 0 | 0 | 0 | 1 | 0 | 0 | 0 |
| aaBB | 0 | 0 | 0 | 0 | 0 | 0 | 1 | 0 | 0 |
| aaBb | 0 | 0 | 0 | 0 | 0 | 0 | 0 | 1 | 0 |
| aabb | 0 | 0 | 0 | 0 | 0 | 0 | 0 | 0 | 1 |

$$y_j = \sum_{i=1}^{9} X_{ij} g_i + \varepsilon_j \tag{4.32}$$

where the Xs are the coefficients for the nine marker genotypes; $g_i$ is the expected trait value for marker genotypic class i as shown in Table 4.8; r, $r_1$, and $r_2$ were defined as the recombination frequencies between the two markers, the putative QTL and marker A, and the putative QTL and marker B, respectively; and $\varepsilon_j$ is experimental error associated with the individual j. As for the backcross progeny, the recombination frequencies can be reparameterized. The recombination frequency between marker A and the putative QTL is defined as $\rho r$, then we have $\rho = r_1/r$ and $0 \le \rho \le 1$. Four unknown parameters are involved in the model, the three QTL genotypic means and the parameter for relative QTL position. The least-square estimates of the unknown parameters in Equation 4.32 can be solved using an iterative Gauss-Newton algorithm.

#### 4.3.6.3 Linear Regression

As shown by Knapp and Bridges,[36] model 4.32 can be written as a linear model,

$$y_j = \sum_{i=1}^{9} \theta_i X_i + \varepsilon_j \tag{4.33}$$

where the θs are the means for each of the nine marker genotypic means and by setting them equal to the expected trait values, we have

$$\begin{cases} \theta_1 = \mu_1 \\ \theta_2 = (1-\rho)\mu_1 + \rho\mu_2 \\ \theta_3 = (1-\rho)^2\mu_1 + 2\rho(1-\rho)\mu_2 + \rho^2\mu_3 \\ \theta_4 = \rho\mu_1 + (1-\rho)\mu_2 \\ \theta_5 = \dfrac{r^2\rho(1-\rho)}{(1-r)^2 + r^2}(\mu_1 + \mu_3) + \dfrac{\left[(1-r)^2 + r^2(1-2\rho+\rho^2)\right]}{(1-r)^2 + r^2}\mu_2 \\ \theta_6 = (1-\rho)\mu_2 + \rho\mu_3 \\ \theta_7 = \rho^2\mu_1 + 2\rho(1-\rho)\mu_2 + (1-\rho)^2\mu_3 \\ \theta_8 = \rho\mu_2 + (1-\rho)\mu_3 \\ \theta_9 = \mu_3 \end{cases} \tag{4.34}$$

So $\hat{\theta}_1$ and $\hat{\theta}_9$ can be used to estimate $\mu_1$ and $\mu_3$, respectively. The unbiased estimators of $\mu_2$ and ρ are

$$\begin{cases} \hat{\mu}_2 = 0.5\left(\hat{\theta}_2 + \hat{\theta}_4 + \hat{\theta}_6 + \hat{\theta}_8 - \hat{\theta}_1 - \hat{\theta}_9\right) \\ \hat{\rho} = \left(\hat{\theta}_2 + \hat{\theta}_8 - \hat{\theta}_1 - \hat{\theta}_9\right) \Big/ \left(\hat{\theta}_2 + \hat{\theta}_4 + \hat{\theta}_6 + \hat{\theta}_8 - 2\hat{\theta}_1 - 2\hat{\theta}_9\right) \end{cases} \tag{4.35}$$

### 4.3.7 Composite Interval Mapping (CIM)

#### 4.3.7.1 Model and Likelihood Function

As stated by Zeng,[6,7] CIM is a combination of simple interval mapping and multiple linear regression. For CIM analysis on a segment between markers i and i + 1 using backcross progeny, the statistical model is

$$y_j = b_0 + b_i X_{ij} + \sum_{k \neq i, i+1} b_k X_{kj} + \varepsilon_j \tag{4.36}$$

where $y_j$ is the trait value for individual j, $b_0$ is the intercept of the model, $b_i$ is the genetic effect of the putative QTL located between markers i and i + 1, $X_{ij}$ is a dummy variable taking 1 for marker genotype AABB, 0 for AaBb, 1 with a probability of $1 - r_l/r = 1 - \rho$ and 0 with a probability $r_l/r = \rho$ for marker genotype AaBB, 1 with probability of $\rho$, and 0 with a probability of $1 - \rho$ for marker genotype AABb, r is recombination frequency between the two markers, $r_l$ is the recombination between the first marker and the putative QTL, $b_k$ is the partial regression coefficient of the trait value on marker k, $X_{kj}$ is dummy variable for marker k and individual j, taking 1 if the marker has genotype AA and 0 for Aa, and $\varepsilon_j$ is a residual from the model. If $\varepsilon_j$ is normally distributed with mean zero and variance $\sigma^2$, the likelihood function for the CIM is

$$
\begin{aligned}
L = & \frac{1}{\left\{\sqrt{2\pi\sigma}\right\}^N} \exp \frac{1}{-2\sigma^2}\left[\sum_{j=1}^{n_1}(y_{1j}-\mu_1)^2 + \sum_{j=1}^{n_4}(y_{4j}-\mu_2)^2\right] \\
& \prod_{j=1}^{n_2}\left\{(1-\rho)\exp\left[-\frac{(y_{2j}-\mu_1)^2}{2\sigma^2}\right] + \rho\exp\left[-\frac{(y_{2j}-\mu_2)^2}{2\sigma^2}\right]\right\} \\
& \prod_{j=1}^{n_3}\left\{\rho\exp\left[-\frac{(y_{3j}-\mu_1)^2}{2\sigma^2}\right] + (1-\rho)\exp\left[-\frac{(y_{3j}-\mu_2)^2}{2\sigma^2}\right]\right\}
\end{aligned}
\tag{4.37}
$$

### 4.3.7.2 Regression Models

The CIM can be implemented using the likelihood approach[6] and the nonlinear regression model for interval mapping and the multiple linear model for controlling the residual genetic effects. Here, I will discuss the regression model. The model can be written as

$$
\begin{aligned}
y_j = & X_{i1j}\mu_1 + X_{i2j}\left[(1-\rho)\mu_1 + \rho\mu_2\right] + X_{i3j}\left[\rho\mu_1 + (1-\rho)\mu_2\right] \\
& + X_{i4j}\mu_2 + \sum_{k\neq i,i+1} b_k X_{kj} + \varepsilon_j
\end{aligned}
\tag{4.38}
$$

where the $X_i$s, $\mu_1$, $\mu_2$, and $\rho$ were defined in Table 4.6 and Equation 4.26. Combining with Equation 4.27 and treating $\rho$ as known constant, Equation 4.38 can be rearranged as

$$
\begin{aligned}
y_j = & \theta_1\left[X_{i1j} + (1-\rho)X_{i2j} + \rho X_{i3j}\right] \\
& + \theta_4\left[\rho X_{i2j} + (1-\rho)X_{i3j} + X_{i4j}\right] + \sum_{k\neq i,i+1} b_k X_{kj} + \varepsilon_j
\end{aligned}
\tag{4.39}
$$

This is a simple linear multiple regression model. So we have the estimates of the two genotypic means for the putative QTL located at position $\rho$ between the two markers

$$
\begin{cases}
\hat{\theta}_1 = \mu_1 \\
\hat{\theta}_4 = \mu_2
\end{cases}
\tag{4.40}
$$

Under the null hypothesis $\mu_1 = \mu_2$ or $\theta_1 = \theta_4 = \theta$, Equation 4.38 becomes

$$y_{j(\theta_1=\theta_4=\theta)} = \theta\left[X_{i1j} + X_{i2j} + X_{i3j} + X_{i4j}\right] + \sum_{k \neq i,i+1} b_k X_{kj} + \varepsilon j$$
$$= \theta + \sum_{k \neq i,i+1} b_k X_{kj} + \varepsilon j \quad (4.41)$$

So the hypothesis test can be implemented using the log likelihood ratio test statistic

$$G^2 = \frac{SSE_{reduced} - SSE_{full}}{SSE_{full}/df_{Efull}} \quad (4.42)$$

where SSEs are residual sum of square for the full (Equation 4.38) and reduced models (Equation 4.41). The test statistic can be estimated for each of the biologically meaningful positions on the genome.

### 4.3.8 Mapping Populations

In quantitative and population genetics mating design is developed in a manner to simplify the partitioning and interpretation of genetic variance components. The purpose of mapping population design is also for clear genetic interpretation and genomic data analysis. The basic observations for genomic data are genotype of genetic marker, fingerprint of an individual (genotype or a clone), sequence of a DNA segment, trait value, known gene genotype, etc. The basic conclusions of genomic analysis are usually linkage relationships and physical locations of genes of interest, and relationships between genes and trait. Mating design for mapping population establishment is for making the relationships among the polymorphic markers and traits of interest detectable and tractable. Commonly used mapping populations can be classified into controlled crosses and natural populations.

#### 4.3.8.1 Population of Controlled Cross

To create a population used for genomic research involves choosing parents and determining mating types. To make decisions on parents and mating types, type of markers and objectives of the experiment should be taken into consideration. Parents of a mapping population must have sufficient variation at the DNA sequence level and at phenotype level for the traits of interest. The variation at the DNA level is the basis for tracing recombination events using genetic markers. The more variation the easier to find polymorphic and informative markers. When the objective of the experiment is to search for genes controlling a particular trait, the genetic variation of the trait between the parents is also essential. If the parents have a great variation at phenotypic level for a trait then there is a large chance that genetic variation exists. However, no phenotypic variation among the parents does not mean that there is certainly no genetic variation. Different sets of genes could result from the same phenotype.

Different types of DNA markers may have different resolutions on detecting genome variation. For some species, little genome variation exists in natural populations. This genome variation may be not sufficient for detection using some marker systems. However, technology is being developed to detect even a single base change, for example SSCP (single-strand conformational polymorphism).

Regarding mating types, if the parents are inbred lines, progeny of the cross between the two parents is called F1 and will be uniformly heterozygous without segregation. Another generation of mating is needed to have the genes and the trait segregating. Commonly used mating types are F2 and backcross. F2 progeny is produced by selfing the F1 individuals. Backcross is produced by a cross between the F1 and one of the parents. The disadvantage of the F2 and backcross progenies is that true replications of the experiment can only be obtained for the species which can be clonally reproduced. However, for most species, the true replications cannot be done because the progenies cannot be reproduced identically. To solve this problem, recombinant inbred line (RIL) and doubled haploid line (DHL) are used for some plant species. RIL is produced by selfing the F2 for a large number of generations, say 10 generations, using single seed descent approach. DHL is produced by doubled haploid gametes produced by the F2 plants using tissue culture. RIL and DHL can reproduce themselves for repeated experiments. However, RIL and DHL are only available for a limited number of plant species.

If the parents are heterozygous, progeny of crosses between the two parents will segregate at some of the loci and a portion of the segregating loci have information on linkage. Some of the locus combinations are not informative for linkage. At a single locus level, the F1 progeny between two heterozygous parents is a mixture of F2 and backcross. At the whole genome level, linkage phase is a mixture of coupling and repulsion. To determine linkage phase, a three generation pedigree may be needed.

The backcross and F2 populations were used to illustrate the rationale and methodology for QTL mapping using experimental populations (see Figure 4.1). Commonly used mapping populations obtained by controlled matings usually can be classified as these two population types at whole genome or individual genome segment model levels. For example, doubled-haploid lines and the recombined inbred lines can be treated as backcross model in terms of the data analysis because the expected genotypic frequencies are corresponding to each other. However, the interpretations of the QTL mapping results may be different. QTL effect in backcross population is a mixture of additive and dominance effects. QTL effects in the doubled haploid and recombinant inbred lines are a purely additive genetic effect. When mapping using hybrids of two heterozygous populations the progeny is a mixture of the backcross and F2. When mapping using open-pollinated populations the progeny is a mixture of F2, backcross, and random mating.

If the marker used is dominance, F2 is not recommended because dominance markers in repulsion linkage phase in F2 have a low information content on linkage. If the recurrent parent is recessive for the dominance loci, the backcross progeny is same using dominance and codominance markers in terms of genomic analysis. In the F1 of two heterozygous parents, a pseudo-backcross approach has been used to avoid the problem.[65]

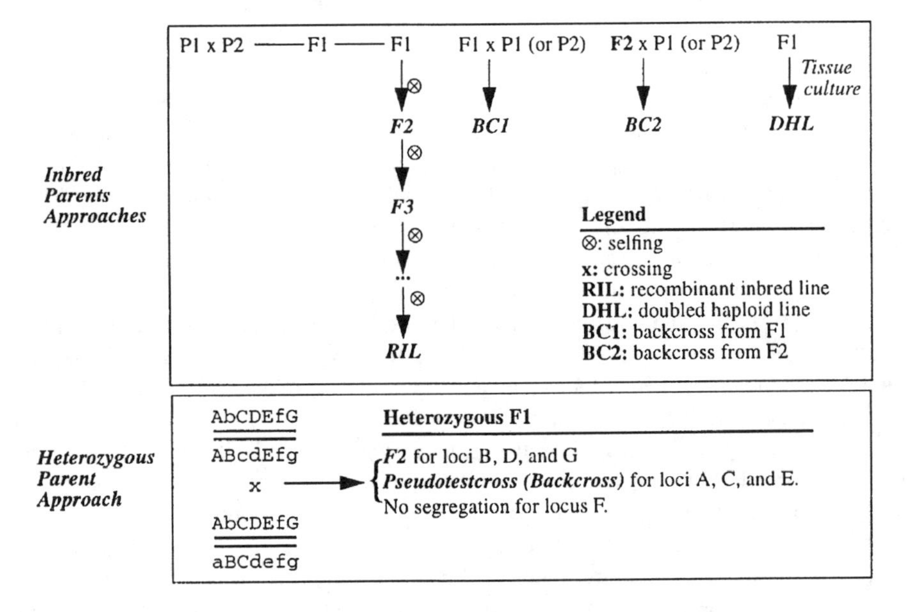


**FIGURE 4.1** Commonly used mating schemes for QTL mapping.

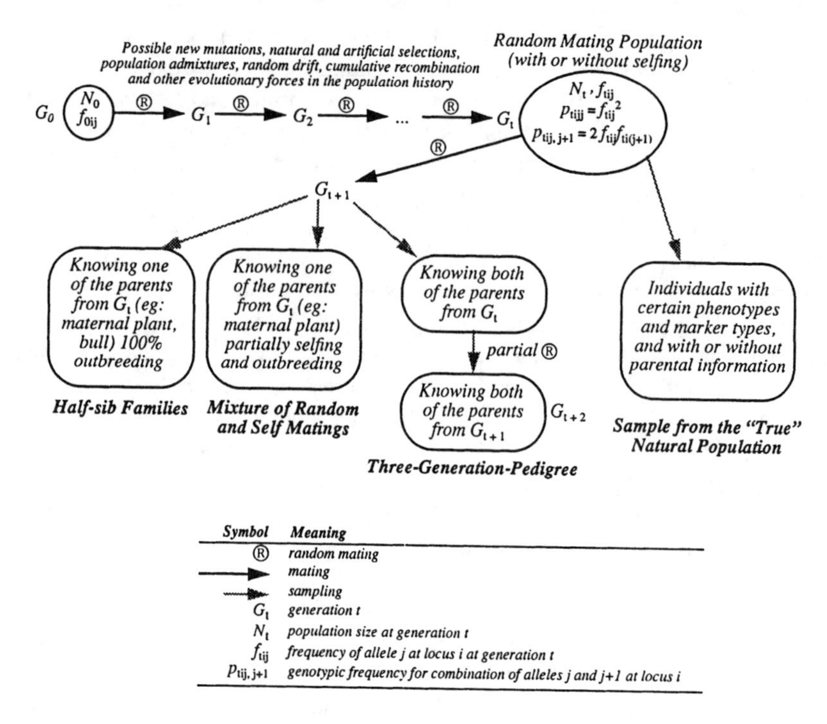


**FIGURE 4.2** "Natural" populations used for genomic research for typical outbred plant and animal species.

### 4.3.8.2 Natural Population

It is difficult to make a clear distinction between natural and artificial (experimental) populations. In the context of genomics, the populations obtained using controlled crosses between selected parents can be considered as experimental populations. The populations produced by naturally occurring matings (without artificial control) can be considered as "natural populations" (Figure 4.2). Population parameters, such as allelic frequencies, genotypic frequencies, and disequilibrium at different levels, are commonly used to characterize genetic architecture of the population. The evolutionary forces, such as mutations, natural and artificial selection, population admixtures (migration), random drift, cumulative recombination events, and others, may play important roles in the population history (Figure 4.2). For some species, natural populations may be generated by assortative mating or even complete self mating instead of random mating, such as wheat and other self-pollinated crop species. However, natural populations of self-pollinated crop species are seldom as populations for genomic research because genetic relationships at genomic level are difficult to trace among individuals within the population. The natural populations referred to here are for the naturally outbred species.

The samples from the "natural populations" can be half-sib families, mixture of random and self matings, three- (or more or less) generation pedigree, and the sample with characteristics of the "true natural population" (Figure 4.2). Genetic variation among half-sib families are commonly used for classical quantitative genetics in many plant and animal species. For examples, populations generated by a single plant pollinated by many unknown or partially known pollen resources, or populations generated by using semen of a bull to inseminate many female animals are typical ways to produce half-sib families. In genomics, genetic variation within a half-sib family is a common resource for searching genes controlling traits of interest. In some way, the matings to produce the half-sib families is controlled instead of completely random. However, in genetic terms the pollen sources and the collection of the female animals are random subsets of the population. Another type of sample from natural population is generated by matings of mixture of random and self. This type occurs commonly in plant kingdoms. For example, seeds on a tree could be results of outcrossing or selfing. Another commonly used sample from the natural population has pedigree structure. The pedigree can be recorded for two, three, or even more generations.

In terms of genomic analysis, what is essential for a mating type is that the genomic relationships among genes and markers and among genes and phenotypes can be detected and traced in the population using means of the genomic DNA assays, such as hybridization, PCR amplification, etc. As new technologies for the detection and tracing are developed, different mating types may become appropriate.

## 4.4 COMPUTER SOFTWARE

### 4.4.1 Specific Packages

Compared to general statistical analysis of biological data, statistical analysis for the study of genes controlling complex traits has the following characteristics: (1) many repeated analysis in one task, (2) lack of standard distribution for some test statistics, and (3) complexity of models used in QTL mapping. Some specialized software package to perform some specific genetic and statistical models have been developed mainly by scientists working in the area of statistical genetics, such as MAPMAKER/QTL,[13] QTLSTAT,[14] QTL Cartographer,[15] PGRI,[16] MAPQTL,[17] Map Manager QTL,[18] and QGENE,[19] etc. These are all public domain packages except for MAPQTL. These packages have some similarities, such as (1) interface is not user friendly compared to some commercial software; (2) user support is also limited due to their noncommercial status; (3) statistical models which can be built using the software are limited; and (4) speed of model building is high for the models which the software can build. Because of limitation (3), these software packages usually cannot handle data with complex experimental designs. In practice, means or least square means of the genotypes are used as input data for these packages. These packages can usually perform simple t-test, linear and nonlinear regressions, interval mapping using the likelihood approach, and the composite interval mapping (Table 4.10).

Most of the available models for QTL mapping can be implemented using statistical software packages, such as SAS (SAS Institute, 1990). The advantages of using the general statistical packages are (1) they are commercially available, (2) user interface are usually friendly, (3) user support is available (with or without charge), and (4) user can specify models. However, general statistical software packages are usually not efficient for a large number of repeated analyses and data manipulation. Software, such as SAS, is flexible to build any kind of models. Knapp[45] listed a suite of SAS programs for QTL mapping data with experimental designs.

In the following section, I will discuss implementation of QTL analysis using SAS and list the specific software packages in Table 4.10. For QTL analysis using SAS, I only give the regression approaches here. For ANOVA-based analysis, please see Knapp.[45] The introduction to the packages is very limited. Please refer to their user manuals for details.

For using the software packages in Table 4.10, a known linkage map is needed for either running the programs or interpreting results. Companion packages, such as MAPMAKER/EXP, GMENDEL, PGRI, MapManager, and JoinMap, for linkage map construction are available for MAPMAKER/QTL, QTLSTAT, PGRI, Map Manager QT, and MAPQTL. It is common to analyze marker data and obtain linkage map before QTL analysis. For packages QTL Cartographer and QGENE, linkage maps obtained using MAPMAKER/EXP can be incorporated in the analysis.

### 4.4.2 QTL Analysis Using SAS

#### 4.4.2.1 Interval Mapping Using Nonlinear Regression

The regression approach can be carried out using commonly used statistical packages, such as SAS and also using QTLSTAT, developed by the author at Oregon State University (Liu and Knapp;[14]

## TABLE 4.10
## Some Available Software Packages for QTL Mapping

| | | |
|---|---|---|
| MAPMAKER/QTL | Models | Interval mapping, multiple QTL modeling |
| | Mating type | F2, backcross, RIL, DH |
| | Computer platform | SUN SPARCstation |
| | Graphic interface | No |
| | Graphic output | Postscript |
| | Contact | Eric Lander (mapmaker@genome.wi.mit.edu) |
| QTLSTAT | Models | Interval mapping using nonlinear regression |
| | Mating type | F2, backcross, RIL, DH |
| | Computer platform | SUN SPARCstation |
| | Graphic interface | No |
| | Graphic output | No |
| | Contact | Steve Knapp (sknapp@helix.css.orst.edu) |
| PGRI | Function | t-test, conditional t-test, linear regression, multiple QTL modeling, permutation test |
| | Mating type | F2, backcross, RIL, DH, heterozygous F1, OP |
| | Computer platform | SUN SPARCstation |
| | Graphic interface | No |
| | Graphic output | No |
| | Contact | Ben Liu (benliu@unity.ncsu.edu) |
| QTL Cartographer | Model | t-test, composite interval mapping, permutation test |
| | Mating type | F2, backcross |
| | Computer platform | SUN SPARCstation, Mac, PC Windows |
| | Graphic interface | No |
| | Graphic output | Gunplot (public domain software) |
| | Contact | Christopher Basten (basten@esssjp.stat.ncsu.edu) |
| MAPQTL | Model | Interval mapping, MQM, nonparametric mapping |
| | Mating type | F2, backcross, RIL, DH, heterozygous F1 |
| | Computer platform | Vax, Unix, Mac, and PC |
| | Graphic interface | No |
| | Graphic output | No |
| | Contact | Johan Van Ooijen (J.W.vanOOIJEN@cpro.dlo.nl) |
| Map Manager QTL | Model | Interval mapping using regression, multiple-QTL |
| | Mating type | F2, backcross |
| | Computer platform | MAC OS |
| | Graphic interface | Yes |
| | Graphic output | Yes |
| | Contact | Kenneth Manly (kmanly@mcbio.med.bufflo.edu) |
| QGENE | Model | Linear regression |
| | Mating type | F2, backcross |
| | Computer platform | MAC |
| | Graphic interface | Yes |
| | Graphic output | Yes |
| | Contact | James C. Nelson (jcn5@cornell.edu) |

*Note:* (Information in this table may not be accurate due to the update of the packages. Contact the authors of the packages for the latest information.) RIL = recombinant inbred line; DH = doubled haploid, OP = open-pollinated population.

contact Dr. Knapp for availability). For using SAS, the data set required should have a certain format. For example, for a single-environment experiment the data should be arranged as:

| | Marker Name | | Marker Genotype | | line | trait |
|---|---|---|---|---|---|---|
| segment | marker1 | marker2 | Marker1 | Marker2 | number | value |
| 1 | WG622 | ABG313B | 1 | 1 | 1 | 72.90 |
| 1 | WG622 | ABG313B | 1 | 2 | 2 | 70.70 |
| 1 | WG622 | ABG313B | 2 | 1 | 3 | 72.90 |
| 1 | WG622 | ABG313B | 2 | 2 | 4 | 74.75 |
| ... | ... | ... | ... | | | |

Then the nonlinear analysis can be carried out using SAS codes similar to:

```
data a;
  infile 'mayqtl.dat';
  input seg marker1 $ marker2 $ g1 g2 line y;
  x1 = 0; x2 = 0; x3 = 0; x4 = 0;
  if g1 = 1 and g2 = 1 then x1 = 1;
  if g1 = 1 and g2 = 2 then x2 = 1;
  if g1 = 2 and g2 = 1 then x3 = 1;
  if g1 = 2 and g2 = 2 then x4 = 1;
proc nlin data = a noprint method = gauss convergence =
  0.0000001 outest = output;
  by seg;
  parms m1 = constantA m2 = constantB r = constantC;
  bounds r< = 1.0 r> = 0;
  model y = x1*m1+x2*((1-r)*m1+r*m2)+x3*(r*m1+(1-r)*m2)+x4*m2;
  der.m1 = x1+x2*(1-r)+x3*r;
  der.m2 = x2*r+x3*(1-r)+x4;
  der.r = x2*(m2-m1)+x3*(m1-m2);
proc print data = output;
run;
```

The constants (`parms m1 = constantA m2 = constantB r = constantC`) are initial values for the parameters. SAS does not provide the hypothesis test for the contrasts nor the matrix $\hat{C}$ needed for the computation. To obtain the matrix, a linear regression procedure using the estimates of the parameters can be used. For example, for segment 18 of a barley data the following SAS codes were used to generate the matrix:

```
data b; set a; if seg = 18;
  t1 = x1+x3;
  t2 = x2+x4;
  t3 = x2*1.359-x3*1.359;
  y1 = y-(x1*72.447+x2*73.806+x3*72.447+x4*73.806);
proc reg all; model y1 = t1 t2 t3;
run;
```

In this SAS codes, 72.447, 73.806, and 0 are the estimated values for the three parameters; the estimated difference between the two means is –1.359; t1, t2, and t3 are the first derivatives of the model with respect to the three parameters evaluated at the estimated values; and y1 is the residual of the predicted value using the estimates from the observed values. The following SAS output is the matrix $\hat{C}$:

X′X Inverse, Parameter Estimates, and SSE

| | T1 | T2 | T3 | Y1 |
|---|---|---|---|---|
| T1 | 0.0135031624 | −0.001422537 | 0.0057376333 | 0.0274427155 |
| T2 | −0.001422537 | 0.0187884359 | −0.006239866 | −0.054301096 |
| T3 | 0.0057376333 | −0.006239866 | 0.0251677508 | −1.107825897 |
| Y1 | 0.0274427155 | −0.054301096 | −1.107825897 | 328.40021666 |

The test statistic can be easily obtained using a hand calculator when the contrasts are simple. When the contrasts are complicated with several degrees of freedom, the computation may be complicated. For example, the contrasts for the environmental effect and the genotype by environment interaction contains three degrees of freedom each. For those cases, specialized software, such as QTLSTAT are recommended.

#### 4.4.2.2 Composite Interval Mapping Using Regression

In practical data analysis, the original composite interval mapping using the ECM algorithm can be performed by computer software, QTL Cartographer.[17] For using the linear regression approach for the composite interval mapping, commercial software, such as SAS, and specialized software PGRI[16] can be used. For using SAS, data should be manipulated into a format which SAS can read. For using SAS, the data (called markerdata1) should be arranged as

| Genome Segment | Marker1 | Marker2 | Genome Positions | Lines | Z1 | Z2 | m1 | m2 | m3 | m4... | |
|---|---|---|---|---|---|---|---|---|---|---|---|
| ... | | | | | | | | | | | |
| 1 | WG622 | ABG313B | 0.00 | 147 | 1.00 | 0.00 | 2 | 2 | 1 | 1 | 1 |
| 1 | WG622 | ABG313B | 0.00 | 148 | 0.00 | 1.00 | 1 | 1 | 1 | 1 | 1 |
| 1 | WG622 | ABG313B | 0.00 | 149 | 0.00 | 1.00 | 1 | 1 | 1 | 1 | 1 |
| 1 | WG622 | ABG313B | 0.00 | 150 | −1.00 | −1.00 | 1 | 2 | 0 | 0 | 1 |
| 1 | WG622 | ABG313B | 0.01 | 1 | 1.00 | 0.00 | 1 | 1 | 1 | 1 | 1 |
| 1 | WG622 | ABG313B | 0.01 | 2 | −1.00 | −1.00 | 1 | 0 | 1 | 0 | 0 |
| 1 | WG622 | ABG313B | 0.01 | 3 | 1.00 | 0.00 | 1 | 1 | 1 | 1 | 1 |
| 1 | WG622 | ABG313B | 0.01 | 4 | 0.00 | 1.00 | 1 | 1 | 2 | 0 | 2 |
| ... | | | | | | | | | | | |

where the genome segment is flanked by the two markers and each of the segments can be divided into number of genome positions, such as every percent of recombination frequency. For each of the genome positions, there are N corresponding re-coded variables for the position (Z1 and Z2) and the predetermined marker genotypes for controlling the residual genetic background (m1, m2, m3, …). The Zs are the coefficients for the two parameters and they are

$$\begin{cases} Z1 = X_{i1j} + (1-\rho)X_{i2j} + \rho X_{i3j} & \text{for} \quad \theta_1 \\ Z2 = \rho X_{i2j} + (1-\rho)X_{i3j} + X_{i4j} & \text{for} \quad \theta_4 \end{cases} \tag{4.43}$$

For determining which markers needed to control the residual genetic background, a stepwise regression can be used to obtain markers linked to potential QTLs and a preconstructed linkage map is needed to find out the relative positions of the target interval and the markers. For stepwise regression using SAS, a predetermined variance and covariance matrix for the markers and trait phenotype is recommended if there are missing values for the marker data and the trait data. For this case, if the original data is used for the regression analysis, most likely SAS only uses a portion of the data (SAS only uses the observations without any missing values for all the markers and the trait).

Another two data sets are needed for using SAS. They are the data set (called traitdata) which contains trait values corresponding to the lines, and a data set (called markerdata2) containing the data of markers on the other chromosomes (will be used in the model). The following SAS codes can be used for the CIM using the linear regression approach:

```
options ps = 60 ls = 80 nocenter;
/* Read trait data */
data trait;
 infile 'yourtrait.dat';
 input line y;
proc sort; by line;
/* Read marker data for the linkage group */
data markerdata1;
 infile 'yourmarker1.dat';
 input segment name1 $ name2 $ position line z1 z2 m1 m2 m3 ...;
proc sort; by line;
/* Read marker data for the rest of the linkage groups */
data markerdata2;
 infile 'yourmarker2.dat';
 input line l1 l2 l3 ...;
data all; merge trait markerdata1 markerdata2; by line;
proc sort data = all; by segment name1 name2 position;
/* Full model */
proc glm data = all noprint outstat = fullmodel;
by segment name1 name2 position;
model y = z1 z2 m1 m2 m3 ... l1 l2 l3 .../solution noint;
data fullmodel; set fullmodel; if _type_ = 'ERROR';
 keep segment name1 name2 position df ss;
data fullmodel; set fullmodel; rename ss = ssfull;
/* Reduced model */
proc glm data = all noprint outstat = redumodel;
by segment name1 name2 position;
model y = m1 m2 m3 ... l1 l2 l3 .../solution noint;
data redumodel; set redumodel; if _type_ = 'ERROR';
 keep segment name1 name2 position df ss;
data redumodel; set redumodel; rename ss = ssredu;
/* Merge the two data sets and compute the statistic and p-value */
data model; merge fullmodel redumodel;
by segment name1 name2 position;
gstatistic = df*(ssreduc-ssfull)/ssfull;
if g<0 then g = 0.0;
pvalue = 1.0-probchi(gstatistic,1.0);
proc print data = cd;
run;
```

Some modifications may be needed for your computer and the version of the software.

## 4.5 DISCUSSION

### 4.5.1 Commercial Quality Software Is Needed

Available public domain software packages for studying genes controlling complex traits are not adequate for development of genomic research on complex traits in terms of user interface, flexibility, and user support. Software packages with commercial quality are needed to accommodate the growing needs of data analysis and management in genomic research. This is especially true for study of genes controlling complex traits.

Private and public organizations have been working on automated devices for DNA sequencing and marker genotyping during the last decade. Technologies, such as sequencing using DNA chips, capillary array electrophoresis (CAE), multiplex PCR, whole genome shotgun sequencing, and transposon-facilitated direct sequencing, in combination with automated devices and new molecular biology techniques, reduce cost of genome research and increase the speed for DNA sequencing and fragment sizing. Despite rapid advances in technology, applications of genome information are small scale and limited within research laboratories. The technology is still expensive and inefficient relative to its potential. Those technologies are limited, not by the speed of biochemical reactions and laser beams, but by the speed of processing and integrating the data, which is a function of software design, computer speed, and algorithms. The currently available BIAM systems for genome research are not adequate for real world applications because (1) there are limitations in the mathematical and statistical algorithms used for analyzing DNA sequence data and genomic mapping, (2) current BIAM systems are designed for gene discovery and not for practical applications, (3) the software packages are not user friendly and not widely available because they are mostly developed by biologists and are not on the commercial market, and (4) the packages are not complete in terms of functionality and are difficult to incorporate into an automated and complete system. Only a few institutes with in-house informatics teams are able to semi-automate their genomic analysis process. Data entry, assembly, and analysis take more time than generating the data. The error and inaccuracy of genome research are largely due to human handling of the data. BIAM is becoming a bottleneck for practical applications of genomic research.

Methodology for DNA sequence polymorphism detection, such as SSCP, genomic mismatch scanning (GMS), high-density multiplex PCR, Cleavase fragment length polymorphism (CFLP), genetic bit analysis (GBA), etc. are becoming routine biochemical assays for many laboratories. Automated devices for carrying out those processes are being developed. An integrated system which can carry out BIAM from initial biochemical assays to final practical applications is needed. The system can provide a bridge from molecular biologists to practical users, such as clinical physicians, attorneys, plant and animal breeders, environmentalists, etc. A multi-disciplinary collaboration among molecular biologists, physicians, plant and animal biologists, forensic scientists, mathematicians, and computer scientists is needed to develop the system (Figure 4.3).

### 4.5.2 Interpretation of QTL Analysis Results

Common mistakes for interpreting results of the QTL analysis are (1) too confident on QTL position, (2) not an adequate significance level, and (3) wrong conclusions because of an over-simplified model.

#### 4.5.2.1 Where Are QTLs?

It is common to present results of QTL analysis by plotting lod score or other test statistics for detecting QTL against genome position. For example, Figure 4.4 shows an F statistic plotted against genome locations on the chromosome IV of barley using the nonlinear regression approach for simple interval mapping. Genome locations corresponding to peaks of the test statistic have been commonly considered as genome segments where the potential QTLs are located. However, this is true for some limited situations (Figure 4.5A and C). The peaks of the curve may not be the results of QTLs under the peaks. When there are more than one QTL located near the group of linked markers, the peak may be very wide or the peak does not correspond to the QTL location. The peak may correspond to the QTL location when a single QTL contributed to the peak. Figure 4.4 may have a very small amount of information on precise QTL locations. Even though the plot shows the whole genome, the results are still based on single-QTL analysis.

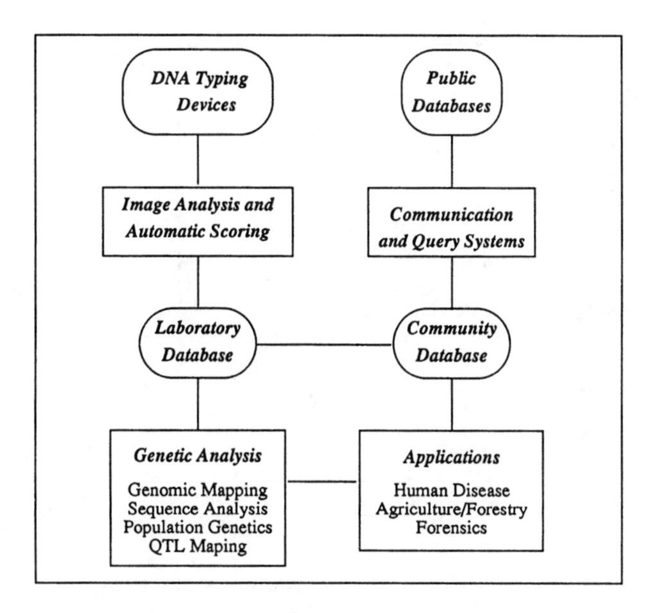


**FIGURE 4.3** Bioinformation analysis and management (BIAM) system and its relationship with automated DNA typing devices and public genome databases.

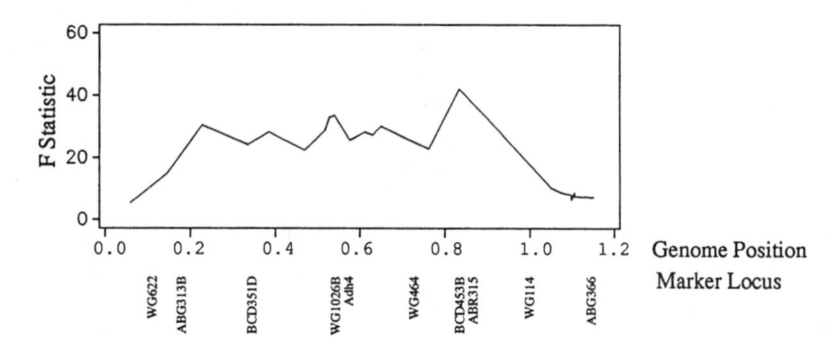


**FIGURE 4.4** F statistics are plotted against genome locations on the chromosome IV of barley using the nonlinear regression approach for simple interval mapping. Trait is malt extract. For information on experimental materials, see Hayes et al.[28] (Data is provided by the North American Barley Genome Mapping Project.)

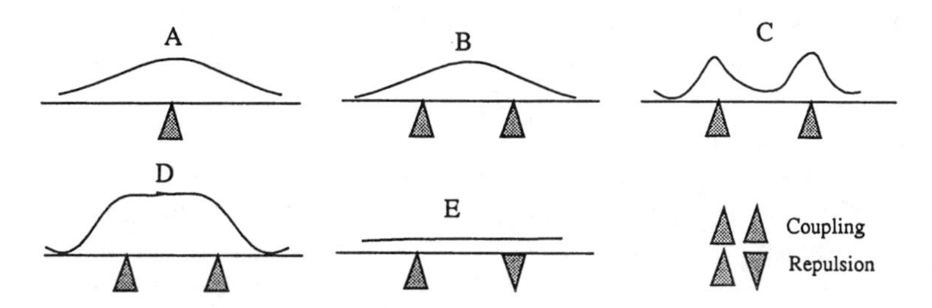


**FIGURE 4.5** Some possible patterns of test statistic plot for one-QTL and two-QTL models in QTL mapping. (A) peak corresponds to a single QTL; (B) peak corresponds to genome location between two linked QTLs in coupling phase; (C) two peaks correspond with two QTLs; (D) a wide plateau corresponds to two linked QTLs in coupling; and (E) there is a low peak or no peak for two linked QTLs in repulsion phase.

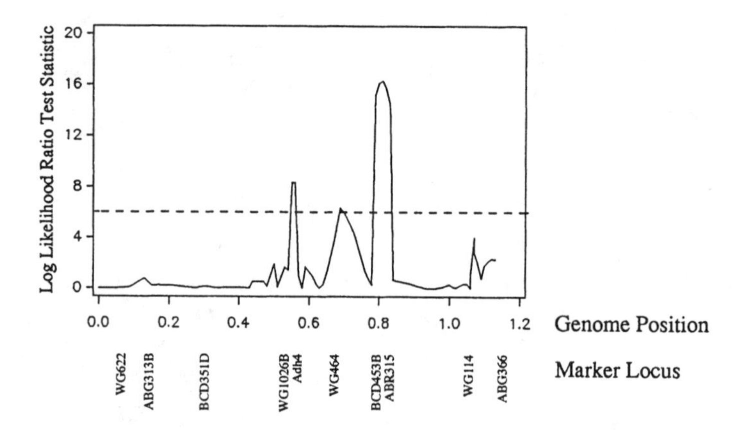


**FIGURE 4.6** The log likelihood ratio test statistic are plotted against genome locations on chromosome IV of barley using the composite interval mapping approach.

Composite interval mapping (CIM) approach increases the resolution of QTL location by using markers other than the markers flanking the segment to control residual noise in the model. Figure 4.6 shows the results for the same chromosome IV and the same trait using linear regression model 39. The analysis was done for every 1% recombination point by treating the putative QTL position parameters as known constants. The data contains 292 markers for the seven chromosomes of barley. It is impossible to fit all the markers in the models because the mapping population only contains 150 DHL. A subset of markers on different chromosomes (other than chromosome IV) were picked using stepwise regression. Markers Adh7, ABC165, ABR337, and ABR337 (on chromosomes I, II, V, and VII, respectively) were found to be significantly associated with malt extract and selected as markers to be fitted in the models. On chromosome IV, markers were selected based on their significance of association with the trait and their position relative to each of the 25 segments on the genome.

Compared with the simple interval mapping (Figure 4.4), the CIM approach gives higher resolution on QTL location. Both approaches identify the peak near marker BCD453B. However, a peak near marker WG1026B and a peak between these markers were identified differently. The simple interval mapping identifies the peaks located near markers WG1026B and ABR315 (Figure 4.4). However, the CIM approach identifies the peak near markers Adh4 (closely linked with WG1026B with recombination 3% away) and WG464 (the peak is about 8% recombination units from ABR315).

The difference for a peak near WG1026B and Adh4 can be explained as the QTL resolution differences among these methods. The single marker analysis and the simple interval mapping cannot resolve precise location for the potential QTL near that region of genome. The composite interval mapping approach can efficiently use the markers as a boundary for locating QTLs. For this case, Adh4 is a marker at the boundary point. The potential QTL is located between Adh4 and ABA3 a region about 2 cM long.

For the difference on the peak in the middle, the explanation is that the peak of ABR315 using the single marker analysis and the simple interval mapping is a result of linked markers Adh4 and BCD453B, which are significantly associated with potential QTLs. When the analysis for segment ABR315 and WG464 conditional on Adh4 and BCD453B (put Adh4 and BCD453B in the model to control the genetic background) the peak disappears. For the segment between WG464 and BCD453B, the peak shows up when the analysis is conditional on the adjacent possible QTLs. This is a sign of repulsion linkage phase among the possible QTLs. However, a marker or several markers between WG464 and BCD453B and more individuals in the mapping population are needed (depending on where exactly the QTL located in the segment) because the next potential major QTL are located in the next segment between markers BCD453B and ABG472. WG464 is about

10% recombination from BCD453B of the next peak and there is no marker between WG464 and the marker (BCD453B) at the boundary between two possible QTLs.

#### 4.5.2.2 How Significant Are the QTL?

Adding to confusion in interpreting results of a QTL analysis is an undetermined significance level due to non-independent multiple test problem. QTL analysis is a multiple test problem. Number of repeated analyses can range in number of markers in the data and in number of small segments (e.g., 1 cM) of the genome depending on methodology of the QTL analysis. The repeated analyses are not independent because some of the markers or the segments are linked to each other. For the two extremes, all markers and segments are linked completely and all markers and segments are independent, the significance level can be simply determined using standard statistical distributions, such as chi-square, t-distribution, F-distribution, or simple load score test statistic. However, situations are always between the two extremes in practical QTL analysis. The markers and segments used in QTL analysis are linked in some degree.

Three basic approaches have been proposed to solve the problem. A Monte Carlo simulation approach was suggested to build a distribution of test statistic under a null hypothesis.[3,7,43] It was also suggested to use a conservative threshold value to reduce a false positive.[36,40] Churchill and Doerge[20] discussed threshold value for declaring a significant QTL and proposed a permutation approach to obtain empirical threshold value. They proposed two levels of threshold values, comparisonwise and experimentwise. A test statistic is obtained by permuting trait values. N test statistic for each analysis point (a marker or a small segment) can be obtained by repeating the permutation N times. A comparisonwise critical value is the $100(1 - \alpha)$ percentile in distribution of the test statistic at the analysis point ($\alpha$ is the significant level). For each permutation, a maximum test statistic is obtained over all the analysis points. The $100(1 - \alpha)$ percentile in the distribution of the N maximum test statistic is the experimentwise critical value.

Intuitively, the permutation approach should provide an adequate significance level for a test statistic in QTL analysis. However, this approach is computational intensive and has not been implemented in all computer software for QTL analysis. To determine a relative precise significance for a test statistic in QTL analysis a large computational effort is needed.

#### 4.5.2.3 Are the QTLs Real?

The resolution of QTL location is commonly low and the significance level of a test statistic is not precise. A logical question to ask is are the QTL real? Are the QTL real in statistical terms? Are the QTL real in biological terms? Do the estimated gene effects mean anything statistically? I will try to answer the statistical questions using the following example and answer the biological question in the next section.

For the case in point, malt extract means for Steptoe and Morex genotypes of combinations of markers Adh4, BCD453B, and WG464 are shown in Table 4.11. If the overall means are compared for WG464, the Morex genotypes are 1.0899 units higher than the Steptoe genotype. However, if the conditional means on Adh4 and BCD453B are compared to WG464, the Steptoe genotype is 1.7083 and 0.5843 higher than the Morex genotype for the two classes observed, respectively.

When QTLs are linked in the repulsion phase, peaks of the test statistic can be lowered or limited. However, when QTLs are linked in the coupling phase, the estimates of the QTL effects can be overly estimated. In practical data analysis, these cases can occur frequently. For the barley data, Morex is considered a high malt-extract variety and a commonly used malting barley whereas Steptoe is a low extract variety and a feeding barley. Common sense would say Morex has all the high malt extract alleles and Steptoe has all the low malt extract alleles. Only the conditional analysis can resolve the problem of repulsion linkage phase for linked QTLs. Using the QTL mapping for assisting in plant and animal breeding, by finding the bad alleles in the good families or varieties and the good alleles in the bad families or varieties is more meaningful than finding the good alleles in good varieties and the bad alleles in bad varieties.

**TABLE 4.11**
**Malt-Extract Means and Frequencies for Steptoe and Morex Genotypes for Marker Combinations among Adh4, BCD453B, and WG464**

| Adh4 | BCD453B | WG464 Steptoe | WG464 Morex | Mean difference |
|---|---|---|---|---|
| Steptoe | Steptoe | 72.3673 (57) | — (0) | — |
| | Morex | 74.5250 (4) | 72.8167 (6) | 1.7083 |
| Morex | Steptoe | 73.1143 (7) | 72.5300 (10) | 0.5843 |
| | Morex | — (0) | 74.0044 (46) | — |
| Mean | | 72.5732 (71) | 73.6631 (65) | −1.0899 |

*Note:* The number of observations is in parentheses.

From the barley example, it is obvious that the QTL may not be real and the estimated QTL effects may not mean anything if the underlying genetic or statistic model is wrong. A genetic model may be wrong due to assumptions on number of genes and linkage relationships among the genes. A statistical model may be wrong due to the imbalance of the data. To have a high confidence level on a QTL, extensive analyses under many possible genetic and statistical models are needed. However, most of the computer software packages for QTL analysis are not suitable for extensive model screening.

### 4.5.3 High Resolution QTL Mapping

To increase the resolution of QTL location, a multiple-step strategy may be effective (Figure 4.7). It is important to initially analyze traits of interest quantitatively. The quantitative analysis has a potential to screen for families with major and useful gene segregating and eliminate some experimental errors. Traditionally, attention has been paid to family mean and variance among families in plant and animal breeding and quantitative trait evolution study. Within-family variation has been considered secondary in plant and animal improvement and evolutionary process of biology. This is not because within-family variation is not important but rather lack of strategies to study within-family variation. QTL mapping mainly explores within-family variation. Linkage disequilibrium among genes and markers is best studied within family. For some cases, existence of segregating major genes within family can be determined by studying trait value distribution and variation within family. These segregating genes are targets of QTL mapping. For a lot of cases, gene segregating in the family may not show any effect on trait distribution because of small genetic effects, large number of genes, and large environmental and experimental errors. However, it is still a good start to analyze the trait quantitatively because usually only a small effort is needed relative to the future work.

For organisms, in which controlled crosses are commonly used for genomic research, quantitative trait analysis can be implemented for parents of the crosses and the progeny. Certainly, large trait difference between the parents may increase the chance to detect QTL in their progeny. However, this may not always be desirable depending on the objectives of the experiment. For example, progeny of a cross between a high and a low yield crop variety may increase the chance to find yield QTLs. However, the QTLs found in this cross may have little use if the objective of the genomic experiment is to assist breeding for yield because there is a big chance that all the desirable QTL alleles come from the same parent and are already fixed. Only QTLs with desirable alleles from the low yield parent are useful for further yield improvement. In this case, two high-yield parents may be desirable. Any QTL found in the cross between two high-yield parents will be useful for further yield improvement. This is especially true for organisms with a long history

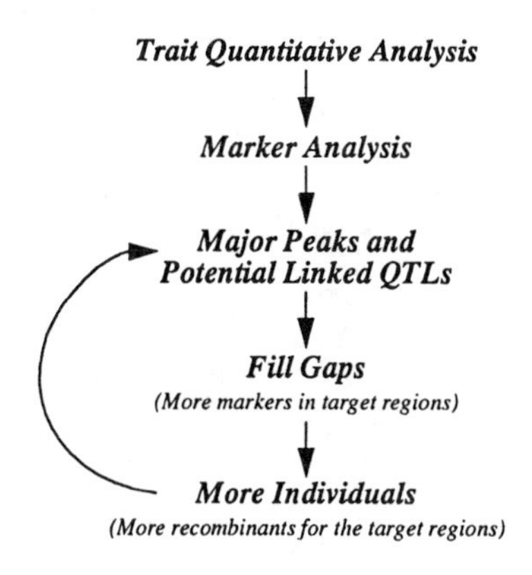


**FIGURE 4.7** Steps to obtain a high resolution QTL mapping.

of breeding. In most cases, genetic variation within progeny of cross between two parents with similar trait value may be small. The chance to find QTL in the progeny may be small. So it is essential to screen crosses of parents with desirable trait value and within-population genetic variation to have a successful and useful QTL mapping experiment. Within-population genetic variation is the single most important factor for the quantitative analysis for obtaining high-resolution QTL analysis. Characters of both parents are important for a useful QTL mapping experiment.

For organisms, in which controlled crosses are difficult to make, quantitative trait analysis should focus on existing populations, e.g., most forest tree species. For practical applications of QTL mapping in forest-tree breeding, it is critical to use the populations that exist and are ready for use in breeding. Theoretically, results of a QTL mapping using designed experiments should be useful in tree breeding. In practice, however, the results may have little use because of time needed for making crosses, evaluating the trait, and transferring QTL. A quantitative screening among available materials will ensure a high resolution and useful QTL analysis.

Besides quantitative analysis on within-population variation, another important aspect of quantitative analysis is to control environmental and experimental errors. Contrary to traditional quantitative genetics, QTL analysis depends more on single genotypic value. Precision of measurement of the trait at an individual genotype level is key for a high resolution QTL analysis. Environmental heterogeneity within population should be eliminated before QTL analysis. It is common to obtain least square means for each of the genotypes and to use them as trait values for QTL analysis.

The initial QTL analysis on the data should find the major test statistical peaks and potential linked QTL. For finding the major peaks, simple linear model approach (t-test, linear regression, or ANOVA) or interval mapping should be adequate. To narrow down the QTL locations, the composite interval mapping approach can be used.[6,7] An approach using markers as boundary conditions should be an alternative due to the principal that QTL effect cannot pass the marker which is used as a conditional boundary. In this way, precise QTL locations may not be determined. However, a large portion of genome can be excluded from target QTL locations. The next step would be searching for more markers in the regions which are not excluded from the conditional analysis, e.g., the regions between markers WG464 and BCD453B and between BCD453B and ABR315 for the barley malt extract QTL in Figure 4.6. A commonly used approach to fill saturate map of a region, such as bulk segregating, can be used to efficiently fill the gaps.

For many situations, more individuals in the mapping population may be needed to observe recombinants when more markers are genotyped in a region. It is also logical to genotype only the recombinant individuals. For example, only individuals with genotypes AAbb or aaBB should be used to fill more markers within region between markers A and B if AAbb and aaBB are known as recombinants. By repeating the process of QTL analysis (fill gaps — more individuals), a relatively high QTL resolution can be reached.

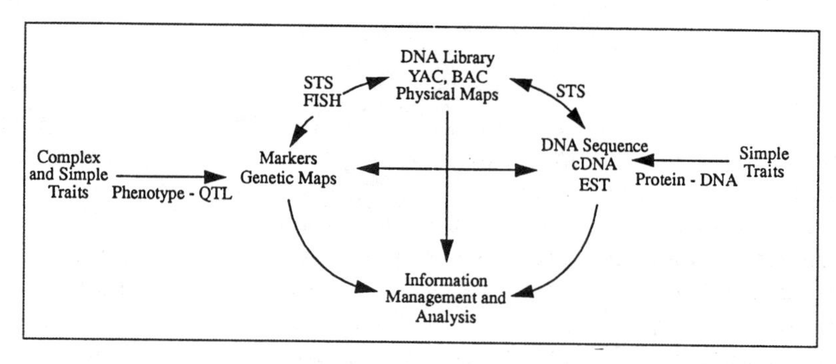


*Some Concepts for Making the Connections among Traits – Maps – Sequences:*

1. STS, EST and cDNA can be mapped on both genetic and physical maps and sequenced.
2. Genes for simple traits can be cloned by using a protein to DNA approach. Genes controlling complex traits can be located coarsely on the genetic map using statistically oriented QTL mapping.
3. FISH brings classical cytogenetics and genetic and physical maps together.
4. Physical map using overlapped cloned DNA fragments is an efficient way for getting high resolution genetic map within a short region and is a powerful tool for isolating the target genes.
5. Genomic information goes to information management and analysis systems (IMAS). Ultimately, Genomics prevails when the IMAS can resemble the cell where the information comes from.

**FIGURE 4.8** A prospective of genomics. STS = sequence tagged sites; FISH = fluorescent *in situ* hybridization; YAC = yeast artificial chromosomes; BAC = bacterial artificial chromosomes; EST = expressed sequence tag.

## 4.5.4 Integration of Genetic and Physical Maps

Resolution of QTL mapping can be increased by picking a mapping population, using improved statistical procedures, and employing the multiple-step strategy. However, there is a limit on how high the resolution can be reached using those approaches because those approaches are based on number of recombinants between marker and gene and genetic variation of the trait. The resolution may still be insufficiently high for some applications. The number of recombinants between marker and gene is limited by number of polymorphic markers which can be obtained in the genome region and population size. Once mapping population is chosen there is a limited genetic variation which can be worked within QTL mapping. Alternatives, which are not based on recombination, are needed for further improvement of QTL resolution. Approach using QTL mapping as an intermediate step to connect trait and physical map may an alternative (Figure 4.8).

### 4.5.4.1 Genetic and Physical Maps

Both genetic and physical maps have been obtained for several organisms with small genomes, such as *E. coli*, yeast, fruit fly, *C. elegans*, *Arabidopsis thaliana*, rice, etc. High density genetic maps have been developed for organisms with relatively large genomes, such as human, barley, maize, tomato, and loblolly pine. Physical maps have been constructed for some of the human chromosomes.

In principle, genetic and physical maps should resemble same genome constitution, such as order of genes or markers and equivalent distances among the loci. However, the relationship between the distances in units of recombination and physical length may not be straight forward due to nonrandomness of the crossing-over events. Moreover, the gene orders on genetic map and physical map may not agree with each other due to random sampling errors or inappropriate laboratory (including data analysis) procedures. Based on this logic, caution should be paid in making the connection between the genetic and the physical maps. Unnecessary connections should not be made between the maps; contradiction between the maps is only one side of the problem. Instead, advantages of both maps should be taken to serve the needs of genomic analysis.

A genetic map is built based on meiotic recombination between homologous chromosomes. Polymorphic markers are needed to identify the crossing-over event which generates the meiotic recombination. Resolution of a genetic map is limited by the crossing-over events. If no crossing-over event can be detected within a region of genome then the genetic mapping cannot resolve the precise genome positions for the genes and markers in the region. To increase the resolution for genetic map usually means that a relatively large mapping population is needed. Physical length of a genome has a relatively small impact on the size of genetic mapping. For a small genome, a recombination unit may represent a small physical segment of the genome. For a large genome, a recombination unit may represent a relatively large segment of the genome. For example, a recombination unit represents on average approximately 200 kb in *Arabidopsis thaliana* and 15 to 20 Mb in loblolly pine.

On the other hand, physical mapping is based on the overlapped DNA fragments. Polymorphism and recombination are not essential for physical mapping. The limitation on resolution is not applied to physical mapping. Instead, physical mapping is limited by the size of the cloned DNA fragments. If size of the DNA fragments is small then a large number of clones are needed to cover even a small region of genome. As genome size increases, the number of clones needed to cover certain portions of genome increases.

Genetic map and physical map can be related to each other by placing genetic markers on the physical map and the DNA fragments on the genetic map. By doing this the genetic map and the physical map can be cross validated with each other. Closely linked genes and markers on the genetic map usually having little confidence on gene order can be checked with the physical map at the region. Gaps are usually common for low coverage physical map and can be filled using a genetic map.

Most importantly, the genetic map is a bridge between traits and physical map. A coordination between the genetic and physical map is an efficient way to identify and isolate genes of interest.

#### 4.5.4.2 Trait — Maps — Sequence

It is not logical (at least for now) to relate a complex trait with a massive DNA sequence data directly. However, DNA sequence is the ultimate information resource for the trait development. Reverse genetics, from protein to DNA, is an elegant way to isolate genes with simple functions. Many genes have been identified and cloned using the reverse-genetics approaches. These genes control mostly the single-gene traits. The coordination among genetic and physical maps and DNA sequences may provide more efficient ways to isolate genes with simple functions and possibly to isolate genes controlling complex traits. Genetic and physical maps are bridges between complex traits and DNA sequences (Figure 4.9).

Based on genetic mapping, an association between a marker or a group of markers segregated and a trait variation is the sign that a gene controlling the trait is located near the marker. To increase the resolution of the gene location, approaches such as increasing mapping population size, improving statistical methodology, using linkage equilibrium accumulated from generations, etc. have been proposed. However, gene location using those improvements is usually still not precise enough for isolating the gene. For example, a disease gene may be located within a centiMorgan region flanked by two markers (Figure 4.10). In human, 1 cM still represents 1 Mb of DNA. This is not precise enough to clone the disease gene efficiently. A physical map for the region can be obtained. For example, 12 BAC clones will cover the 1-Mb region. We can use the BAC clones to identify disease haplotypes and sample the disease haplotypes from population. The precise location of the disease gene can be located relative to the overlapped BAC clones.

Statistics has limited power. This is especially true when statistics is inappropriately used. The biology will ensure the statistics appropriately used. When complicated statistics and intensive computing cannot solve the problem, biological evidences from different aspects may solve the problem efficiently. There are many difficult statistical and computational problems associated with genetic mapping, physical mapping and DNA sequence analysis, and multiple QTL mapping. For

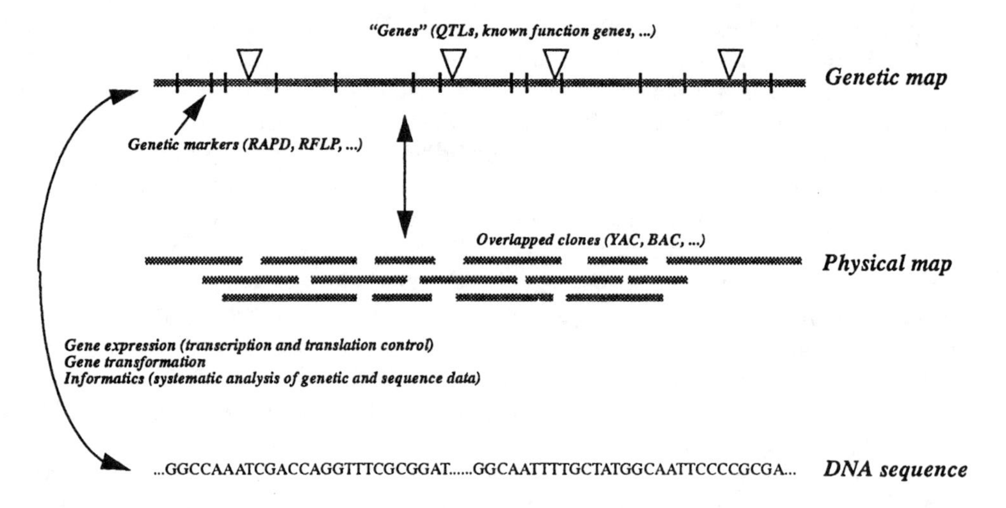


**FIGURE 4.9** Relationship among genetic map, physical map, and DNA sequence. RAPD = random amplification of polymorphic DNA; RFLP = restriction fragment length polymorphism; YAC = yeast artificial chromosomes; BAC = bacterial artificial chromosomes.

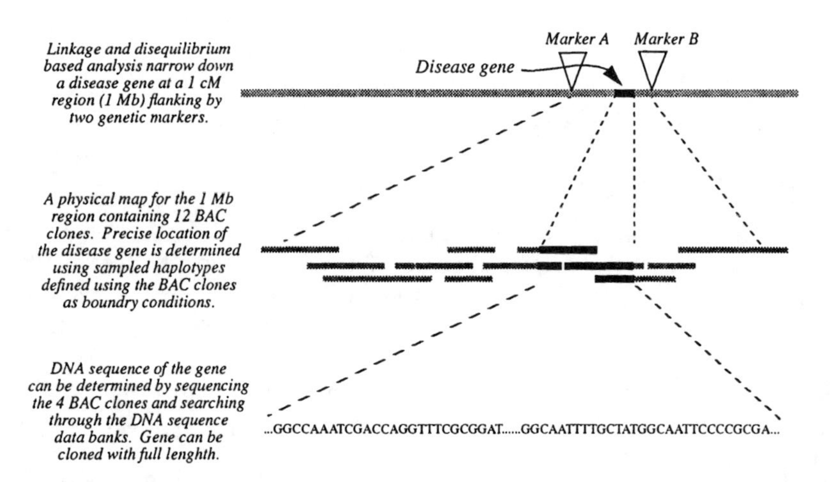


**FIGURE 4.10** Study of a disease gene using a combination of genetic mapping (linkage and disequilibrium), physical mapping, and DNA sequencing. BAC = bacterial artificial chromsomes.

example the gene ordering in genetic mapping and the double digest problem belong to a mathematically difficult problem, NP-complete. There is no general solution for the NP-complete problem. However, if we put the genetic mapping and physical mapping together, those problems may be solved without touching the NP-complete problem.

### 4.5.5 Integration of Metabolic Pathway with QTL Information

#### 4.5.5.1 What Are QTLs?

In recent years, great progress has been made in both molecular and quantitative genetics. In molecular genetics many specific genes and proteins have been identified and characterized affecting growth, metabolism, development and behavior of plants and animals. New methods of quantitative genetics have similarly identified genetic regions regulating important functions through the genetic dissection of quantitative traits. QTL analysis has fundamentally changed the conventional view of polygenic inheritance and has led to the identification of a new set of loci, with important biological roles, but usually lacking in molecular identification. Most genes identified and characterized at the molecular level have no known function, and most genes of presumably known function have not been subjected to direct tests of their biological roles through loss of function-gain of function experiments. In most cases, the quantitative effects of genes of known functions have not been explored. Few cases exist, such as the human hemoglobins, where the underlying molecular events are well understood and the consequences of these changes in metabolism and development have a reasonable molecular basis. Where some individual genes and their effects on complex phenotypic traits are understood, much less is known about the genetic and biochemical basis of the regulation and expression of coordinately and differentially controlled gene expression and protein regulation of complex biochemical pathways.

Lander and Schork[22] wrote in their review paper on genetic dissection of complex traits:

> ... one can systematically discover the genes causing inherited diseases without any prior biological clue as to how they function. The method of genetic mapping, by which one compares the inheritance pattern of a trait with the inheritance patterns of chromosomal regions, allows one to find where a gene is without knowing what it is....

What they wrote is true if one compares the gene mapping approach with traditional quantitative approaches. The gene mapping approach has revolutionized the ways of finding genes from the traditional approaches. However, without knowing what QTLs are biologically, the QTL mapping approach does not provide much more information in terms of fundamental biology than the traditional quantitative approaches. Definition of QTL, such as QTLs are genes located on the genome with significant additive, dominance, or epistatic genetic effects, is superficial. Without knowing what QTLs are may promote wrong genetic models and imprecise definitions. Most importantly, without knowing what QTLs are, complete genetic models are difficult to construct in practical experimentations and QTL mapping cannot be very efficient. In this section, I will discuss limitations of QTL models, the relationship among the traditional quantitative approach, QTL mapping approach and the molecular biology, and the possibility to construct the metabolic genetic models.

#### 4.5.5.2 What Are QTL Effects?

There are four possible reasons for an unsuccessful multiple-locus model for a quantitative trait. (1) QTLs are hypothetical genes based on statistical inference; (2) genetic models on which the QTL mapping are based are not accurate; (3) the amount of genetic information contained is not adequate; and (4) the methodology of statistical analysis is not powerful enough. It is generally believed that the current statistical tools are not adequate for dealing with high levels of epistatic interactions. Another limitation is that the definitions of gene effects used in QTL mapping have very little biological meaning.

In molecular biology, genes are defined as segments of DNA on genomes and include structural genes encoding enzymes on biosynthetical pathways and regulatory genes regulating gene functions. Definitions of gene used in the QTL mapping and the traditional quantitative genetics are not precise, such as the unit of inheritance and a functional genome segment. QTL is a statistical entity instead of a biological one. In terms of gene effects, additive, dominance, and epistatic effects have been used for QTL mapping and quantitative genetics. How these definitions can be related to each other and how those approaches can be integrated with each other will be important for the development of modern biology.

Figure 4.11 shows a hypothetical relationship among different approaches for modeling QT inheritance. The molecular approach models QT as gene regulation, one-gene (structural)

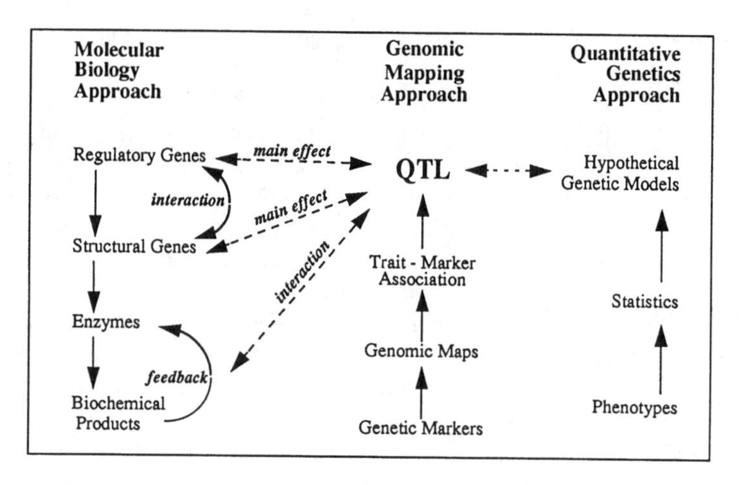


**FIGURE 4.11** The metabolic genetic model (MGM) will validate molecular, QTL mapping and quantitative genetic approaches for a better understanding of quantitative trait inheritance.

-one-enzyme, and possible feedback or feed-forward between products and enzyme activities. Quantitative genetics models QT in the context of the properties of population using terms, such as additive effect, dominance effect, and epistatic interactions. Genomic mapping approach models QT using hypothetical QTLs obtained by analysis of trait and genetic marker association or so-called QTL mapping. QTL mapping has shown a greater power in finding possible genome positions of genes controlling QTs over the classical quantitative approaches. It is logical to say that the QTL mapping approach will serve as a bridge between the quantitative genetics and molecular biology. To make the connection between "gene effects" of quantitative genetics and biochemical processes, QTLs should have their physical definitions instead of "genes". So the question is: what are QTLs.

What is an adequate approach for modeling QT? A combination of molecular and QTL approaches may be a logical alternative. Let's call the model which integrates metabolic pathway with QTL mapping MGM (metabolic genetic model). MGM has at least the following advantages: (1) cDNAs encoding enzymes or regulatory proteins can be mapped on the genome in the same way as for regular genetic markers. The QTL mapping based on this prior information begins searching for other structural and regulatory genes. MGM can be built based on predetermined trait-gene relation. The QTL mapping models are based purely on statistical association between traits and molecular markers. (2) For MGM, a complicated biological relationship among the components of the trait is determined by biochemical experiments and the information can be integrated with the analysis. (3) Genetic models and metabolic pathways can be cross validated with each other. The genetic models will help in searching for additional genes on the pathways and the pathway will help in building more accurate and precise genetic models.

## ACKNOWLEDGMENTS

Thanks to Dr. Andy Kleinhofs at Washington State University and Dr. Pat Hayes at Oregon State University for providing the barley data, to Drs. Ron Sederoff, David O'Malley, Ross Whetten, Bruce Weir, and Zhaobang Zeng at North Carolina State University for stimulating discussion on QTL mapping, to Dr. Andy Paterson for critical comments on the manuscript and to my graduate student, Ling Li, for useful comments. This research was partially supported by USDA Plant Genome Program awards 93-37300-8839 and 94-37300-0281, NIH grant GM32518, Pioneer Hi-Bred International, and NCSU Forest Biotechnology Industrial Associates Consortium.

## REFERENCES


1. Stuber, C. W., Lincoln, S. E., Wolff, D. W., Helentjaris, T., and Lander, E. S., Identification of genetic factors contributing to heterosis in a hybrid from two elite maize inbred lines using molecular markers, *Genetics,* 132: 823–839, 1992.
2. Weller, J. I., Soller, M., and Brody, T., Linkage analysis of quantitative traits in an intraspecific cross of tomato (*Lycopersicon esculentum* × *Lycopersicon pimpinellifolium*) by means of genetic markers, *Genetics,* 118, 329–339, 1988.
3. Lander, E. S., Botstein, D., Mapping Mendelian factors underlying quantitative traits using RFLP linkage maps, *Genetics,* 121, 185–199, 1989.
4. Knapp, S. J., Bridges, W. C., and Liu, B.-H., Mapping quantitative trait loci using nonsimultaneous and simultaneous estimators and hypothesis tests, in *Plant Genomes: Methods for Genetic and Physical Mapping,* Beckmann, J. S. and Osborn, T. S., Eds., Kluwer Academic Publishers, Dordrecht, The Netherlands, 1992, 209–237.
5. Lande, R. and Thompson, R., Efficiency of marker-assisted selection in the improvement of quantitative traits, *Genetics,* 124, 743–756, 1990.
6. Zeng, Z. B., Theoretical basis of separation of multiple linked gene effects on mapping quantitative trait loci, *Proc. Natl. Acad. Sci. U.S.A.,* 90, 10,972–10,976, 1993.
7. Zeng, Z. B., Precision mapping of quantitative trait loci, *Genetics,* 136, 1457–1466, 1994.
8. Haseman, J. K. and Elston, R. C., The investigation of linkage between a quantitative trait and a marker locus, *Beh. Genet.,* 2, 3–19, 1972.
9. Lange, K., A test statistic for the affected-sib-set method, *Am. J. Hum. Genet.,* 40, 283–290, 1986.
10. Weeks, D. E. and Lange, K., The affected-pedigree-member of linkage analysis, *Am. J. Hum. Genet.,* 42, 315–325, 1988.
11. Fulker, D. W. and Cardon, L. R. A sib-pair approach to interval mapping of quantitative trait loci, *Am. J. Hum. Genet.,* 54, 1092–1103, 1994
12. Cardon, L. R. and Fulker, D. W., The power of interval of quantitative trait loci, using selected sib pairs, *Am. J. Hum. Genet.,* 55, 825–833, 1994.
13. Lander, E. S., Green, P., Abrahamson, J., Barlow, A., Daly, M. J., Lincoln, S. E., and Newburg, L., MAPMAKER: an interactive computer package for constructing primary genetic linkage maps of experimental and natural populations, *Genomics,* I, 174–81, 1987.
14. Liu, B. H. and Knapp, S. J., QTLSTAT1.0, a software for mapping complex trait using nonlinear models, Oregon State University, 1992.
15. Basten, C. J., Zeng, S. B., and Weir, B. S., ZMAP-A QTL cartographer. Proceedings of the 5th World Congress on Genetics Applied to Livestock Production, 22, 65–66, 1994.
16. Lu, Y. Y., and Liu, B. H., PGRI, a software for plant genome research. Plant Genome III conference abstract, p. 105, San Diego, CA, 1995.
17. Manly, K. F. and Cudmore, E. H., Jr., New versions of MAP Manager genetic mapping software, Plant Genome IV (Abst.), 1996, 105.
18. van Ooijen, J. W. and Maliepaard, C., MAPQTL version 3.0: Software for the calculation of QTL position on genetic map, Plant Genome IV (Abst.), 1996, 105.
19. Tanksley, S. D. and Nelson, J. C., Advanced backcross QTL analysis: a method for the simultaneous discovery and transfer of valuable QTLs from unadapted germplasm into elite breeding, *Theor. Appl. Genet.,* 92, 191–203, 1996.
20. Churchill, G. A. and Doerge, R. W., Empirical threshold values for quantitative trait mapping, *Genetics,* 138, 963–971, 1994.
21. Tanksley, S. D., Mapping polygenes, *Annu. Rev. Genet.,* 27, 205–233, 1993.
22. Lander, E. S. and Schork, N. J., Genetic dissection of complex traits, *Science,* 265, 30, 1994.
23. Mackay, T. F. C., The genetic basis of quantitative variation: number of sensory bristles of *Drosophila melanogaster* as a model system, *Trends Genet.,* 11, 464–470, 1995.
24. Frankel, W. N., Taking stock of complex trait genetics in mice, *Trends Genet.,* 11, 471–477, 1995.
25. Haley, C. S., Livestock QTLs — bringing home the bacon?, *Trends Genet.,* 12, 488–492, 1995.
26. McCouch, S. R. and Doerge, R. W., QTL mapping in rice, *Trends Genet.,* 11, 482–487, 1995.
27. Hayes, P. M., Liu, B.-H., Knapp, S. J., Chen, F., Jones, B., Blake, T., Franckowiak, J., Rasmusson, D., Sorrells, M., Ullrich, S. E., Wesenberg, D., and Kleinhofs, A., Quantitative trait locus effects and environmental interaction in a sample of North American barley germplasm, *Theor. Appl. Genet.,* 87, 392–401, 1993.

28. Sturtevant, A. H., The linear arrangement of six sex-linked factors in Drosophila, as shown by their mode of association, *J. Exp. Zool.,* 14, 43–59, 1913.
29. Sax, K., The association of size differences with seed-coat pattem and pigmentation in *Phaseolus vulgaris, Genetics,* 8, 552–560, 1923.
30. Penrose, L. S., The detection of autosomal linkage in data which consist of pairs of brothers and sisters of unspecified parentage, *Ann. Eugen.,* 6, 133–38, 1935.
31. Soller, M., Brody, T., and Genizi, A., On the power of experimental design for the detection of linkage between marker loci and quantitative loci in crosses between inbred lines, *Theor. Appl. Genet.,* 47, 35–39, 1976.
32. Stuber, C. W., Edwards, M. D., and Wendel, J. F., Molecular-marker facilitated investigations of quantitative trait loci in maize. II. Factors influencing yield and its component traits, *Crop Sci.,* 27, 639–648, 1987.
33. Edwards, M. D., Stuber, C. W., and Wendel, J. F., Molecular marker facilitated investigations of quantitative trait loci in maize. 1. Number, genomic distribution, and type of gene action, *Genetics,* 116, 113–125, 1987.
34. Stuber, C. W., Mapping and manipulating quantitative traits in maize, *Trends Genet.,* 11, 477–481, 1995.
35. Weller, J. I., Maximum likelihood techniques for the mapping and analysis of quantitative trait loci with the aid of genetic markers, *Biometrics,* 42, 627–640, 1986.
36. Knapp, S. J. and Bridges, W. C., Using molecular markers to estimate quantitative trait locus parameters: power and genetic variances for unreplicated and replicated progeny, *Genetics,* 126, 769–777, 1990.
37. Knott, S. A. and Haley, C. S., Aspects of maximum likelihood methods for the mapping of quantitative trait loci in line crosses, *Genet. Res.,* 60, 139–151, 1992.
38. Martinez, O. and Curnow, R. N., Estimation of the locations and the sizes of the effects of quantitative trait loci using flanking markers, *Theor. Appl. Genet.,* 85, 480–488, 1992.
39. Jansen, R. C., A general mixture model for mapping quantitative trait loci by using molecular markers, *Theor. Appl. Genet.,* 85, 252–260, 1992.
40. Jansen, R. C., Interval mapping of multiple quantitative trait loci, *Genetics,* 135, 205–211, 1993.
41. Jansen, R. C., Maximum likelihood in a generalized linear finite mixture model by using the EM algorithm, *Biometrics,* 49, 227–231, 1993.
42. Lou, Z. W. and Kearsey, M. J., Maximum likelihood estimation of linkage between a marker gene and a quantitative trait locus, *Heredity,* 66, 117–124, 1989.
43. Knott S. A. and Haley, C. S., Maximum likelihood mapping of quantitative trait loci using full-sib families, *Genetics,* 132, 1211–1222, 1992.
44. Rodolphe, F. and Lefort, M., A multi-marker model for detecting chromosomal segments displaying QTL activity, *Genetics,* 134, 1277–1288, 1993.
45. Knapp, S. J., Mapping quantitative trait loci, in *DNA-Based Markers in Plants,* Phillipes, R. L., and Vasil, K., Eds., Kluwer Academic Publishers, Dordrecht, The Netherlands, 1994, 58–96.
46. Moreno-Gonzalez, J., Genetic models to estimate additive and non-additive effects of marker-associated QTL using multiple regression techniques, *Theor. Appl. Genet.,* 85, 435–444, 1992.
47. Cockerham, C. C. and Weir, B. S., Linkage between a marker locus and a quantitative trait of sibs, *Am. J. Hum. Genet.,* 35, 263–73, 1983.
48. Weeks, D. E. and Lange, K., A multilocus extension of the affected-pedigree-member method of linkage analysis, *Am. J. Hum. Genet.,* 50, 859–868, 1992.
49. Fulker, D. W., Cherny, S. S., and Cardon, L. R., Multipoint interval mapping of quantitative trait loci, using sib pairs, *Am. J. Hum. Genet.,* 56, 1224–1233, 1995.
50. Jiang, C. and Zeng, Z.-B., Multiple trait analysis of genetic mapping for quantitative trait loci, *Genetics,* 140, 1111–1127, 1995.
51. Rebai, A., Goffinet, B., and Nangin, B., Approximate thresholds of interval mapping tests for QTL detection, *Genetics,* 138, 235–240, 1994.
52. Rebai, A., Goffinet, B., and Mangin, B., Comparing power of different methods for QTL detection, *Biometrics,* 51, 87–99, 1995.
53. Boehnke, M., Limits of resolution of genetic linkage studies: implications for the positional cloning of human disease genes, *Am. J. Hum. Genet.,* 55, 379–390, 1994.

54. Jansen R. C. and Stam, P., High resolution of quantitative traits into multiple loci via interval mapping, *Genetics,* 136, 1447–1455, 1994.
55. Kruglyak, L. and Lander, E. S., High-resolution genetic mapping of complex traits, *Am. J. Hum. Genet.,* 1212–1223, 1995.
56. Terwilliger, J. D. and Ott, J., *Handbook of Human Genetic Linkage,* The Johns Hopkins University Press, Baltimore, 1994.
57. Nienhuis, J., Helentjaris, T., Slocum, M., Ruggero, B., and Schaefer, A., Restriction fragment length polymorphism analysis of loci associated with insect resistance in tomato, *Crop Sci.,* 27, 797–803, 1987.
58. Paterson, A. H., Lander, E. S., Hewitt, J. D., Peterson, S., Lincoln, S. E., and Tanksley, S. D., Resolution of quantitative traits into Mendelian factors by using a complete RFLP linkage map, *Nature,* 335, 721–726, 1988.
59. Paterson, A. H., Damon, S., Hewitt, J. D., Zamir, D., Rabinowitch, H. D., Lincoln, S. E., Lander, E. S., and Tanksley, S. D., Mendelian factors underlying quantitative traits in tomato: comparison across species, generations, and environments, *Genetics,* 127, 181–197, 1991.
60. Groover, A., Devey, M., Fiddler, M., Lee, T., Megraw, J., Mitchell-Olds, T., Sherman, T., Vujcic, B., Williams, C., and Neale, D., Identification of quantitative trait loci influencing wood specific gravity in an outbred pedigree of loblolly pine, *Genetics,* 138, 1293–1300, 1994.
61. Bradshaw, H. D., Jr. and Stettler, R. F., Molecular genetics of growth and development in Populus. IV. Mapping QTLs with large effects on growth, form and phenology traits in a forest tree, *Genetics,* 139, 963–973, 1995.
62. Grattapaglia, D., Bertolucci, F. L., and Sederoff, R., Genetic mapping of QTLs controlling vegetative propagation in *eucalyptus grandis* and *eucalyptus urophylla* using a pseudo-testcross: mapping strategy and RAPD markers, *Theor. Appl. Genet.,* 90, 933–947, 1995.
63. Cockerham, C. C., An extension of the concept of partitioning hereditary variance for analysis of covariance among relatives when epistasis is present, *Genetics,* 39, 859, 1954.
64. Falconer, D. S., *Introduction to Quantitative Genetics,* Longman, New York, 1981.
65. Grattapaglia, D. and Sederoff, R., Genetic linkage maps of *eucalyptus grandis* and *eucalyptus urophylla* using a pseudo-testcross: mapping strategy and RAPD markers, *Genetics,* 137, 1121–1137, 1994.

# 5 QTL Mapping in Outbred Pedigrees

*Claire G. Williams*

**CONTENTS**



## 5.1 INTRODUCTION

Outcrossing, perennial plants include a wide range of taxa important for commodity and horticultural purposes as well as undomesticated plant species: tubers, some turf, forage and range grasses, many fruit and nut perennials, agroforestry species, horticultural ornamentals and timber species. For brevity and familiarity, I have focused on quantitative trait loci (QTL) mapping in timber species as a case in point.

0-8493-7686-6/98/$0.00+$.50

QTL mapping for outcrossed pedigrees is complicated by the absence of complete homozygosity in the parents. The use of highly heterozygous parents alters QTL mapping by: (1) redefining mating type at a locus level rather than all loci in parental cross; (2) requiring the use of a single family to create linkage disequilibrium; and (3) allowing the detection of multiple QTL alleles within a single outcrossed pedigree using separate maps for each parent.

The distinctions between QTL mapping for inbred lines vs. outcrossed pedigrees can be described using standardized terminology and concepts. A single nuclear family is sampled to generate linkage disequilibrium. Terminology based on filial generations (F1, F2) refers to the use of inbred or related matings; if the parents are unrelated, it is preferable to use GP for grandparents, P for parents and F1, for the first filial generation of segregating individuals. Pedigrees for QTL mapping which do not use inbred lines are outcrossed pedigrees although they vary in the degree of relatedness between parents of the segregating progeny population. Outcrossed pedigrees for QTL mapping are either inbred-like with some relatedness between parents or outbred-like where the parents have no shared coancestry.

QTL mapping in outcrossing plants is divided into four parts. Part 1 discusses how life history attributes shape QTL mapping with outcrossing plants. Part 2 defines terminology and main differences between QTL mapping in outcrossed and inbred pedigrees. Part 3 is a general discussion of pedigree designs for outcrossed pedigrees. Part 4 discusses analytical approaches for detecting QTL in outbred pedigrees.

## 5.2 LIFE HISTORY ATTRIBUTES SHAPE QTL MAPPING IN OUTCROSSING PERENNIAL PLANTS

### 5.2.1 Genetic Load and High Heterozygosity

Genetic load tends to be higher for outcrossing perennial plants than for other plants and animals, making related matings, inbred lines or nearly isogenic lines unavailable.[1] Because genetic load is high, population-level improvement is common for forest tree improvement programs. Large breeding populations coupled with recent domestication also make it likely that QTLs with large phenotypic effects are still segregating in these populations.[2]

High levels of heterozygosity are common in outcrossing perennials.[3] In forest trees, restriction fragment length polymorphism (RFLPs), isozymes, and microsatellites commonly show multiple alleles per locus.[4–6] The higher heterozygosity can compensate for the loss of power of QTL mapping due to the lack of inbred parental lines. These highly heterozygous populations are usually in linkage equilibrium[7] so sampling a single family for QTL mapping rather than a population is necessary.

### 5.2.2 Longevity

The longevity of perennials means that all three generations required for some QTL mapping designs may be living at the time that QTL mapping is conducted. Phenotypic and DNA samples may be available for all three generations of a single pedigree and there is often the possibility of collecting more phenotypic data as needed.

Similarly, phenotypic data often accumulate on each plant over decades, adding a temporal complexity to QTL mapping. In the case of forest trees, QTL effects must be detected in mature trees if the trait has economic importance at harvest. This would include traits such as wood quality valued for manufacturing timber, pulp, or paper. Measurement at harvest is not diffficult for short-rotation species such as poplars or tropical eucalypts. Phenotypic measurements can be taken at harvest at 4 to 8 years after planting. At the other extreme, boreal conifers have harvest ages from 40 to 100 years which leaves QTL detection at mature ages an open question. QTL detection at maturity often requires decades of waiting before measuring within the segregating progeny population.

### 5.2.3 Slow Onset of Reproduction, High Rate of Reproduction

Many perennial plants have delayed onset of reproduction, slowing the creation of large, informative pedigrees and advanced generations of crosses. Slow domestication and large population sizes also means that few useful pedigrees for QTL detection already exist. Until recently, a three-generation forest tree pedigree was quite rare.[8]

Full-sib family sizes are virtually unlimited and typically large (N = 100 to 1000) due to the high rate of reproduction in both males and females. Individual trees sustain high annual reproductive rates for decades. Multiple matings for both male and female parents are possible, so that extended pedigree structure such as circular mating or diallels are common.[2] Reciprocal matings can also be made to test for gender-specific recombination rates.

## 5.3 QTL MAPPING IS UNIQUE FOR OUTBRED PEDIGREES

A comparison between outbred and inbred pedigrees for QTL mapping is summarized in Table 5.1.

**TABLE 5.1**
**Comparison of QTL Mapping for Inbreds vs. Outcrossed Species**

| Criterion | Inbred | Outcrossed pedigrees |
|---|---|---|
| Parental genotypes | Maximum of two alleles per locus | Maximum of four alleles per locus |
| | Each line is 100% homozygous for alternative alleles | Segregating, highly heterozygous for alternative alleles |
| Progeny genotypes | 100% Heterozygous at all loci | Segregating 1:2:1, 1:1, or fixed at different loci |
| Linkage phase | Same for all founders | Must be obtained within each family for each pair of linked loci; also varies widely among families in a reference population |
| | Same for all founders; no inference needed | Inference via grandparent genotypes, multi-allelic markers or haplotyping gametes |
| Segregating progeny population | All markers equally informative | Markers vary in information content |
| | No among-family variance; can ignore family structure | Among-family variance quite high; must consider reference population inferences |

### 5.3.1 Mating Type Refers to a Locus

With highly heterozygous parents, mating type is used in the same broad sense as Kempthorne.[9] In general, mating type for outcrossing pedigrees refers to individual marker genotypes of the parents (Table 5.2). Mating type for inbred crosses refers to all loci for a particular mating; outcrossing pedigrees have different mating types within a single parental cross (Table 5.2). There may be backcross mating types at marker loci A, D, and F and intercross mating types at marker loci B, C, and E. This heterogeneity of mating types makes QTL detection more complex and less efficient compared to inbreds.[10-14]

For a single cross $P_1 \times P_2$, there are five possible mating types at a marker locus (Table 5.2). Only two of the possible five mating types are informative for QTL mapping. For a mating type to be informative, one or both parents must be heterozygous at the marker locus. Informative mating types are the backcross with either a heterozygous maternal parent, i.e., $A_1 A_2 \times A_2A_2$ or a heterozygous paternal parent $A_1A_1 \times A_1A_2$ and the intercross where both $P_1$ and $P_2$ are heterozygous $A_1A_2 \times A_1A_2$. The mating types are standardized according to Haseman and Elston[15] in Table 5.2.

**TABLE 5.2**
**Comparison between Mating Types for Different Outcrossed Pedigree Designs and Inbred Crosses**

| Marker type | Abbreviation | Informativeness | Segregation Pattern | | Inbred | | Inbred-like | Outbred F1 | Outbred 3-gen |
|---|---|---|---|---|---|---|---|---|---|
| | | | Codominant | Dominant | F2 | BC | | | |
| I: $A_iA_i \times A_iA_i$ | — | NI | 1 | 1 | — | — | √ | √ | √ |
| II: $A_iA_i \times A_jA_j$ | — | NI | 1 | 1 | — | — | √ | √ | √ |
| III: $A_iA_i \times A_iA_j$ | BC1 | PI | 1:1 | 1:1 | — | √ | √ | √ | √ |
| III: $A_iA_j \times A_iA_i$ | BC1 | MI | 1:1 | 1:1 | — | √ | √ | √ | √ |
| IV: $A_iA_i \times A_jA_k$ | MBC | FI | 1:1 | 1:1 | — | — | √ | √ | √ |
| V: $A_iA_j \times A_iA_j$ | IC | BI | 3:1 | 1:2:1 | √ | — | √ | √ | √ |
| VI: $A_iA_j \times A_iA_k$ | MIC-3 | FI | 3:1 | 1:1:1:1 | — | — | √ | √ | √ |
| VII: $A_iA_j \times A_kA_l$ | MIC-4 | FI | 3:1 | 1:1:1:1 | — | — | — | √ | √ |

*Note:* Marker types is standardized according to Haseman and Elston[15] and informativeness from Groover et al.[16] BC1 = backcrossed one generation, 2-allele marker locus; IC = intercross, 2- allele marker locus; MBC = backcross, multiple (n = 3) allele locus; MIC = intercross, multiple (n = 4) allele locus; NI = not informative; PI = informative for the paternal map; MI = informative for the maternal map; BI = informative for both maternal and paternal maps (two alleles per locus); FI = fully informative for both maternal and paternal maps (multiple alleles per locus).

### 5.3.2 Concept of a Reference Population

In an outbreeding population in linkage equilibrium the same marker allele will not be associated with the same QTL allele in all families. Evidence for a marker-QTL linkage cannot be obtained at the population level using mean differences between marker genotypes.[14] Thus a single nuclear family is selected to create linkage disequilibrium (see Chapter 7, this volume, for approaches to QTL mapping near linkage equilibrium). This parallels the use of inbred lines and can be best explained by considering a three-generation outbred pedigree. Ideally, there are two sets of unrelated grandparents chosen to maximize divergence from one another, so that each GP pair is AABB × aabb. Each pair produces only two types of gametes: AB and ab. The absence of gamete types aB and Ab creates linkage (gametic) disequilibrium.

This single family is a sample used to make inference about the larger reference population which is assumed to be in linkage equilibrium; this is a reasonable assumption for forest trees.[7] Thus, linkage phase must be determined within each family sampled from the reference population. Unlike inbred lines per se, information about the single family is not useful for QTL applications. For forest trees, there are no breeds or sire testing in the sense of plant and animal breeding because domestication is so recent.

The reference population is the source of the single family pedigree for QTL mapping. It is not synonymous with the population targeted for marker-based applications. For example, if an intraspecific full-sib cross is used for QTL mapping then the reference population is the population from which the family was sampled. QTL mapping results may well have broader application to other populations or even to other species or taxa.

Allelic frequency at the population level strongly influences how informative a single family will be for QTL detection. There will be polymorphic loci in the reference population which are monomorphic in both parents of a single mating. These markers will yield no segregation data, requiring screening of a large number of DNA markers in order to assemble a complete genetic map. Unequal allele frequency in the population will shift the proportion of intercross to backcross mating types. With low-frequency alleles, more than 95% of the loci will be the backcross mating type rather than the more informative intercross type (Figure 5.1).

Multiple alleles per locus in the population have the opposite effect: there will be more intercross matings due to the higher proportion of heterozygotes (Table 5.3).[13] Allelic frequency and number at the population level determines the heterozygosity levels in the single family sampled for QTL

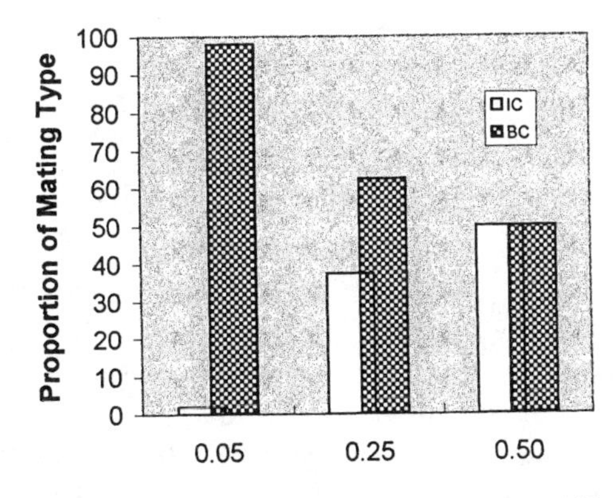


**FIGURE 5.1** Unequal marker allele frequencies in a reference population shifts the probability of informative mating types in a single-family outcrossed pedigree assuming two alleles per locus. (Calculations based on Beckmann, J. S. and Soller, M., *Theor. Appl. Genet.*, 76, 228, 1988.) For low-frequency alleles, the backcross mating type is prevalent; at equal allele frequencies there are equal numbers of loci with backcross and intercross configurations.

**TABLE 5.3**
**The Effect of Multiple Alleles in the Reference Population Affects Proportion of Mating Types within a Single Family Assuming Equal Allele Frequencies**

| Genotypes or matings within each class | No. of different types 2 alleles/locus | Frequency of each type $p = 0.5$ | No. of different types 3 alleles/locus | Frequency of each type $p = 0.33$ | No. of different types 4 alleles/locus | Frequency of each type $p = 0.25$ |
|---|---|---|---|---|---|---|
| Grandparental genotypes | | | | | | |
| Homozygotes | 2 | 0.250 | 3 | 0.109 | 4 | 0.063 |
| Heterozygotes | 1 | 0.500 | 3 | 0.218 | 6 | 0.125 |
| Parental genotypes | | | | | | |
| Homozygotes | 2 | 0.250 | 3 | 0.109 | 4 | 0.063 |
| Phase-known heterozygotes | 2 | 0.188 | 6 | 0.097 | 12 | 0.059 |
| Mixed heterozygotes | 2 | 0.010 | 6 | 0.012 | 12 | 0.004 |
| Informative mating types | | | | | | |
| Intercross | 2 | 0.035 | 30 | 0.009 | 132 | 0.003 |
| Backcross | 8 | 0.026 | 36 | 0.011 | 96 | 0.004 |

Based on Beckmann, J. S. and Soller, M., *Theor. Appl. Genet.*, 76, 228, 1988.

mapping. Without heterozygosity, a family is not informative for any type of segregation analysis (Table 5.3).

### 5.3.3 Multiple Alleles in Outcrossing Perennial Plants

Multiple-allele markers (n > 2) are a unique asset to QTL mapping with outbred pedigrees, in contrast to inbred pedigrees which are constrained to two alleles per locus. A frequently-asked question is how can there be more than two alleles per locus in a QTL mapping pedigree? A diploid individual has two alleles at a locus but in the segregating population there can be multiple (n > 2) alleles per locus. In the three-generation case, each of four unrelated grandparents contributes two alleles per locus. Each allele can be different so a total of eight alleles can be present. Of these eight alleles, only four alleles are transmitted to the offspring generation if parents are unrelated. The segregating progeny (full-sib) population will have a maximum of four alleles per locus.

### 5.3.4 Generating Separate Maternal and Paternal Maps

The marker genotypes in the F1 progeny population result from independent meioses and crossovers in the maternal and paternal parents. Thus individual maps are often constructed for each parent if progeny numbers are sufficiently large.[16-18]

With codominant markers, the maternal map includes segregation data for the following: (1) maternal informative loci; (2) fully informative loci recoded to contain only maternal segregations (i.e., the paternal parent marker data were recoded to be homozygous); (3) both-informative loci, excluding linkages between pairs of both-informative loci. The paternal map is constructed similarly. Partitioning of data from both-informative loci and recoding fully informative loci results in statistical independence of the two parental maps which are then joined into a consensus map.

With dominant markers such as random amplification of polymorphine DNA (RAPDs), informative backcross marker configurations are searched *a posteriori* in an F1 cross between two heterozygous parents. If one parent is heterozygous and the other is homozygous null, then the segregation pattern will be 1:1. Separate genetic maps are then generated for each parent based on backcross marker configurations only.[18,19]

### 5.3.5 Linkage Phase Must Be Determined for Each Outbred Family in the Reference Population

Providing linkage phase information increases the number of informative F1 double heterozygotes for two-allele marker loci. When a pair of loci are scored, it is not possible to distinguish between two types of double heterozygotes for any two pairs of alleles, i.e., $A_1 - B_1/A_2 - B_2$ cannot be distinguished from $A_1 - B_2/A_2 - B_1$ unless linkage phase is known. The haplotype data for the parents makes it possible to determine whether the parental genotypes are in coupling ($A_1 - B_1/A_2 - B_2$) or in repulsion phase ($A_1 - B_2/A_2 - B_1$). Additional segregating progeny become informative with phase information from the grandparents, further increasing the effficiency of an outbred pedigree design. Phase information is also important for marker-QTL linkages.

There are several ways to deduce the parental haplotypes in outbred pedigrees: (1) use grandparent genotype data; (2) track multiple alleles per locus from the parental genotype through to the segregating progeny population; and (3) haplotype the parents directly using PCR-based marker technology and DNA from parental gametes. Single pollen grain genotyping has been demonstrated for *Pinus sylvestris*.[20] The *Pinus* female gametophyte provides ideal tissue for haplotyping segregating progeny because it is the identical, haploid genetic complement to the egg nucleus.[21] This approach has been used subsequently to detect QTL in pines.[22,23]

## 5.4 OUTCROSSED PEDIGREE DESIGNS

To generalize, these are classified as "inbred-like" or "truly outbred" pedigrees. This is a broader distinction than the "F2-like" or "BC-like" (BC = backcrossed) strategies also used to describe outcrossed pedigrees.[2] "F2-like" and "BC-like" refers to "inbred-like" pedigrees, excluding the common case in forest trees where parents of an intraspecific are completely unrelated. Here we use the symbols GP for grandparents, P for parents and F1 for the first filial generation of segregating individuals in the case where parents are truly unrelated.

### 5.4.1 Inbred-Like Pedigrees

#### 5.4.1.1 Inbred-Like Pedigree: Interspecific F2 Intercross or BC1 Backcross Design

Two highly heterozygous parents (P) from separate species are mated to produce the first filial (F1) generation. Two F1 full-sibs are mated (or backcrossed) to produce a true second filial generation (F2). Map construction and QTL analysis are computationally straightforward for codominant or dominant markers using methods, algorithms, and software written for inbred pedigrees.

This design is well-suited to dioecious species if genetic load is low and hybrids are heterotic (e.g., *Populus deltoides* × *P. tremuloides*[24]). In the case of the *Populus* hybrid, there are some advantages. QTL effects are enhanced by the contrast between recessive mutant and wild-type alleles and QTL effects are further enhanced by cloning each F2 genotype, separating environmental variation from estimates of true QTL effects.[25] The prevalence of multiple QTL alleles can be tested in the single pedigree and across other pedigrees due to short generation intervals, ease of crossing and mass cloning, thus making QTL detection in this *Populus* system more expedient than with other forest tree species. Genetic mapping efforts for other woody perennial species with similar designs include walnut,[26] citrus,[27] and *Prunus* spp.[23,29]

### 5.4.2 Outbred Pedigree

#### 5.4.2.1 Outbred Pedigree: Interspecific F1 Design

Two heterozygous, unrelated parents ($P_1$, $P_2$) from different species are mated to produce a full-sib F1 family which is subsequently replicated through cloning. QTL mapping is conducted using phenotypic measurements on these F1 clones. This design is well-suited to species where full-sib crossing is diffficult, vegetative propagation is easy, and hybrids are heterotic (e.g., *Eucalyptus grandis* × *E. urophylla*[19]). This pedigree is also quite effficient with the "pseudo-testcross" approach as shown using two tropical eucalyptus species.[18]

The F1 interspecific design used with dominant markers becomes a pseudo-testcross analysis. This is a case where QTL mapping is defined by the use of the dominant marker system rather than the pedigree design itself. With dominant markers, the number of mating types can be reduced to the maternal and paternal backcross mating types hence the term "pseudo-testcross". The pseudo-testcross mating type has two genotypic classes *Aa, aa* which can be discerned with dominant markers. The intercross mating type cannot be discerned because two of its three genotypic classes, *AA* and *Aa* are indistinguishable. The main advantage of the dominant markers is that QTL detection is expedient for species which are not widely studied as genetic models or have insuffficient pedigree records.[19] The main drawback is that failed polymerase chain reactions (PCR) cannot be distinguished from a null allele.

The same F1 interspecific pedigree used with codominant markers has four informative mating types: intercross, maternal- and paternal-backcrosses, and multiallelic fully-informative types (Table 5.2). Also, the computational ease of QTL analysis is considerable; data can be analyzed using maximum likelihood algorithms written for backcrossed inbred lines. Multiple-allele markers or QTL can be detected with the F1 interspecific outbred pedigree using codominant markers, increasing the power of QTL mapping.

#### 5.4.2.2 Outbred Pedigree: Three-Generation Full-Sib Pedigree Design

In this design, there are two unrelated parents ($P_1$ and $P_2$) and their progeny, the segregating population as well as four grandparents $GP_{11}$, $GP_{12}$, $GP_{21}$, $GP_{22}$. Three-generation pedigrees are uncommon with long-lived perennial species so the need for grandparent marker data is often questioned.

The grandparents' phenotypic and genotypic data maximize the efficiency of the QTL mapping effort. The phenotypic data are useful for increasing the chances of a highly informative pedigree, especially if the trait heritability is low. For each pair of grandparents, one is selected for a high phenotypic value for the trait of interest and the other is selected for a low value, increasing genetic divergence between the two grandparents. Genetic divergence maximizes the probability that their offspring, the two unrelated parents, are highly heterozygous at marker loci.[8]

Grandparents' marker genotyping data also provide linkage phase information, increasing the number of informative F1 double heterozygotes for two-allele marker loci. Additional F1 progeny become informative with phase information from the grandparents, making the outbred pedigree more effficient. Phase information is also increased for marker-QTL linkages in the same fashion.

## 5.5 QTL ANALYSIS FOR OUTCROSSING PEDIGREES

The basics of QTL analysis are illustrated using a single marker approach. In practice, the single-marker approach has some serious drawbacks and other analytical methods are more commmonly used. Three alternatives to the single marker approach for outcrossing pedigrees are as follows:

1. Alter the pedigree design and use dominant markers to simplify the analysis, to fold an outbred pedigree into an inbred design. This has been done with inbred-like pedigree

designs and dominant marker systems (i.e., the pseudo-testcross approach). There is some loss of information for multiple-allele marker and QTL loci.

2. Develop maximum likelihood programs which parallel the algorithms used for inbred pedigrees. This is computationally demanding and biased for small samples or extreme genotyping in the segregating population; it can also be quite diffficult to add fixed effect parameters (i.e., site, gender, treatment) to the model.
3. Use flanking markers, or simultaneously search all markers along a chromosome rather than single-marker analyses[30-32] and use highly heterozygous markers to maximize marker information content.[13,31]

### 5.5.1 Basics of QTL Analysis in Outbred Pedigrees

For an intercross mating type, phenotypic values are compared at one codominant marker M with two alleles $M_1$ and $M_2$. The null hypothesis $H_0$ is the absence of a QTL linked to marker M. The alternative hypothesis $H_1$ is the presence of a QTL linked to M. The contrast effects for the QTL are defined as QQ = +a, Qq = d, qq = –a where +a, –a are substitution effects at locus Q. The term d is the deviation from additivity. Estimates for d or dominance deviations are obtained by regressing the phenotypic values of the heterozygote marker $M_1M_2$ class upon the values at the two homozygote classes, $M_1M_1$ and $M_2M_2$.

$$Y_{M_{11}} - Y_{M_{22}} = (1-2r)\,2a \tag{5.1}$$

The contrasts +a, –a are tested independently as $Y_{M_{11}} - Y_{M_{22}}$ where $Y_{M_{ij}}$ is the phenotypic mean of the trait at marker locus M and r is the recombination fraction between marker and QTL. This assumes no other fixed effects in the model. If all three QTL genotypes have the same variance, i.e., all are equal to $\sigma^2$ then this is considered a homoscedastic model. Under this model, the statistics for a single-marker test for marker M is as follows:

$$t = \frac{Y_{M_{11}} - Y_{M_{22}}}{\sqrt{2(\sigma^2/n)}} \tag{5.2}$$

A more general assumption is that the variances of the three QTL genotypes are unequal or heteroscedastic. Computer simulation studies show that this assumption affects the accuracy of regression analysis.[33] If the variances are heteroscedastic, the accuracy of QTL detection is more sensitive to the recombination distance (r) between two linked loci[33] although the power of any linkage analysis declines rapidly when r exceeds 0.3.[34]

In both models, these statistics are distributed under $H_0$ as standard normal distributions and their powers for the altemate hypothesis can be computed for a given significance level and a family size N. The single-marker method can be extended to an analysis of variance (ANOVA), adjusting phenotypic values for each individual in the segregating population on the basis of treatment, site or genotype × site.

The ANOVA model for a large segregating population at a single location is as follows:

$$Y_{ij} = \mu + G_j + \varepsilon_{ij} \tag{5.3}$$

where $Y_{ij}$ = phenotypic value, $\mu$ = progeny mean, $G_j$ = the marker genotype i, and $\varepsilon_{ij}$ = the error term associated with the kth progeny with the ith genotype. Genotypic effects are generally assumed to be fixed.[14,16] For forest trees, QTL effects are often expressed as a percentage of the total phenotypic variance. If the QTL effect estimated from $Y_{M_{11}} - Y_{M_{22}}$ is 10 U and the total phenotypic variance of the segregating population is 100 U then the QTL at marker M accounts for 10% of the phenotypic variance.

There are better ways to analyze QTL data than using single markers because single-marker analyses (1) produce estimates of QTL effects which are biased by the recombination distance from the marker so that loose linkage between the marker and QTL reduces the estimated magnitude of the QTL effect; (2) cannot be used to infer QTL location as the size of the QTL effect is confounded with its recombination distance from the marker; and (3) thus have relatively low power for detecting QTL.

### 5.5.2 Flanking Markers and Maximum Likelihood

Later QTL mapping studies with inbred and inbred-like pedigrees have used interval mapping, a combination of flanking marker pairs and maximum likelihood analysis as advocated by Lander and Botstein.[35] The criterion for detecting the presence of a QTL is the log-odds ratio or the LOD score:

$$\mathrm{LOD} = \log_{10}\left(\mathrm{L}_1/\mathrm{L}_0\right) \tag{5.4}$$

The LOD score is the logarithm to the base 10 of the ratios of two likelihoods: $L_1$ is the likelihood that the QTL is linked to the marker and $L_0$ is the likelihood that there is no QTL in the interval (i.e., the null hypothesis or $H_0$). If the LOD score threshold exceeds 3.0, representing odds of 1000:1,[36] then the null hypothesis of no QTL in the interval is rejected. If recombination is gender-specific as suggested for *Pinus*[37,38] then appropriate LOD threshold increases to 3.5.[39]

For inbred-like pedigrees or the F1 interspecific pedigree with a pseudo-testcross marker scoring, interval mapping is straightforward because data can be analyzed similar to inbred lines. Interval mapping is appealing because it is insensitive to the assumptions about heteroscedastic within-genotype variances.[33] It becomes too computationally complex for outbred pedigrees with codominant markers. This is problematic for outbred pedigrees. For outbred pedigrees, single-marker analysis has too many drawbacks yet there are no computational tools for maximum likelihood.

Using flanking or multiple markers based on a regression approach is a logical answer. A comparative analysis of flanking and single-marker QTL mapping using maximum likelihood vs. regression revealed similar results for inbred pedigrees.[30,40,41] Thus the added information from pairs of markers is more important for regression analysis than the assumptions regarding the variance distribution among QTL genotypic classes.[40] This result corroborates Lander and Botstein's[35] suggestion that there is little difference in power between regression and maximum likelihood for single marker analyses.

The criterion for detecting the presence of a QTL using the regression approach is the likelihood ratio test. This test statistic allows comparison between regression and maximum likelihood methods and can be easily converted to LOD scores for a threshold test of significance:[31,32,40]

$$\text{Test statistic } = 2\log_e\left(\mathrm{L}_1/\mathrm{L}_0\right) \text{ for maximum likelihood}$$

becomes (5.5)

$$= n\log_e\left(\text{RSS Reduced/RSS Full}\right)$$

and this can be converted back for testing LOD thresholds:

$$\mathrm{LOD} = n/2\log_{10}\left(\text{RSS Reduced/RSS Full}\right)$$

where n is the number of observations, RSS Full is the residual sum of squares for the full model fitting the regression, RSS Reduced is the residual SS for the reduced model, omitting the regression.

The likelihood test ratio is asymptotically distributed as $\chi^2$ with p degrees of freedom where p represents the estimated regression parameters.[31] Like the LOD score, it is plotted at regular intervals along the chromosome. The peak value for the likelihood test ratio represents the most likely QTL position. It can be equated to products of mean squares and their degrees of freedom:[31]

$$\begin{aligned} \text{Test statistic} &\approx \text{p MS regression/MS residual} \\ \text{Test statistic} &\approx \text{p F test for regression} \end{aligned} \tag{5.6}$$

A likelihood ratio test value of 1.38 approximates a LOD score of 3.0.[32] The flanking marker analysis based on regression offers similar results to the more computationally difficult maximum likelihood. However, application of the likelihood ratio test to outbred pedigrees presents another difficulty: the uneven information content of paired markers along a chromosome generates bias in detecting QTL.[32]

### 5.5.3 Flanking Markers and Simultaneous Search with Regression Analysis

A simultaneous search along a chromosome is an improvement over flanking markers in outbred pedigrees. Outbred pedigrees have different mating types at each locus and these mating types have different information content. For example, an intercross marker is more informative than a back-cross marker and neither type is as informative as a fully informative marker (Table 5.2). The different information content of paired, linked markers results in a bias in QTL detection in outbred pedigree designs.[32]

Simulation models suggested three ways to reduce this serious source of bias unique to outbred pedigrees:[32] (1) increase the density of the markers so that less informative markers can be discounted in favor of markers with higher information content; (2) use multiple-allele markers to increase the proportion of heterozygous parents, thus increasing the proportion of intercross and fully informative markers in the segregating progeny population; or (3) use all markers along a chromosome simultaneously rather than pairwise.

Simultaneous search methods are the most useful for removing bias and increasing the power by reducing the residual variance.[32] Applying simultaneous search methods to outbred pedigrees also favors the use of multiple regression rather than maximum likelihood because regression is easier to analyze and to add other parameters such as fixed (treatment, site, gender) effects to the analysis; maximum likelihood is too computationally complex to extend to this type of a simultaneous search. Simultaneous search methods have been extended to outbred pedigrees in forest trees.[33]

### 5.5.4 A Special Case: Detecting Multiple QTL Alleles per Locus

Only the fully informative mating types with $n > 2$ alleles per locus (Table 5.1) are useful for detecting the number of segregating QTL alleles and for estimating their total intralocus interaction. Multiple alleles increase the number of genotypic classes; for n alleles per locus there are n homozygote and $n(n - 1)/2$ heterozygote classes. In the case of three alleles per locus, there are three homozygote and three heterozygote classes; for four alleles per locus there are four homozygote and six heterozygote classes. Unlike the two-allele case, the effect of individual QTL alleles and their specific interactions cannot be quantified because there are more terms to solve than there are equations.[16,17]

Multiple QTL analysis is demonstrated for an outbred pedigree of *Pinus taeda*.[16] In this case, marker locus S6a is fully informative with a large phenotypic effect which accounts for 5.6% of the total phenotypic variance. An ANOVA was used to test the effects of each parent's QTL alleles and the interaction among them (Table 5.3).[16] Both parents were heterozygous for alternate alleles, posing the possibility of a 4-allele QTL model. A test of the four-allele QTL model indicated the presence of at least three segregating QTL alleles as well as a significant interaction among them.[16]

In the preceding case, three rather than four QTL alleles may have been detected because the fourth QTL allele is masked or "hidden" by a more dominant QTL allele segregating in the other parent. Alternatively, there may be only three QTL alleles. One maternally segregating QTL allele might have been masked by the joint effect of the paternal QTL alleles. To dissect specific interactions among multiple QTL alleles, one may use either very tight linkage with a fully-informative marker or develop nearly isogenic lines in order to hold genetic background constant.[42] Testing allelism in the latter way is a formidable obstacle for many perennial plant species because it requires development of near-isogenic lines by four to six generations of backcrossing despite high genetic loads and long generation intervals.[17]

Dissecting multiple QTL alleles is an emerging area of research unique to outcrossing pedigrees although theory in quantitative variance in population genetics is based upon the concept of multiple alleles at a locus.[9] Multiple QTL alleles have profound consequences for extending QTL results from a single family to the reference population. If prevalent, then we would expect to find fewer QTLs common across families.[43] QTL mapping in outcrossed pedigrees is beginning to require powerful analytical methods for detection of multiple-allele loci in extended family pedigrees.

## 5.6 SUMMARY

QTL mapping in outcrossing perennial plants is typically based on highly heterozygous parents and on large families. A single family is sampled from the larger reference population to create linkage disequilibrium and linkage phase must be established for each family. Mating types are assigned at each locus; a single parental cross will have loci in backcross, intercross, and fully informative multi-allelic configurations. The heterogeneity of mating types biases QTL detection using single-marker or flanking-marker QTL analyses. Simultaneous searches per linkage group, highly heterozygous markers on a saturated map, and a large segregating progeny population are the preferred tools for QTL mapping in outcrossing perennial plants. Up to four QTL alleles per locus can be detected in outbred pedigrees, contrary to the two-allele models implicit to the use of inbred lines. The prevalence of multiple alleles per locus in the larger reference population is an emerging area of research interest for outbred QTL mapping. If prevalent, multiple QTL alleles will decrease the probability of detection of the same QTL allele across extended pedigrees, unrelated families, and the reference population itself. This chapter reviews QTL mapping in outbred pedigrees: the influence of life history attributes, terminology and concepts unique to outbred QTL mapping, general pedigree designs, and the basics of QTL analyses. For genomic mapping applications, the reader is referred to reviews by Strauss et al.[44] and O'Malley and McKeand.[45]

## ACKNOWLEDGMENTS

Special thanks to Dr. Jerry Taylor and Dr. Floyd Bridgwater for helpful discussions.

## REFERENCES

1. Williams, C. G. and Savolainen, O., Inbreeding depression in conifers: implications for breeding strategy, *For. Sci.*, 42(1), 102, 1996.
2. Muranty, H., Power of tests for quantitative trait loci detection using full-sib families in different schemes, *Heredity*, 76, 156, 1995.
3. Hamrick, J. L. and Godt, M. J. W., Allozyme diversity in plant species, in *Plant Populations Genetics, Breeding and Genetic Resources*, Brown, A. H. D., Clegg, M., Kahler, A., and Weir, B. S., Eds., Sinauer Associates, Sunderland, MA, 1990, 43–63.

4. Devey, M. E., Fiddler, T., Liu, B., Knapp, S., and Neale, D., An RFLP linkage map for loblolly pine based on a three-generation outbred pedigree, *Theor. Appl. Genet.*, 88, 273, 1994.
5. Conkle, M. T., Genetic diversity: seeing the forest through the trees, *New Forests*, 6, 5, 1992.
6. Echt, C. S., May-Marquardt, P., Hsieh, M., and Zahorchak, R., Characterization of microsatellite markers in eastern white pine, *Genome*, 36(6), 1102–1108, 1996.
7. Epperson, B. K. and Allard, R. W., Linkage disequilibrium between allozymes in natural populations of lodgepole pine, *Genetics*, 115, 341, 1987.
8. Williams, C. G. and Neale, D. B., Conifer wood quality and marker-aided selection: a case study, *Can. J. For. Res.*, 22, 1009, 1992.
9. Kempthorne, O., An introduction to genetic statistics, John Wiley, New York, 1957, 545.
10. Jakayar, S. D., On detection and estimation of linkage between a locus influencing a quantitative character and a marker locus, *Biometrics*, 26, 451, 1970.
11. Hill, A. P., Quantitative linkage: a statistical procedure for its detection and estimation, *Ann. Hum. Genet.*, 38, 439, 1975.
12. Soller, M. and Genizi, A., The efficiency of experimental designs for the detection of linkage between a marker locus and a locus affecting a quantitative trait in segregating populations, *Biometrics*, 34, 47, 1978.
13. Beckmann, J. S. and Soller, M., Detection of linkage between marker loci and loci affecting quantitative traits in crosses between segregating populations, *Theor. Appl. Genet.*, 76, 228, 1988.
14. Knott, S. A., Prediction of the power of detection of marker-quantitative trait locus linkages using analysis of linkage, *Theor. Appl. Genet.*, 89, 318, 1994.
15. Haseman, J. K. and Elston, R. C., The investigation of linkage between quantitative trait and a marker locus, *Beh. Genet.*, 2(1), 3, 1972.
16. Groover, A. T., Devey, M., Fiddler, T., Lee, J., Megraw, R., Mitchell-Olds, T., Shemman, B., Vujcic, S., Williams, C. G., and Neale, D. B., Identification of quantitative trait loci influencing wood specific gravity in an outbred pedigree of loblolly pine, *Genetics*, 138, 1293, 1994.
17. van Eck, H. J., Jaconbs, J., Stam, P., Ton, J., Stiekema, W. J., and Jacobsen, E., Multiple alleles for tuber shape in diploid potato detected by qualitative and quantitative genetic analysis using RFLPs, *Genetics*, 137, 303, 1994.
18. Grattapaglia, D. and Sederoff, R. R., Genetic linkage maps of *Eucalyptus grandis* and *E. urophylla* using a pseudotestcross mapping strategy and RAPD markers, *Genetics*, 137, 1121, 1994.
19. Grattapaglia, D., Bertolucci, F. L., and Sederoff, R. R., Genetic mapping of QTLs controlling vegetative propagation in *Eucalyptus grandis* and *E. urophylla* using a pseudo-testcross strategy and RAPD markers, *Theor. Appl. Genet.*, 90, 933, 1995.
20. Kostia, S., Varvio, S.-L., Vakkari, P., and Pulkkinen P., Microsatellite sequences in a conifer, *Pinus sylvestris*, *Genome*, 38, 1244, 1996.
21. Carlson, J. E., Tulsieram, L. K., Glaubik, J. C., Luk, V. W. K., Kauffeldt, C., and Rutledge, R., Segregation of random amplified DNA markers in F1 progeny of conifers, *Theor. Appl. Genet.*, 83, 194, 1991.
22. Plomion, C., O'Malley, D. M., and Durel, C. E., Genomic analyses in maritime pine *(Pinus pinaster)* comparison of two RAPD maps using selfed and open-pollinated seeds of the same individual, *Theor. Appl. Genet.*, 90(7–8), 1028, 1995.
23. Wilcox, P. L., Amerson, H. V., Kuhiman, E. G., Liu, B., O'Malley, D. M., and Sederoff, R., Detection of a major gene for resistance to fusiform rust resistance in loblolly pine by genomic mapping, *Proc. Natl. Acad. Sci.*, 93(9), 3859, 1996.
24. Bradshaw, H. D. and Stettler, R. F., Molecular genetics of growth and development in *Populus*. IV. Mapping QTLs with large effects on growth, form and phenology traits in a forest tree, *Genetics*, 139, 963, 1995.
25. Bradshaw, H. D. and Foster, G. S., Marker-aided selection and propagation systems in trees: advantages of cloning for studying quantitative inheritance, *Can. J. For. Res.*, 22, 1044, 1992.
26. Fjellstrom, R. G. and Parfitt, D. E., RFLP inheritance in walnut, *Theor. Appl. Genet.*, 89, 665, 1994.
27. Durham, R. E., Lion, P. C., Gmitter, F. G., and Moore, G. A., Linkage of restriction fragment length polymorphisms and isoenzymes in *Citrus*, *Theor. Appl. Genet.*, 84, 39, 1992.
28. Chaparro, J. X., Wemer, D. J., O'Malley, D., and Sederoff, R. R., Targeted mapping and linkage analysis of morphological, isozyme and RAPD markers in peach, *Theor. Appl. Genet.*, 87, 805, 1994.
29. Foolad, M. R., Arulsekar, S., Becerra, V., and Bliss, F. A., A genetic map of *Prunus* base on an interspecific cross between peach and almond, *Theor. Appl. Genet.*, 91, 262, 1995.

30. Haley, C. S. and Knott, S. A., A simple regression method for mapping quantitative trait loci in line crosses using flanking markers, *Heredity,* 69, 315, 1992.
31. Haley, C. S., Knott, S. A., and Elsen, J.-M., Mapping quantitative trait loci in crosses between outbred lines using least squares, *Genetics,* 136, 1195, 1994.
32. Knott, S. A. et al., *Theor. Appl. Genet.,* in press.
33. Luo, Z. W. and Woolliams, J. A., Estimation of genetic parameters using linkage between a marker gene and a locus underlying a quantitative character in $F_2$ populations, *Heredity,* 70, 245, 1993.
34. Risch, N., A note on multiple testing procedures in linkage analysis, *Am. J. Hum. Genet.,* 48, 1058, 1991.
35. Lander, E. S. and Botstein, D., Mapping Mendelian factors underlying quantitative traits using RFLP linkage maps, *Genetics,* 121, 185, 1989.
36. Morton, N. E., Sequential tests for the detection of linkage, *Am. J. Hum. Genet.,* 7, 277, 1955.
37. Moran, G. F., Bell, J. C., and Hilliker, A. J., Greater meiotic recombination in male vs. female gametes in *Pinus radiata, J. Hered.,* 74, 62, 1983.
38. Groover, A. T., Williams, C. G., Devey, M. E., Lee, J. M., and Neale, D. B., Sex-related differences in meiotic recombination frequency in *Pinus taeda, J. Hered.,* 86(2), 157, 1995.
39. Lander, E.S. and Lincoln, S. E., The appropriate threshold for declaring linkage when allowing sex-specific recombination rates, *Am. J. Hum. Genet.,* 43, 396, 1988.
40. Knott, S. A. and Haley, C. S., Aspects of maximum likelihood methods for the mapping of quantitative trait loci in line crosses, *Genet. Res. Camb.,* 60, 139, 1992.
41. Martinez, O. and Cumow, R. M., Estimating the locations and the sizes of the effects of quantitative trait loci using flanking markers, *Theor. Appl. Genet.,* 85, 480, 1992.
42. Kowyama, Y., Takahashi, H., Muraoka, K., Tani, T., Hara, K., and Shiotani, I., Number, frequency and dominance relationships of S-alleles in diploid *Ipomoea trifida, Heredity,* 73, 75, 1994.
43. Beavis, W. D., Grant, D., Albertsen, M., and Fincher, R., Quantitative trait loci for plant height in four maize populations and their associations with qualitative genetic loci, *Theor. Appl. Genet.,* 83, 141, 1991.
44. Strauss, S. H., Lande, R., and Namkoong, G., Limitations of the molecular marker-aided selection in forest tree breeding, *Can. J. For. Res.,* 22, 1050, 1992.
45. O'Malley, D. M. and McKeand, S. E., Marker-assisted selection for breeding value in forest trees, *For. Genet.,* 1(14), 207, 1994.

# 6 Mapping QTLs in Autopolyploids

*Sin-Chieh Liu, Yann-Rong Lin, James E. Irvine, and Andrew H. Paterson*

## CONTENTS



## 6.1 INTRODUCTION

DNA markers such as restriction fragment length polymorphisms (RFLP) enable the development of large numbers of genetic markers useful for the construction of genetic linkage maps, and for systematic analysis of quantitative trait loci (QTLs). Based on the idea of detecting the association of quantitative traits with monogenic traits as first reported by Sax,[1] many methods have been developed for systematically detecting genetic linkage between QTLs and DNA markers.[2-7]

However, methods based on the segregation of diploid populations are not readily applicable for QTL mapping in polyploids, due to the unique genetic characteristics of polyploids. Specifically, segregating populations of polyploids have more genotypes than those of diploids; DNA markers may not be able to identify all these genotypes; and the genome constitution of polyploids is often indeterminate, or a mixture of autopolyploid, allopolyploid, and/or aneuploid.[8,9] In this chapter, we use sugarcane as an example to discuss the special considerations for QTL detection in autopolyploids by means of RFLP linkage maps.

Autopolyploids are composed of multiple basic sets of chromosomes from within one species. Normally, each basic set of chromosomes contains one representative from each homologous group. Somatic cells of an autopolyploid have a chromosome number of 2n = mx, where n = the number of chromosomes in a gamete; m = ploidy number indicating the number of chromosomes in a homologous group; and x = monoploid number of chromosomes in a basic set. Cultivated species of alfalfa (2n = 32) and coffee (2n = 44) are examples of autotetraploids, i.e., m = 4.[10] Some species of sweet potato (2n = 90) are considered as autohexaploids, i.e., m = 6.[10] Higher levels of autopolyploidy among crop plants become progressively more difficult to characterize. Sugarcane is a crop species with elevated ploidy levels and cytogenetic complexity contributed in part by

0-8493-7686-6/98/$0.00+$.50

**TABLE 6.1**
**Expected Gamete Segregation Ratios in Autopolyploids and Allopolyploids for Different Ploidy Levels and RFLP Marker Dosages**

| Dosage number of RFLP marker (corresponding genotype of dominant alleles) | Expected segregation ratios of presence: absence in gametes | | | | |
|---|---|---|---|---|---|
| | Autodiploid (m = 2) | Autotetraploid (m = 4) | Autohexaploid (m = 6) | Autooctoploid (m = 8) | Allopolyploid (m > 2) |
| 1 (Simplex) | 1:1 | 1:1 | 1:1 | 1:1 | 1:1 |
| 2 (Duplex) | All present | 5:1 | 4:1 | 11/3 | All present or 3:1 |
| 3 (Triplex) | — | All present | 19:1 | 13:1 | All present or 15:1 |
| 4 (Quadruplex) | — | — | All present | 69:1 | All present or 64:1 |

interspecific hybridization. Autopolyploidy in sugarcane has been indicated by RFLP segregation data of progeny derived from a cultivar[11] and a wild sugarcane species,[12,13] necessitating linkage analysis using single dosage restriction fragments.[8,17]

## 6.2 CONSTRUCTING RFLP LINKAGE MAPS IN AUTOPOLYPLOIDS

### 6.2.1 Segregation of DNA Markers

Segregation of DNA markers in autopolyploids can be analyzed based on the presence and absence of the DNA fragments. A DNA fragment behaves like a dominant allele in autopolyploids: absence of the fragment represents nulliplex; and presence of the fragment represents all other genotypes (simplex, duplex, triplex, up to *m*-plex). A single dose restriction fragment is equivalent to the dominant allele of a simplex genotype that segregates in a single-dose ratio (presence:absence = 1:1) in the gametes. Gamete segregation ratios of DNA markers with a higher dosage number *k* can be calculated as

$$\text{presence}:\text{absence} = \left[\binom{m}{m/2} - \binom{k}{o}\binom{m-k}{m/2}\right] : \left[\binom{k}{o}\binom{m-k}{m/2}\right]$$

(Table 6.1).[17] DNA markers with dosage numbers higher than m/2 do not segregate in the gametes, because all the gametes carry at least one copy of the RFLP fragment.

In a segregating population in linkage disequilibrium (i.e., a "mapping population"), the number of progeny scored as presence (P) for a particular DNA fragment is a binomial random variable. DNA markers with different dosages (single-dose restriction fragment (SDRF), double-dose (DDRF), triple-dose (TDRF), etc.) will yield specific expected values of P with corresponding binomial probability distributions. Statistical methods distinguishing these binomial probability distributions enable the determination of the most likely marker dosage yielding the observed P. Confidence intervals of P for different marker dosages can be defined based on the specific population, the probability distributions of P, and the a-level of significance. Ripol[17] has discussed the statistical aspects of several methods for assigning dosages for segregating DNA markers. However, segregation ratios of DNA fragments may deviate from the expected values due to preferential chromosome pairing and/or segregation distortion. Any scoring errors may complicate the segregation ratios. Many highly polymorphic DNA probes will detect multiple polymorphic bands in autopolyploids. Some of these bands may not be well distinguished, because the small size differences between the fragments cannot be resolved by agarose gel electrophoresis or the specific exposure conditions used. Caution needs to be exercised during scoring and assigning dosages, to avoid ambiguous DNA markers.

### 6.2.2 Linkage Analysis

Most linkage maps of autopolyploids are based on coupling-phase linkages between SDRFs, together with identification of homologous relationships between these linkage groups.

Wu et al.[8] presented a method for mapping polyploids based on the segregation of SDRFs. Use of SDRFs resembles mapping of diploid backcross populations, i.e., SDRFs segregate in a 1:1 ratio, a population size of 75 is efficient for identifying SDRFs and detecting linkages in coupling phase. A much larger population size is required for detecting SDRF linkages in repulsion. For example, a population of more than 750 progeny is needed for detecting repulsion-phase linkages of 20 cM in auto-octoploids. Therefore, SDRFs are usually only effective for detecting linkages in coupling.

Homologous groups can be determined based on different SDRFs generated by the same highly polymorphic DNA probes.[12] Identification, or verification of homologous relationships[12-16] can be achieved by mapping DDRFs and TDRFs as demonstrated by Ripol et al.[17]

SDRFs in coupling-phase linkages are the most informative markers for constructing linkage maps of autopolyploids.[8,17] Ripol et al.[17] have discussed the pitfalls of mapping linkages in repulsion, including the low probabilities of the specific chromosome pairing required for recombination and the possible confusion with recombination of coupled fragments for higher dosage markers. Mapping populations with a high frequency of SDRFs can be derived from a modified backcross formation in which only one informative parent contributes to the polymorphism. The informative parent should be a highly heterozygous individual (e.g., an outcrossed heterozygote or an individual from a BC1 population). An F1 from two homozygous plants cannot be used as an informative parent for SDRFs, since all the RFLP markers of the F1 between pure lines have a dosage number of $m/2$. Selfing progeny of an informative parent will show a 3:1 segregation ratio for SDRFs. However, this modified F2 formation is less efficient for constructing linkage maps.[8] It will be more effective to first screen a subset of progeny to determine dosage of RFLP markers. A primary linkage map can then be built by mapping SDRFs. Linkage groups of homologous chromosomes are then associated by connecting SDRFs generated by highly polymorphic DNA probes and/or mapping higher dosage markers.

## 6.3 DETECTING QTLs IN AUTOPOLYPLOIDS

As a fundamental requirement for detecting QTLs, the mapping population must demonstrate segregation at the QTLs, as evidenced by showing significant genetic variation for the quantitative trait. Quantitative traits are contributed by the collective effects of individual quantitative trait alleles at many different loci. In autopolyploids, a parent with a simplex genotype for a dominant quantitative trait allele will contribute highest variation to the progeny population with 50% of the progeny possessing the dominant allele (in modified backcross formation). If the parent carries a higher number of alleles, lower variation is expected in the progeny population, with the majority of progeny having at least one copy of the dominant allele.

### 6.3.1 Detecting Quantitative Trait Alleles

The association between the segregation of an SDRF and the variance of a quantitative trait can be tested by procedures used for diploids such as maximum likelihood, analysis of variance (ANOVA), or a t-test. The inference of linkage between an SDRF and a quantitative trait allele is based on statistically significant difference in the means of simplex vs. nulliplex progeny for the SDRF. To minimize the effect of experiment-specific factors conflicting with the standard assumptions of QTL analysis,[4] an empirical threshold value estimated by permutation tests may be employed for declaring statistical significance.[19]

We studied two populations of F1 progeny from the crosses between heterozygous sugarcane clones: *Saccharum officinarum* Green German × *S. spontaneum* Ind 81-146 (GI population) and

*S. spontaneum* Pin 84-1 × *S. officinarum* Muntok Java (PM population). DNA extraction, southern hybridization were performed largely as described in Chittenden et al.[20] The DNA fragments present in one parent but absent in the other parent were scored as those in a modified backcross population. A subset of 85 progeny was randomly chosen from the total of approximately 250 individuals in each population, for assigning dosages for the RFLP markers. The SDRFs of each parent are identified by chi-squared tests for an expected segregation ratio of 1:1.[8] Linkage of the SDRFs is analyzed as backcross data for constructing linkage groups in coupling for each parent. Since the subsets and the entire populations have similar distributions for the traits of interest, initial tests on the subsets were performed for detecting the association of SDRFs with quantitative trait alleles. A one-way ANOVA was used to test the difference in the means of a quantitative trait between the two classes of progeny (presence vs. absence for a marker fragment). The effect of a SDRF was estimated from a least-squares linear model:

$$\mathrm{Trait}_i = \mathrm{Mean} + (a \times \mathrm{Marker}) + e_i,$$

where $\mathrm{Trait}_i$ is the trait value of the ith individual in the progeny population; Mean is the mean value of the quantitative trait component not affected by the presence of the marker fragment (equivalent to the mean quantitative trait value of the progeny scored as absence); a is the effect of presence of the marker fragment; Marker is 1 if the marker fragment is present, and 0 otherwise; and $e_i$ is a normal random variable representing the variation in the quantitative trait not controlled by the segregation of the SDRF.

As examples of mapping QTLs in autopolyploids, we describe the identification of two QTLs associated with variation in sugar content of sugarcane. In the GI population, a SDRF generated by a sorghum genomic DNA probe *pSB604* shows association with sugar content estimated from field refractometer readings (brix). Presence of the *S. spontaneum* Ind 81-146 fragment contributes a decrease of 1.8 in brix value ($F = 8.54$ and $p = 0.005$). In the PM population, a SDRF generated by an oat cDNA probe *CDO87* also shows association with the brix value. Presence of the *S. officinarum* Muntok Java fragment contributes an increase of 1.8 in brix ($F = 9.82$ and $p = 0.002$). To evaluate the statistical significance of the association of these two SDRFs with the brix value, permutation tests were performed with 1000 times random shuffling of the brix values within the population. We found that the two SDRFs generated by *pSB604* and *CDO87* are significantly associated with the brix value, each with adjusted $p$ values of 0.004 based on the distribution of the 1000 $F$ values from the permutation tests.

The SDRFs identified are associated with alleles having major effects on the quantitative trait. However, the tests only account for the segregation of one half of the genome that is inherited from one of the two parents. To further evaluate the variation of the trait that is explained by the segregation of the marker locus, a linkage map with completed homologous groups and an extended mathematical model with marker terms to account for the segregation of the whole genome will be needed.

### 6.3.2 Modified Approaches for Detecting Quantitative Trait Alleles

Tests can also be performed for detecting major alleles associated with traits measured on discrete scales. For example, photoperiodic response of flowering in the GI and PM sugarcane populations were recorded as floral development stages observed on a particular date. Stages of floral development were recorded as "no flower", "sheath elongation", "boot", "emerging", and "full flower". A chi-squared test was used to test the association of the segregation of the SDRFs (presence vs. absence) with the phenotypes of flower initiation (no flower observed vs. incipient flowering or flower observed).

One of the two SDRFs of *S. officinarum* Muntok Java generated by sorghum genomic DNA probe *pSB188* shows association with flower initiation in the PM population with a p value of

**TABLE 6.2**
**Numbers of GI Population Plants Sorted by Floral Development Stages and RFLP Phenotypes of Three SDRFs Generated by *pSB188***

| pSB188al | pSB188bl | pSB188cl | Dose | No flower | Sheath elongation | Boot | Emerging | Full flower |
|---|---|---|---|---|---|---|---|---|
| – | – | – | 0 | 9 | 0 | 2 | 0 | 2 |
| + | – | – | 1 | 5 | 0 | 0 | 0 | 2 |
| – | – | + | 1 | 2 | 0 | 1 | 0 | 2 |
| – | + | – | 1 | 4 | 0 | 0 | 2 | 3 |
| + | – | + | 2 | 2 | 1 | 0 | 1 | 3 |
| – | + | + | 2 | 3 | 0 | 0 | 3 | 7 |
| + | + | – | 2 | 4 | 3 | 0 | 3 | 7 |
| + | + | + | 3 | 2 | 1 | 1 | 0 | 10 |

*Note:* The floral development stages are recorded in progressive stages of flowering: no flower, sheath elongation, boot, emerging, full flower. The phenotypes of the three SDRFs are indicated as presence (+) and absence (–) on the *S. spontaneum* Ind 81-146 fragment.

0.025. Presence of the *S. officinarum* Muntok Java fragment is associated with a higher number of plants showing incipient or full flowering. This corresponds with the observation of earlier flowering of the *S. officinarum* Muntok Java parent than the *S. spontaneum* PIN 84-1 parent. Comparatively, three of the four SDRFs of *S. spontaneum* Ind 81-146 generated by the probe *pSB188* show association with flower initiation in the GI population with p values of 0.020, 0.024, and 0.026. Presence of the *S. spontaneum* Ind 81-146 fragment is associated with a higher number of plants showing incipient or full flowering. This corresponds with the observation of earlier flowering of the *S. spontaneum* Ind 81-146 parent than the *S. officinarum* Green German parent. When the GI population plants in the subset are sorted according to their flowering phenotypes and RFLP scores of these three SDRFs together, higher dosage of *S. spontaneum* Ind 81-146 fragments detected by pSB188 was associated with more plants in full flower (Table 6.2).

It is very likely that the different SDRFs of *pSB188* mark homologous loci, since they are not linked in coupling phase with each other, and two of the *pSB188* SDRFs of *S. spontaneum* Ind 81-146 are individually linked in coupling with the SDRFs generated by another sorghum probe *pSB314*. A colinear linkage between *pSB188* and *pSB314* is observed in sorghum linkage group D (Figure 6.1). Moreover, a significant association of the *pSB188* locus and the photoperiodic response of flowering was identified in a sorghum interspecific population from *Sorghum bicolor* × *S. propinquum* (Figure 6.1).[21,22] This locus appears to be responsible for photoperiodic flowering response of many cultivated grasses.[21]

## 6.4 SUMMARY

The examples demonstrate that QTLs in autopolyploids can be detected by identifying alleles with large effects on the quantitative trait. Other QTLs may be detected when more SDRFs are identified in the rest of the genome. Alleles with smaller effects on the quantitative trait may also be detected, after larger populations are scored for the SDRFs. The power of QTL detection may be increased by analyzing phenotypic data collected from multiple years or locations to minimize the environmental variance.[4] Further, selective genotyping of the progeny with extreme phenotypes can improve the efficiency of QTL detection.[4] Finally, breeding designs which reduce the level of genetic variation segregating in a population can further increase the power to detect genes with small effects. The example of detecting a flowering-related locus makes evident the application of comparative information from diploid relatives to QTL mapping in polyploids. This approach will

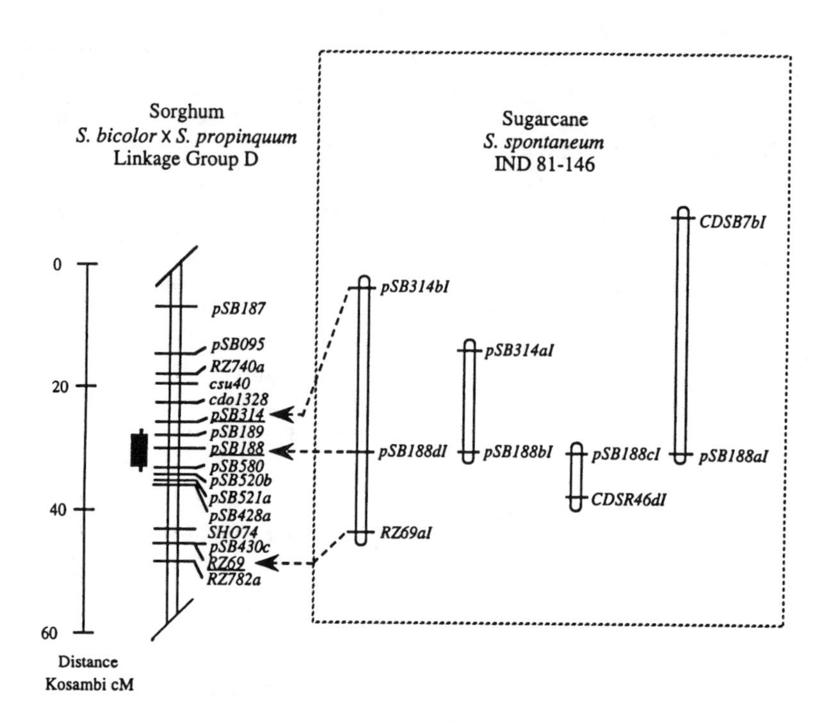


**FIGURE 6.1** RFLP markers associated with photoperiodic response of flowering in sugarcane and sorghum.

not only accelerate QTL mapping in polyploids, but also provide valuable information for further understanding evolutionary processes of polyploids.[21]

## ACKNOWLEDGMENTS

The authors express their appreciation to partners of the International Consortium for Sugarcane Biotechnology for funding, specifically the American Sugar Cane League, Australian Sugar Research and Development Corporation, Cenicaña, Copersucar, Florida Sugar Cane League, Hawaii Sugar Producers Association, and Mauritius Sugar Industry Research Institute.

## REFERENCES

1. Sax, K., The association of size differences with seed-coat pattern and pigmentation in *Phaseolus vulgaris, Genetics,* 8, 552, 1923.
2. Thoday, J. M., Location of polygenes, *Nature,* 191, 368, 1961.
3. Weller, J. I., Maximum likelihood techniques for the mapping and analysis of quantitative trait loci with the aid of genetic markers, *Biometrics,* 42, 627, 1986.
4. Lander, E. S. and Botstein, D., Mapping Mendelian factors underlying quantitative traits using RFLP linkage maps, *Genetics,* 121, 185, 1989.
5. Haley, C. S. and Knott, S. A., A simple regression method for mapping quantitative trait loci in line crosses using flanking markers, *Heredity,* 69, 315, 1992.
6. Jansen, R. C., A general mixture model for mapping quantitative trait loci by using molecular markers, *Theor. Appl. Genet.,* 85, 252, 1993.
7. Krugylak, L. and Lander, E. S., A nonparametric approach for mapping quantitative trait loci, *Genetics,* 139, 1421, 1995.
8. Wu, K. K., Burnquist, W., Sorrells, M. E., Tew, T. L., Moore, P. H., and Tanksley, S. D., The detection and estimation of linkage in polyploids using single-dosage restriction fragments, *Theor. Appl. Genet.,* 83, 294, 1992.

9. de Wet, J. M. J., Origins of polyploids, in *Polyploidy: Biological Relevance*, Lewis, W. H., Ed., Plenum Press, New York, 1980, 3.
10. Simmonds, N. W., *Evolution of Crop Plants*, Longman Scientific & Technical, Essex, 1976.
11. Grivet, L., D'Hont, A., Roques, D., Feldmann, P., Lanaud, C., and Glaszmann, J. C., RFLP mapping in cultivated sugarcane (*Saccharum* spp.): genome organization in a highly polyploid and aneuploid interspecific hybrid, *Genetics*, 142, 987, 1996.
12. Da Silva, J. A. G., Sorrells, M. E., Burnquist, W. L., and Tanksley, S. D., RFLP linkage map and genome analysis of *Saccharum spontaneum, Genome*, 36, 782, 1993.
13. Al-Janabi, S. M., Honeycutt, R. J., McClelland, M., and Sobral, B. W. S., A genetic linkage map of *Saccharum spontaneum* L. 'SES208.' *Genetics*, 134, 1249, 1993.
14. Price S., Cytogenetics of modern sugar canes, *Econ. Bot.*, 17, 97, 1963.
15. Sreenivasan, T. V., Ahloowalia, B. S., and Heinz, D. J., Cytogenetics, in *Sugarcane Improvement through Breeding*, Heinz, D. J., Ed, Elsevier, New York, 1987, 211.
16. Mather, K., Segregation in autotetraploids, *J. Genet.*, 32, 287, 1936.
17. Ripol, M. I., Churchill, G. A., Da Silva, J. A. G., and Sorrells, M. E., Statistical aspects of genetic mapping in autopolyploids, submitted.
18. Ripol, M. I., Statistical aspects of genetic mapping in autopolyploids, Master thesis, Cornell University, 1994.
19. Churchill, G. A. and Doerge, R. W., Empirical threshold values for quantitative trait mapping, *Genetics*, 138, 963, 1994.
20. Chittenden, L. M., Schertz, K. F., Lin, Y. -R., Wing, R. A., and Paterson, A. H., A detailed RFLP map of *Sorghum bicolor* × *S. propinquum*, suitable for high-density mapping, suggests ancestral duplication of *Sorghum* chromosomes or chromosomal segments, *Theor. Appl. Genet.*, 87, 925, 1994.
21. Paterson, A. H., Lin, Y. -R., Li, Z., Schertz, K. F., Doebley, J. F., Pinson, S. R. M., Liu, S.-C., Stansel, J. W., and Irvine, J. M., Convergent domestication of cereal crops by independent mutations at corresponding genetic loci, *Science*, 269, 1714, 1995.
22. Lin, Y. -R., Schertz, K. F., and Paterson, A. H., Comparative analysis of QTLs affecting plant height and maturity across the Poaceae, in reference to an interspecific sorghum population, *Genetics*, 141, 391, 1995.

# 7 QTL Analysis Under Linkage Equilibrium

*Jeremy F. Taylor and Joao L. Rocha*

## CONTENTS



## 7.1 INTRODUCTION

Summarization of the state of the art in quantitative trait loci (QTL) analysis under linkage equilibrium and considering new developments within a historical context is an enormous challenge. We identified nearly 300 manuscripts addressing aspects pertinent to this topic, which precludes a meaningful synthesis of these contributions within the confines of this chapter. Consequently, our strategy will be to: (1) focus on the contributions from animal breeding; (2) place developments within an historical context, considering the excellent reviews of Soller[1,2] and Weller[3,4] as our launch point; (3) focus on the key technical contributions in the areas of mixed linear model analysis, Bayesian analysis, and maximum likelihood approaches; and (4) assume that the target audience possesses a modest knowledge of statistics.

While we consider the major contributions in QTL analysis from technical and statistical perspectives, we emphasize that technology is only as useful as the extent to which it finds application. In this regard, the trend that we detect in the literature is not entirely satisfying. The huge number of publications in the area of QTL detection by far exceeds the number of reported applications, which in turn, greatly exceeds the number of QTL identified and actually utilized in animal improvement programs. The reasons for this are many, and include the cost of implementation of mapping experiments in livestock species and the difficulties associated with the utilization of marker assisted selection (MAS) in populations under conditions of linkage equilibrium.

0-8493-7686-6/98/$0.00+$.50

However, an emerging issue is: Just how sophisticated an analysis is necessary for QTL detection? Beavis[5] coined the expression "statistical responsibility" in reference to the ample opportunity for statistical[4] and genetic design flaws in QTL analysis, and to our scientific obligation to utilize the appropriate statistical tools for the analysis of data from segregating populations. However, in practice, this obligation has to be tempered by our need for "operational simplicity" which stems from our lack of understanding of the underlying genetic model and a limited access to computer software which implement alternative statistical methodologies. The trend towards statistical complexity will be evident in this chapter. However, we acknowledge that conventional breeding is an extremely efficient machine that relies on fairly simple ideas and methods. Tools that are intended to augment conventional breeding, such as MAS, that are complex or technically demanding will meet resistance from this machine. In view of the complexity of QTL detection and utilization, our primary challenge will be to not lose sight of the simplicity that is inherent to efficient improvement programs.

## 7.2 BASIC PRINCIPLES

### 7.2.1 General Considerations

Figure 7.1 illustrates the essential principles underlying QTL mapping. A nonobservable gene (Q) with a quantitative effect on a trait, is assumed to be syntenic with a marker locus (M) at a physical distance that precludes the independent assortment of QTL and marker alleles at meiosis. If alternate QTL and marker alleles are fixed within two inbred lines, the detection of the QTL can be accomplished using an F2 design in which the linkage disequilibrium present in the parental lines is diminished only to the extent that there is a recombination between the marker and QTL alleles in the F1 gametes. Weller[3] provided a thorough review of the statistical approaches applicable to test for the presence of a putative QTL and for the simultaneous estimation of QTL additive ($a$) and dominance ($d$) effects and recombination rate ($r$). However, from Figure 7.1, it is evident that if the trait means among the progeny marker genotype classes differ, as determined by an analysis of variance (ANOVA), the null hypothesis of no QTL may be rejected.

If the assumption of alternate allele fixation in the inbred lines is not violated, the power of this approach depends on the number of F2 progeny, the magnitude of the QTL effect, the genetic distance separating the marker and QTL loci, and on the Type I error rate that is appropriate to the analysis.[4,6] However, the main weakness of least-squares analysis is that the recombination rate and QTL $a$ and $d$ effects are not individually estimable (see Figure 7.1 for estimable functions). Maximum likelihood (ML) approaches overcome this limitation by capturing the information in both the marker genotype means and within marker genotype variances that are functions of $a$, $d$, and $r$ (see Weller[4,7] for the specification of likelihood functions). Thus, ML methods that are based on pairs of flanking markers (interval mapping[8] and composite interval mapping[9,10]) have become the standard for QTL analysis in inbred line-cross designs. Regression approaches that approximate ML analysis have been proposed.[11]

### 7.2.2 Linkage Equilibrium Vs. Disequilibrium

When the experimental design involves a cross between inbred parental lines, the genetic architecture in the F2, advanced $F_n$ generations or recombinant inbred lines (RILs), is basically that illustrated in Figure 7.1. In the absence of selection, the complete linkage disequilibrium between QTL and marker alleles in the parental lines is reduced by $1 - 2r$ in the F2 generation. In the F3 and subsequent generations, recombination continues to erode the disequilibrium, however, considerable disequilibrium will remain even among RILs.

Other than the limited amount of inbreeding that has occurred in the formation of livestock breeds, inbreeding has not routinely been applied in most animal breeding systems. Consequently,

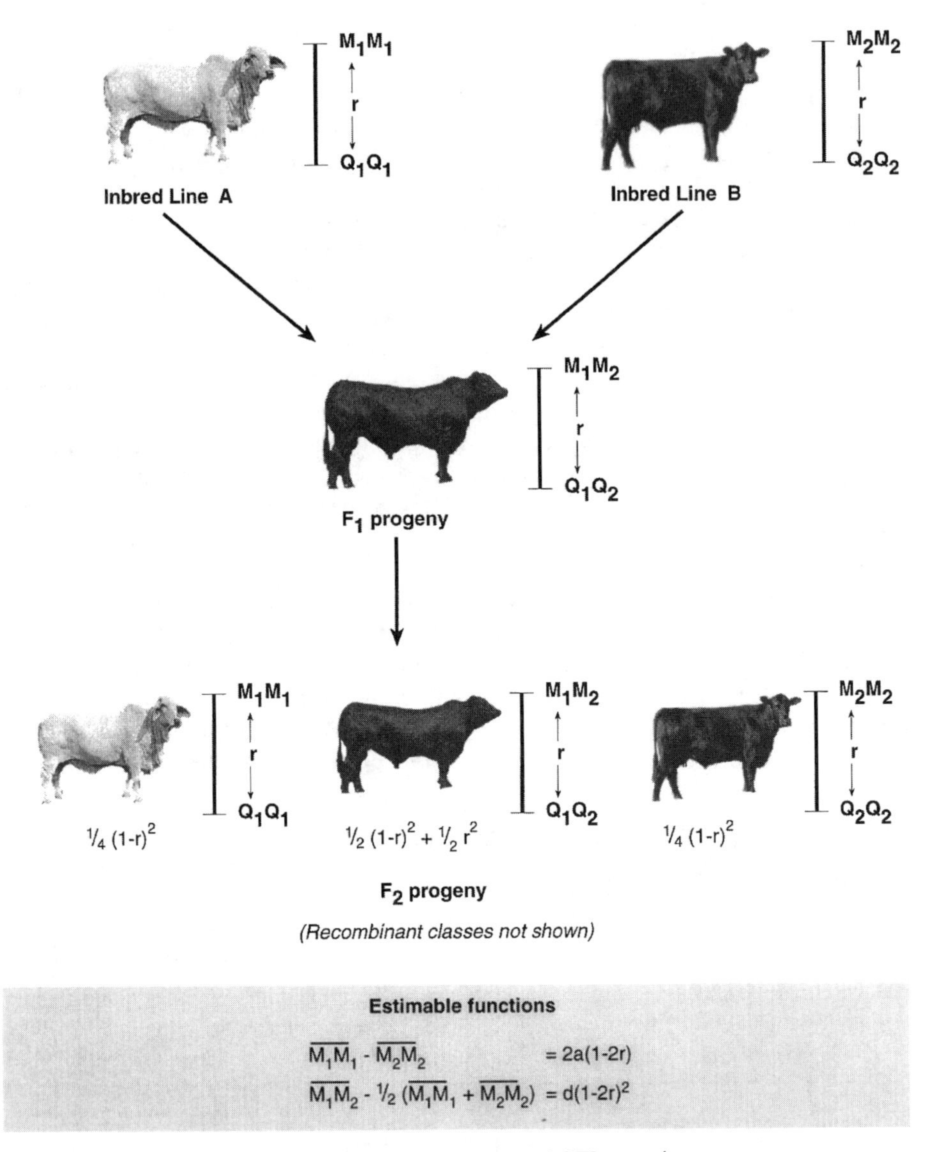


**FIGURE 7.1** Basic principles of QTL mapping.

the majority of livestock populations are characterized by complex and unbalanced pedigree structures in which individuals possess considerable levels of heterozygosity. While the genetic maps for most economically important species are well advanced,[12-14] the degree to which there is linkage disequilibrium among even closely linked markers in these populations is generally unknown.[15] Thus, the presumption that marker and QTL alleles are in linkage equilibrium should probably be made as a conservative basis for QTL mapping (and subsequent MAS) in livestock. The consequence of this assumption is that in a random sample of individuals drawn from a population, a statistical comparison of the trait means of alternative marker genotype classes should not be expected to reveal the existence of a QTL even if the QTL were closely linked to the assayed marker. How then can QTL analysis be performed in populations in which loci are expected to be in linkage equilibrium?

### 7.2.3 The Contribution of Penrose

In 1938, Penrose[16] resolved this problem by capitalizing on the linkage disequilibrium present within pedigrees and set the direction for all subsequent research in this area. As is evident from Figure 7.1, linkage disequilibrium is a necessary condition for the detection of marker-QTL associations, and within outbreeding populations, linkage disequilibrium is guaranteed to exist only among the linked loci segregating within pedigrees. Thus, the apparent paradox of "QTL analysis under linkage equilibrium" is resolved as the consideration of experimental designs and statistical analyses that capture the linkage disequilibrium present within pedigreed population substructures.

Human and animal geneticists have access to data structures that are defined by multigenerational families and marker-QTL phase relationships that differ among the families due to the population level linkage equilibrium. Penrose[16] defined the "reference structure" in which marker and QTL alleles are in linkage disequilibrium to be a family of sibs. Within a group of sibs, the marker-QTL allele phase relationships present in the common parent(s) are disrupted only to the extent that there is recombination between the loci, analogous to the F2 design in Figure 7.1. However, where these designs differ conceptually from a line cross is that not all families may be segregating for a QTL allele, nor be informative for a given marker locus. Thus, any statistical analysis of this type of reference structure must aggregate evidence for the presence of a QTL across families from comparisons of the trait means of sibs with different marker genotypes within each family.

For a reference structure of full-sib pairs, which could be alike or unlike with respect to marker genotype, the quantitative trait difference between the two full-sibs forming a pair is squared and two averages are computed:[16] the average squared difference between all full-sibs (x), and the average squared difference between full-sibs that are unlike with respect to marker-genotype ($x_u$). Penrose's[16] test statistic standardized the ratio of these two averages $[(x_u/x) - 1]$ by its sampling variance under the null hypothesis of independence of the marker and QTL genotypes. Testing for independence is equivalent to testing $r = 0.5$, since the expected value of the test statistic is $(1 - 2r)^2$.

## 7.3 THE LEGACY OF PENROSE IN HUMAN AND ANIMAL GENETICS

From the auspicious contribution of Penrose[16] in 1938, two vigorous branches of the phylogeny of QTL detection research in outbreeding populations emerged: one in 1959,[17] from within poultry breeding which would pave the way for some of the major developments in this field from animal breeding; the other in 1972,[18] from which a considerable series of human genetics contributions in this area would subsequently be derived.

### 7.3.1 Sib-Pairing Approaches in Human Genetics

Haseman and Elston[18] introduced statistical and genetic sophistication to Penrose's[16] method. They based their statistical inference on the regression of the squared sib-pair quantitative trait difference on the estimated proportion of alleles identical by descent (IBD) shared by the two sibs. This required the incorporation of marker data from the sibs' parents, in order to estimate whether any particular sib-pair had 0, 1, or 2 alleles IBD at the trait locus. When sib and parental genotypes are both known, the IBD proportions can easily be calculated for every conceivable mating type and sib-pair.[18] When some of the genotypes are unknown, calculations become more difficult, but an algorithm is provided.[18] This regression coefficient has an expected value of $[-2(1 - 2r)^2\sigma_a^2]$, where $\sigma_a^2$ is the additive genetic variance accounted for by the QTL.[19] A large absolute value of the regression coefficient indicates linkage, since if a QTL is near the marker, there should be an inverse relationship between the sib-pair difference and the proportion of alleles IBD at the marker locus, i.e., the more similar the genotypes of sibs at the marker locus, the smaller should be the difference between them in the metric trait.[18] On the other hand, if there is no QTL near the marker, the squared quantitative difference and the proportion of alleles IBD should be independent, and the regression coefficient should not be significantly different from 0.[18] Haseman and Elston[18] also proposed nonparametric methods to test for linkage, as well as a ML procedure to estimate the recombination rate. Their methods assume random mating and linkage equilibrium, and allow for several loci affecting the quantitative trait, provided that there is no epistasis.

The Haseman and Elston[18] work is the seminal contribution in what have been designated "sib-pairing approaches", and which represent an important component of QTL analysis in human genetics.[20-22] A series of refinements to this analytical approach have subsequently been proposed: the computation of quadrivariate products for groups of three and four sibs;[23-24] extensions to even larger sibships;[25] generalizations to any type of outbred relative pair;[26-27] application of weighted least-squares techniques;[28] and the formulation of multivariate,[29] multiple regression,[30] and variance component[31] strategies. Elston[32] introduced computer software to implement the approach of Haseman and Elston[18] and some of its extensions.

In anticipation of the availability of high resolution genetic maps, Goldgar[33] adapted the Haseman and Elston[18] method to include multiple markers and multiple siblings and to partition the genetic variance of a quantitative trait to include contributions due to loci in specific chromosomal regions.[33] Goldgar's variance component method is based on estimating the expected proportion of genetic material (R) shared IBD by sib pairs in a specified chromosomal region from their genotypes at a set of marker loci spanning the region. The mean and variance of the distribution of R, conditional on IBD status and the recombination rate between two marker-loci, were derived to support the implementation of the method which is based on ML methodology.[33] Goldgar's[33] approach was extended by Schork,[34] and adapted to interval mapping under a random model by Xu and Atchley.[35] An interval mapping extension of the regression approach of Haseman and Elston[18] has also recently been introduced.[36-37]

QTL analysis in human genetics extends well beyond the *sib-pairing approaches*.[20-22] However, regardless of the origin of each branch of QTL analysis in plant, animal, or human genetics, it appears that as they are developed to incorporate general pedigrees these approaches are converging toward a common trunk. The mixed linear model approach of Van Arendonk et al.,[15,38] the variance component formulation of Amos,[31] the random model of Xu and Atchley,[35] the finite polygenic mixed model of inheritance,[39,40] the complex segregation analysis[41] approach, the combined segregation and linkage analysis models of Bonney et al.[42] and Guo and Thompson,[43] and the mixed model likelihood approximations of Hasstedt[44,45] all share common technical and methodological features. Clearly, there is a need for integration of these complex methodologies among the fields.[46,47]

### 7.3.2 Family Designs in Animal Breeding

Animal breeders adapted Penrose's[16] approach to suit the available large half- and full-sib families. In large sib families from heterozygous sires and/or dams, the segregation of both sets of parental alleles can be determined (Figure 7.2). Rather than squaring quantitative differences as proposed by Penrose,[16] animal breeders have used ANOVA in which the trait means of alternative groups of sibs (according to the parental marker allele inherited) are compared within (sire or full-sib) family (Figure 7.2). The sum of squares associated with marker genotype is calculated for each family and these are aggregated across families so that evidence for the segregation of a QTL can be evaluated regardless of the presence of segregating QTL alleles, or of the effects of different phase relationships in different families (Figure 7.2). This approach provides the basis for the "family designs" which have been utilized by animal breeders for the detection of marker-QTL associations. Three primary family designs have emerged: half- and full-sib designs[48] and the granddaughter design[49] which are represented in Figure 7.2. These designs and the statistical methodologies applied to their analysis are familiar to animal breeders and have been the subject of thorough reviews.[1,2,4,46,50,51]

In 1959, Lowry and Shultz[17] were the first animal breeders to utilize these designs for the detection of blood group marker-QTL associations in poultry. Lowry and Shultz[17] concluded that

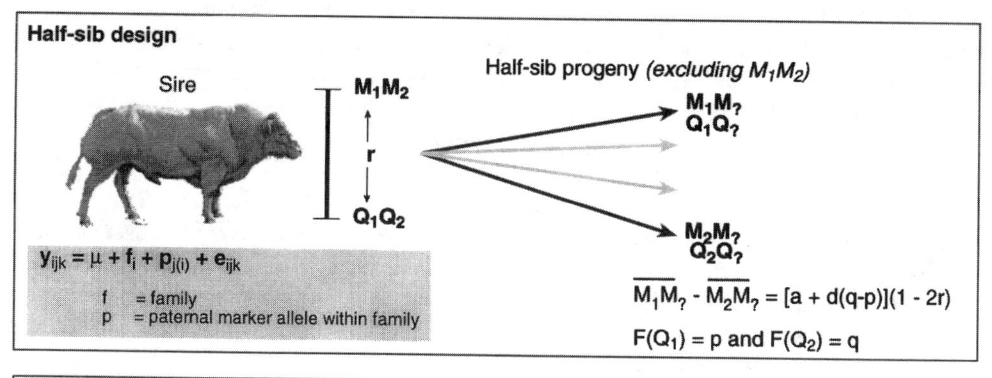


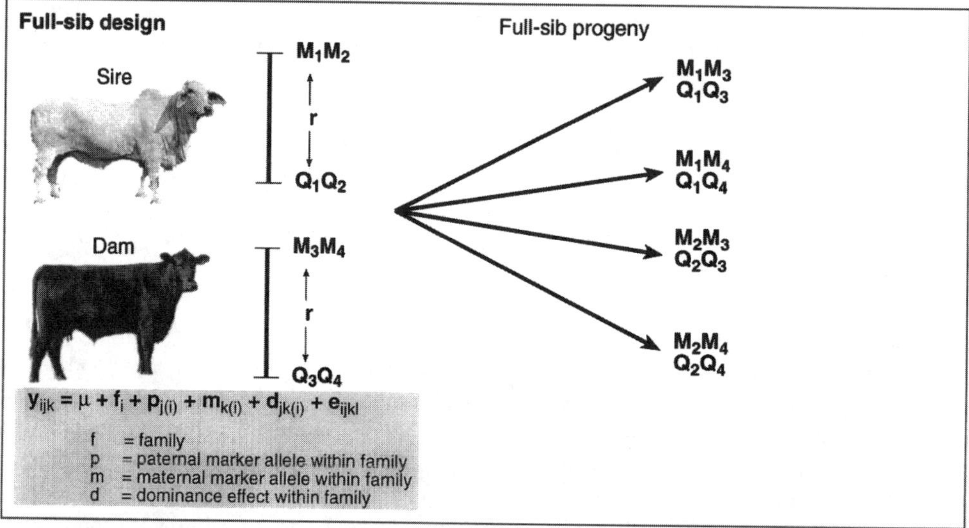


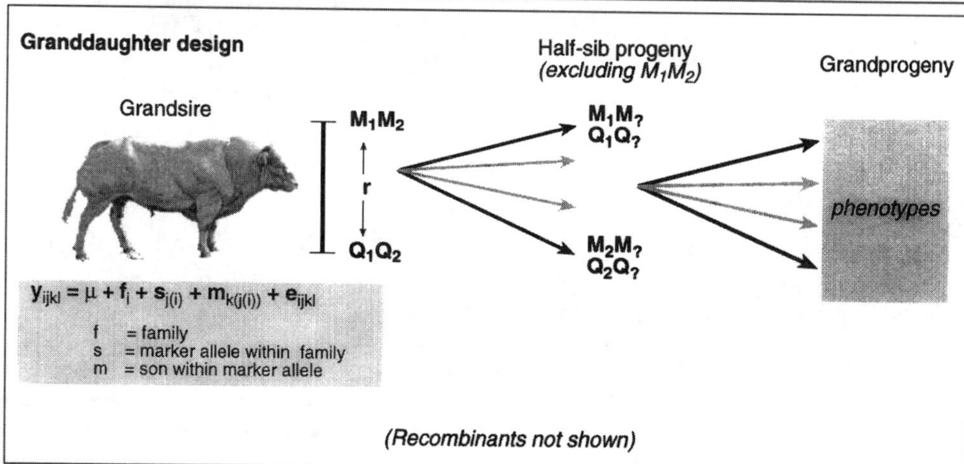


**FIGURE 7.2** Family designs for QTL mapping.

Penrose's[16] method required modification in order to apply to their data structure and proposed an extension which essentially defined the "full-sib design" that would later re-emerge in the contributions of Hill[52] in human genetics and Soller and Genizi[48] in animal breeding. The work of Geldermann in 1975,[53] although apparently an independent development, amounts in essence to the half-sib variant of the Lowry and Shultz[17] model. Rocha[46] provides a detailed chronology of the evolution of family designs, from the model of Lowry and Shultz[17] to the proposal in 1990 of the "granddaughter design" by Weller et al.[49] (Figure 7.2). While the formulation of the granddaughter design represented a considerable innovation for the analysis of sex-limited traits, claims regarding the increased statistical power of this approach relative to a half-sib design have been questioned.[46,54] Another issue that has been raised concerns the appropriate error term to use for hypothesis testing under this design.[46] Likelihood based approaches have also been applied to the granddaughter design to obtain estimates of recombination rates between markers and the detected QTL.[55]

Traditionally, the analysis of family designs has been implemented using least-squares methodology. Heterozygous progeny which share the genotype of the common parent are often excluded from the analysis (see Figure 7.2 and Rocha[46]) because the origin of each parental allele cannot be ascertained. Elegant analytical approaches to circumvent incomplete ascertainment and prevent the exclusion of heterozygous offspring from the analysis have been proposed.[56-58] Family design analyses have been performed utilizing estimates of additive genetic merit as the dependent variable in the statistical model,[59-60] however, Famula and Medrano[61] have recently shown that this approach may lead to biased estimates of QTL effects. Regression-based, multiple-marker extensions applicable to the half-sib design have recently been proposed by Knott et al.[62] and discussed by Haley and Knott.[63] These approaches utilize information from all markers within a linkage group to avoid problems associated with the variability among families in marker locus informativeness and to increase the statistical power and provide better estimates of QTL position and effect.[62,63]

Finally, an analytical approach denoted "trait-based analysis"[64] or "selective genotyping"[8,65] originally proposed in the context of line crosses, can also be implemented as a family design strategy. This experimental design requires the genotyping of only a subsample of the available individuals determined by truncating the distribution of phenotypes within each family. Only a small proportion (1 to 10%) of individuals with the most extreme phenotypes in both tails of the progeny distribution are genotyped. Marker allele frequencies among the two opposing tails of the phenotypic distribution are then compared,[64] or marker genotype means in the pooled selected tails are contrasted.[65] If there is no QTL linked to the marker, marker allele frequencies in both tails should be similar. However, if a QTL is linked to the marker, the divergent selection should result in considerable marker allele frequency differences between the tails, which can be detected by several statistical methodologies.[8,64,65] This is an interesting and flexible approach that can easily be adapted to fit any family design if large half- or full-sib families are available.[46,60,66,67] The method can also be implemented in conjunction with the utilization of pooled DNA strategies[66-71] to maximize its advantages.[46] Haley[72] considered the special problems that the utilization of alternative marker systems [DNA fingerprints/VNTRs (variable number of tandem repeats)] pose to the analysis of family designs.

### 7.3.3 Comparison of Sib-Pairing and Family Designs

Which of these approaches should preferentially be utilized over the others? A recent study[73] concluded that, when highly polymorphic marker systems and large full-sib families are available, the Haseman and Elston[18] sib-pairing method always has a greater statistical power than the "animal breeding" full-sib design (Figure 7.2). The difference in power seems to be important. For families of six full-sibs, a trait heritability of 0.4 and no recombination between the marker and QTL, 500 families provide a statistical power of 37% for the detection of one QTL explaining 4% of the phenotypic variance with the sib-pairing method against a power of 17% with the family design approach.[73] As the size of the full-sib families increases, the advantage in statistical power increases, although some problems remain to be addressed concerning the use of dependent sib-pair comparisons in families larger than six full-sibs.[73-75] This suggests that for full-sib families in animal breeding, the sib-pairing strategy of Haseman and Elston[18] should at least be considered as a powerful alternative for the detection of marker-QTL associations. While Gotz and Ollivier[73] did not perform a comparison between approaches for half-sib families, their opinion was that the family design approach would probably prove to be the most powerful.[73] It would appear to be desirable to verify this opinion. The study of Gotz and Ollivier[73] provided an important first step toward the integration of parallel methodologies that have been independently developed and utilized by animal and human geneticists for QTL detection. Clearly, their results demonstrate the utility of efforts of this kind, and further research is necessary to address the important questions that remain in this area.[46]

## 7.4 CONTRIBUTIONS FROM ANIMAL BREEDING SINCE 1992

### 7.4.1 Marker Genotypes as Fixed Effects in Mixed Linear Models

There are situations where linkage disequilibrium may exist between marker and QTL loci in the data structures available to animal breeders: (1) the marker and the QTL are the same locus, or are extremely closely linked (e.g., candidate gene markers); (2) one of the loci possesses an allele which resulted from a recent mutation event; and (3) genetic drift-selection phenomena due to small effective population sizes.[2,76] In all of these cases, the naturally existing linkage disequilibrium allows the detection of the QTL without the need for the creation of special family structures or statistical analyses which capture phase relationships within families. However, the data analysis must be carefully performed to avoid a confounding of QTL effects with additive genetic family effects or other types of genetic background effects.[77,78] To accomplish this, the analysis must take into account pedigree relationships to eliminate correlations among the phenotypes of related individuals, in order to avoid biases leading to the detection of spurious QTL effects. Kennedy et al.[77] proposed the use of the mixed linear model approach of Henderson[79] to accomplish the partitioning of QTL and polygene effects.

With densely saturated genetic maps, a rapid preliminary genome screen for the existence of marker-QTL associations in unstructured animal breeding data may be accomplished by fitting markers as fixed effects within a mixed linear model.[77,79,80] This strategy has been successfully adopted in a number of recent studies.[61,81,82] In matrix notation, the general mixed linear model can be represented as:

$$y = Xb + Za + e \tag{7.1}$$

where y is a vector of response variable observations; b is a vector of location parameters treated as fixed effects; a is a vector of additive genetic (polygene) effects treated as random; X and Z are design matrices relating elements of b and a to the data y; and e is a vector of random residual errors (nonadditive genetic, permanent, and temporary-environmental effects).

Assumptions inherent to the general mixed model are:[77,79]

$$E\begin{bmatrix} y \\ a \\ e \end{bmatrix} = \begin{bmatrix} Xb \\ 0 \\ 0 \end{bmatrix} \text{ and } Var\begin{bmatrix} y \\ a \\ e \end{bmatrix} = \begin{bmatrix} ZAZ'\sigma_a^2 + I\sigma_e^2 & ZA\sigma_a^2 & I\sigma_e^2 \\ AZ'\sigma_a^2 & A\sigma_a^2 & 0 \\ I\sigma_e^2 & 0 & I\sigma_e^2 \end{bmatrix} \tag{7.2}$$

where A is the numerator relationship matrix[79] (NRM); $\sigma_a^2$ is the additive genetic variance associated with the trait y; I is the identity matrix (residuals are assumed uncorrelated and with homogeneous variance); and $\sigma_e^2$ is the residual variance. The NRM is a symmetric matrix containing all of the pairwise additive genetic covariances (the numerator of Wright's coefficient of relationship) among individuals with additive genetic merits in a.

The best linear unbiased estimator (BLUE) of b and best linear unbiased predictor (BLUP) of a are obtained as the solutions to Henderson's[79] mixed model equations (MME):

$$\begin{bmatrix} X'X & X'Z \\ Z'X & Z'Z + A^{1}\lambda \end{bmatrix} \begin{bmatrix} \hat{b} \\ \hat{a} \end{bmatrix} = \begin{bmatrix} X'y \\ Z'y \end{bmatrix} \tag{7.3}$$

where $\lambda = \sigma_c^2/\sigma_a^2$ and $A^{-1}$ is the inverse of the NRM. The variance components $\sigma_a^2$ and $\sigma_c^2$ are usually estimated by restricted maximum likelihood (REML)[79,83] which requires normality of distribution of polygene and residual effects.

Marker locus genotypes (in either single marker or multiple marker models) are incorporated into the mixed linear model as fixed effects in b. Hypothesis testing for QTL effects is accomplished by the computation of an F-statistic as described by Kennedy et al.,[77] Bovenhuis et al.[81] and Henderson.[79] There are 4 assumptions that are implicit to the parameterization of markers as fixed effects. (1) There exists significant linkage disequilibrium between the marker and the QTL loci. Based on the results of Hill and Robertson,[76] Soller[2] concluded that it would be reasonable to expect linkage disequilibrium between loci with a recombination frequency of less than 5% in many animal populations. However, Van Arendonk et al.[15] discuss evidence which seems to dispute this conclusion. (2) There is a uniform distribution of QTL effects, i.e., QTLs of large and small effect are equally likely. (3) Each marker allele monitors the effect of a single QTL allele (otherwise, the average effect of the QTL genotypes in frequency disequilibrium with each marker genotype is estimated). (4) QTL genotype effects are independent of genetic background (otherwise, an average genotype effect across families is estimated). While these assumptions are probably violated in many circumstances, the widespread availability of mixed linear model computer software which incorporate variance component estimation[84] makes this approach to QTL detection attractive from the perspective of operational simplicity.

### 7.4.2 Marker Genotypes as Random Effects in Mixed Linear Models

To fit markers as fixed effects in mixed linear models fails to capitalize on any of the within-family linkage disequilibrium represented in the data. Theoretically, the mixed model approach should be capable of generalizing the family design concept to incorporate the linkage disequilibrium present within the complete data pedigree reflected in the NRM. Van Arendonk et al.[38] extended important earlier work[85-90] to formulate a mixed linear model at the gametic level which introduced linked marker and QTL loci as random effects. This approach, which recovers all within-family information on marker-QTL allele co-segregation, allows for QTL hypothesis testing and simultaneously provides estimates of recombination rates, magnitudes of QTL effects and of variance components associated with marked QTL. Assumptions implicit to fitting markers and linked QTLs as random effects include: (1) the magnitude of QTL effects is normally distributed; (2) multiple QTL alleles are permitted (QTL effects are not assumed constant across families); (3) a variance component may be parameterized to reflect genetic variance in the population due to each QTL; and (4) markers and QTLs are linked, allowing the recombination rate between the loci to be estimated.

The mixed model introduced by Van Arendonk et al.[38] may be written:

$$y = Xb + Za + Wv + e \tag{7.4}$$

where v is a vector of random gametic effects at the QTL, W is the corresponding design matrix and all other terms are as previously defined. Under this model, the additive genetic value of an individual is partitioned into two components — due to marked QTL effects (v) and due to residual polygenes (a). The additive genetic covariances among elements of a in the NRM are as previously modeled, i.e., for each segregation between a parent and progeny, equal probabilities of 0.5 are assigned, indicating that at each locus either of the two parental alleles could have been inherited with equal probability (unobserved events). However, for the remainder of the marked genome (v), probabilities of 0 or 1 are assigned according to the inheritance of the observed marker alleles.

In this model, $v \sim (0, G_{v|r}\sigma_v^2)$, where $G_{v|r}$ is the gametic relationship matrix (GRM) for the marked QTL, and $\sigma_v^2$ is the additive genetic variance due to gametic effects at the marked QTL. The notation $G_{v|r}$ indicates that the GRM for the QTL is established using linked marker information, and is dependent on the recombination fraction between the marker and QTL. This is accomplished by assigning probabilities of $(1 - r)$ and $(r)$ in $G_{v|r}$ rather than 1 and 0, respectively. The QTL effects are assumed to be uncorrelated with residual polygene effects (Cov(a,v) = 0), an assumption that will fail in the presence of epistasis.

The total additive genetic variance is $\sigma_t^2 = \sigma_a^2 + 2\sigma_v^2$ and the matrix $G_{vlr}$ has twice the number of rows and columns as the corresponding NRM (reflecting that each individual inherited a gamete from each parent), and the MME corresponding to the model of Van Arendonk et al.[38] are

$$\begin{bmatrix} X'X & X'Z & X'W \\ Z'X & Z'Z + A^{1}\lambda & Z'W \\ W'X & W'Z & W'W + G_{v|r}^{-1}\gamma \end{bmatrix} \begin{bmatrix} \hat{b} \\ \hat{a} \\ \hat{v} \end{bmatrix} = \begin{bmatrix} X'y \\ Z'y \\ W'y \end{bmatrix} \quad (7.5)$$

where $\gamma = \sigma_e^2/\sigma_v^2$. Multiple QTL-models can theoretically be fit by this approach (covariances among QTL effects are assumed 0), but a gametic relationship matrix is required for each marker-QTL combination, thus raising the number of equations to be solved per individual from one (in the conventional polygene model) to (2m + 1), where m is the number of markers in the model in Equation 7.4. This creates considerable technical demands and computational problems[91] which currently define the greatest limitation to the application of this approach. Developments such as the Gibbs sampler[43,92-94] and other Markov Chain Monte Carlo (MCMC) approaches[43] may help remediate this limitation. However, once QTL have been identified, the model of Van Arendonk et al.[38] may be reparameterized to include only a single random effect for each animal representing additive genetic merit as the joint effect of marked QTL and residual polygenes. Consequently, the methodology has important consequences for the application of MAS in situations where an appreciable proportion of the additive genetic variance is due to residual polygene effects. The estimation of r and of variance components under this model may be accomplished by the application of derivative-free REML procedures,[38,95] and tests for QTL performed using likelihood ratio tests of hypotheses concerning the QTL variance components.

Wang et al.[96] have recently introduced several refinements to enhance the efficiency of this methodology. Grignola et al.[97] proposed a generalization of the approach based on interval mapping, in which the model is parameterized to incorporate flanking marker information to facilitate the estimation of the QTL position within the interva1.[97] Hoeschele[98] presented a method for reducing the number of equations to be solved by absorbing those for animals with missing marker genotypes.

### 7.4.3 Bayesian Analysis

Bayesian methods offer conceptually different approaches to statistical analysis. Bayesian analysis is often technically and computationally demanding and is not well understood by geneticists. Consequently, the utility of Bayesian analysis for QTL detection is unclear at present. The Bayesian interpretation of likelihood based approaches, such as when marker genotypes are included as fixed effects in a mixed linear model as in Equation 7.1, is that model parameters are estimated as the mode of the posterior distribution with an uniformative (uniform) prior distribution of QTL genotype effects. Beavis[47] has commented that QTL mapping experiments based on a small sample size tend to overestimate both the magnitude of effect and the proportion of the genetic variation attributed to the detected QTL. Results from large experiments now indicate that there may often be many more QTL of small effect than there are of large effect, suggesting that an exponential prior distribution for QTL effects may be appropriate.[92,93] Bayesian estimates of QTL effects are obtained by shrinking the QTL effect estimated from the data toward the mode of the prior distribution,[99] which presumably results in a more conservative test for the presence of QTL. Hoeschele and VanRaden[99,100] and Hoeschele[92,93] have presented and applied models for Bayesian analysis of half-sib family and granddaughter designs in dairy cattle, but developments in this area are best desribed as being at an embryonic stage. This form of analysis may be facilitated using MCMC methods such as the Gibbs sampler.[43,92-94]

### 7.4.4 Likelihood-Based Approaches

ML is an elegant and powerful statistical procedure which is often technically demanding to implement and this has limited its application in livestock QTL mapping. Frequently, the application of ML to family based designs is achieved by partitioning the data into a series of families which are assumed to be independent.[101,102] This approach has the undesirable consequence of omitting relationship information among parents which may have an important consequence for the estimation of QTL effects. In this regard, ML appears to be particularly sensitive to the specification of the underlying genetic model.[21,63] Haley and Knott[63] and Elston[21] discussed the advantages and disadvantages of ML over alternative approaches.

Knott and Haley[103] derived the likelihood for the application of interval mapping[8] to full-sib families in the family design represented in Figure 7.2. Their method incorporates a random component for common family effects due to the presence of additional QTL, residual polygenic variation and/or the environment.[103] Weller,[104] Mackinnon and Weller[105] and Ron et al.[55] introduced likelihood-based models for application to the half-sib[104,105] and the granddaughter[55] designs, however, these approaches have the disadvantage of being single-marker[55,104,105] or single-interval[103] models. The likelihood-based composite interval mapping approach of Jansen and Stam[9] and Zeng,[10] which are now beginning to be used extensively in plant QTL mapping, have yet to be extended for application under the family design paradigm. Likelihood-based QTL analyses in domesticated livestock species have successfully been implemented by Bovenhuis and Weller[101] and Georges et al.[102]

## 7.5 CONCLUSIONS

While initial statistical research into QTL mapping was based on the classical genetic paradigms of inbred line crosses and large kindreds, current research acknowledges the lack of these types of family structure within commercial breeding populations and attempts to capture the disequilibrium that exists within more general pedigrees.[47] While it is likely that the integration of markers into the framework of a mixed linear model will become the rule, particularly as the need eventuates to make selection decisions based on joint QTL and residual polygene information, more complex likelihood-based or Bayesian analysis will grow in popularity only as computer applications are developed and distributed. In the interim, analyses based on family designs and fit by least-squares will continue to be used due to their simplicity.

We should not lose sight of the fact that QTL mapping and utilization are breeding and genetic problems which have a statistical dimension, but they are not statistical problems per se.[80] Opportunities related to the strategic and timely definition of breeding objectives, the careful evaluation and definition of environments (both production and marketing), and the importance of alternative genetic backgrounds[80] will probably provide the framework for success in the utilization of identified QTL, no matter how they were identified. Integration, simplicity, responsibility and utility appear to be four key concepts to be promoted at this stage of QTL research. Efforts to integrate[46] the novel and existing contributions in the fields of human genetics[20-22] and plant[47] and animal breeding into a unified conceptual framework seem to be essential to simultaneously ensure operational simplicity[106] and statistical responsibility[4-6] which are both imperative for utility.[47,106]

## REFERENCES


1. Soller, M., Genetic mapping of the bovine genome using deoxyribonucleic acid-level markers to identify loci affecting quantitative traits of economic importance, *J. Dairy Sci.*, 73, 2628, 1990.
2. Soller, M., Mapping quantitative trait loci affecting traits of economic importance in animal populations using molecular markers, in *Gene-Mapping Techniques and Applications,* Schook, L. B., Lewin, H. A. and McLaren, D. G., Eds., Marcel Dekker, New York, 1991, 21.

3. Weller, J. I., Statistical methodologies for mapping and analysis of quantitative trait loci, in *Plant Genomes: Methods for Genetic and Physical Mapping,* Beckmann, J. S. and Osborn, T. C., Eds., Kluwer Academic Publishers, Dordrecht, The Netherlands, 1992, 181.
4. Weller, J. I. and Ron, M., Detection and mapping quantitative trait loci in segregating populations: theory and experimental results, in *Proc. 5th World Congress on Genetics Applied to Livestock Production*, Vol. 21, Univ. of Guelph, Ontario, Canada, 1994, 213.
5. Beavis, W. D., The power and deceit of QTL experiments: lessons from comparative QTL studies, in *Proc. 49th Annual Corn & Sorghum Industry Research Conference,* 1994, 250.
6. Lander, E. S. and Kruglyak, L., Genetic dissection of complex traits: guidelines for interpreting and reporting linkage results, *Nat. Genet.,* 11, 241, 1995.
7. Weller, J. I., Maximum likelihood techniques for the mapping and analysis of quantitative trait loci with the aid of genetic markers, *Biometrics,* 42, 627, 1986.
8. Lander, E. S. and Botstein, D., Mapping Mendelian factors underlying quantitative traits using RFLP linkage maps, *Genetics,* 121, 185, 1989.
9. Jansen, R. C. and Stam, P., High resolution of quantitative traits into multiple loci via interval mapping, *Genetics,* 136, 1447, 1994.
10. Zeng, Z.-B., Precision mapping of quantitative trait loci, *Genetics,* 136, 1457, 1994.
11. Haley, C. S. and Knott, S. A., A simple regression method for mapping quantitative trait loci in line crosses using flanking markers, *Heredity,* 69, 315, 1992.
12. Bishop, M. D., Kappes, S. M., Keele, J. W., Stone, R. T., Sunden, S. L., Hawkins, G. A., Toldo, S. S., Fries, R., Grosz, M. D., Yoo, J., and Beattie, C. W., A genetic linkage map for cattle, *Genetics,* 136, 619, 1993.
13. Crawford, A. M., Montgomery, G. W., Pierson, C. A., Brown, T., Dodds, K. G., Sunden, S. L., Henry, H. M., Ede, A. J., Swarbrick, P. A., Berryman, T., Penty, J. M., and Hill, D. F., Sheep linkage mapping: nineteen linkage groups derived from the analysis of paternal half-sib families, *Genetics,* 137, 573, 1994.
14. Rohrer, G. A., Alexander, L. J., Keele, J. W., Smith, T. P., and Beattie, C. W., A microsatellite linkage map of the porcine genome, *Genetics,* 136, 231, 1994.
15. Van Arendonk, J. A., Bovenhuis, H., Van der Beek, S., and Groen, A. F., Detection and exploitation of markers linked to quantitative traits in farm animals, in *Proc. 5th World Congress on Genetics Applied to Livestock Production,* Vol. 21, Univ. of Guelph, Ontario, Canada, 1994, 193.
16. Penrose, L. S., Genetic linkage in graded human characters, *Ann. Eugen.,* 8, 233, 1938.
17. Lowry, D. C. and Shultz, F., Testing association of metric traits and marker genes, *Ann. Hum. Genet.,* 23, 83, 1959.
18. Haseman, J. K. and Elston, R. C., The investigation of linkage between a quantitative trait and a marker locus, *Beh. Genet.,* 2, 3, 1972.
19. Elston, R. C., A general linkage method for the detection of major genes, in *Advances in Statistical Methods Applied to Livestock Production,* Hammond, K. and Gianola, D., Eds., Springer, Berlin, 1990, 495.
20. Lander, E. S. and Schork, N. J., Genetic dissection of complex traits, *Science,* 265, 2037, 1994.
21. Elston, R. C., Linkage and association to genetic markers, *Exp. Clin. Immunogenet.,* 12, 129, 1995.
22. Weeks, D. E. and Lathrop, G. M., Polygenic disease: methods for mapping complex disease traits, *Trends Genet.,* 11, 513, 1995.
23. Cockerham, C. C., Sib pairing methodology, in *Genetic Analysis of Common Diseases: Applications to Predictive Factors in Coronary Disease,* Alan R. Liss, New York, 1979, 417.
24. Cockerham, C. C. and Weir, B. S., Linkage between a marker locus and a quantitative trait of sibs, *Am. J. Hum. Genet.,* 35, 263, 1983.
25. Blackwelder, W. C. and Elston, R. C., Power and robustness of of sib-pair linkage tests and extension to larger sibships, *Commun. Stat. Theor. Meth.,* 11, 449, 1982.
26. Amos, C. I. and Elston, R. C., Robust methods for the detection of genetic linkage for quantitative data from pedigrees, *Genet. Epidem.,* 6, 349, 1989.
27. Olson, J. M. and Wijsman, E. M., Linkage between quantitative trait and marker loci: methods using all relative pairs, *Genet. Epidem.,* 10, 87, 1993.
28. Amos, C. I., Elston, R. C., Wilson, A. F., and Bailey-Wilson, J. E., A more powerful robust sib-pair test of linkage for quantitative traits, *Genet. Epidem.,* 6, 435, 1989.

29. Amos, C. I., Elston, R. C., Bonney, G. E., Keats, B. J., and Berenson, G. S., A multivariate method for detecting genetic linkage, with application to a pedigree with an adverse lipoprotein phenotype, *Am. J. Hum. Genet.*, 47, 247, 1990.
30. Fulker, D. W., Cardon, L. R., DeFries, J. C., Kimberling, W. J., Pennington, B. F., and Smith, S. D., Multiple regression analysis of sib-pair data on reading to detect quantitative trait loci, *Reading Writing: Interdiscip. J.*, 3, 299, 1991.
31. Amos, C. I., Robust variance-components approach for assessing genetic linkage in pedigrees, *Am. J. Hum. Genet.*, 54, 535, 1994.
32. Elston, R. C., Segregation and linkage analysis, *Anim. Genet.*, 23, 59, 1992.
33. Goldgar, D. E., Multipoint analysis of human quantitative genetic variation, *Am. J. Hum. Genet.*, 47, 957, 1990.
34. Schork, N. J., Extended multipoint identity-by-descent analysis of human quantitative traits: efficiency, power, and modeling considerations, *Am. J. Hum. Genet.*, 53, 1306, 1993.
35. Xu, S. and Atchley, W. R., A random model approach to interval mapping of quantitative trait loci, *Genetics*, 141, 1189, 1995.
36. Fulker, D. W. and Cardon, L. R., A sib-pair approach to interval mapping of quantitative trait loci, *Am. J. Hum. Genet.*, 54, 1092, 1994.
37. Cardon, L. R. and Fulker, D. W., The power of interval mapping of quantitative trait loci using selected sib pairs, *Am. J. Hum. Genet.*, 55, 825, 1994.
38. Van Arendonk, J. A., Tier, B., and Kinghorn, B. P., Use of multiple genetic markers in prediction of breeding values, *Genetics*, 137, 319, 1994.
39. Fernando, R. L., Stricker, C., and Elston, R. C., The finite polygenic mixed model: an alternative formulation for the mixed model of inheritance, *Theor. Appl. Genet.*, 88, 573, 1994.
40. Stricker, C., Fernando, R. L., and Elston, R. C., Linkage analysis with an alternative formulation for the mixed model of inheritance: the finite polygenic mixed model, *Genetics*, 141, 1651, 1995.
41. Morton, N. E. and MacLean, C. J., Analysis of family resemblance. III. Complex segregation analysis of quantitative traits, *Am. J. Hum. Genet.*, 26, 489, 1974.
42. Bonney, G. E., Lathrop, G. M., and Lalouel, J.-M., Combined linkage and segregation analysis using regressive models, *Am. J. Hum. Genet.*, 43, 29, 1988.
43. Guo, S. W. and Thompson, E. A., A Monte Carlo method for combined segregation and linkage amalysis, *Am. J. Hum. Genet.*, 51, 1111, 1992.
44. Hasstedt, S. J., A mixed model approximation for large pedigrees, *Comput. Biomed. Res.*, 15, 295, 1982.
45. Hasstedt, S. J., A variance components/major locus likelihood approximation on quantitative data, *Genet. Epidem.*, 8, 113, 1991.
46. Rocha, J. L., Blood Group Polymorphisms and Production and Type Traits in Dairy Cattle: After Forty Years of Research, Ph.D. Dissertation, Texas A&M Univ., College Station, 1994.
47. Beavis, W. D., QTL analyses: power, precision and accuracy, *this volume, Chap. 11.*
48. Soller, M. and Genizi, A., The efficiency of experimental designs for the detection of linkage between a marker locus and a locus affecting a quantitative trait in segregating populations, *Biometrics*, 34, 47, 1978.
49. Weller, J. I., Kashi, Y., and Soller M., Power of daughter and granddaughter designs for determining linkage between marker loci and quantitative trait loci in dairy cattle, *J. Dairy Sci.*, 73, 2525, 1990.
50. Soller, M. and Beckmann, J. S., Restriction fragment length polymorphisms in poultry breeding, *Poult. Sci.*, 65, 1474, 1986.
51. Soller, M., Strategies and opportunities for mapping QTL in agriculturally important animals, in *Mapping the Genomes of Agriculturally Important Animals*, Womack, J. E., Ed., The Institute of Biosciences and Technology, Texas A&M Univ., College Station, 1990, 53.
52. Hill, A. P., Quantitative linkage: a statistical procedure for its detection and estimation, *Ann. Hum. Genet.*, 38, 439, 1975.
53. Geldermann, H., Investigations on the inheritance of quantitative characters in animals by gene markers. I. methods, *Theor. Appl. Genet.*, 46, 319, 1975.
54. Mackinnon, M. J. and Georges M. A., The effects of selection on linkage analysis for quantitative traits, *Genetics*, 132, 1177, 1992.
55. Ron, M., Band, M., Yanai, A., and Weller, J. I., Mapping quantitative trait loci with DNA microsatellites in a commercial dairy cattle population, *Anim. Genet.*, 25, 259, 1994.

56. Dentine, M. R. and Cowan, C. M., An analytical model for the estimation of chromosome substitution effects in the offspring of individuals heterozygous at a segregating marker locus, *Theor. Appl. Genet.*, 79, 775, 1990.
57. Hoeschele, I. and Meinert, T. R., Association of genetic defects with yield and type traits: the weaver locus effect on yield, *J. Dairy Sci.*, 73, 2503, 1990.
58. Clamp, P. A., Beever, J. E., Fernando, R. L., McLaren, D. G., and Schook, L. B., Detection of linkage between genetic markers and genes that affect growth and carcass traits in pigs, *J. Anim. Sci.*, 70, 2695, 1992.
59. Andersson-Eklund, L., Danell, B., and Rendel, J., Association between blood groups, blood protein polymorphisms and breeding values for production traits in Swedish Red and White dairy bulls, *Anim. Genet.*, 21, 361, 1990.
60. Rocha, J. L., Taylor, J. F., Sanders, J. O., and Cherbonnier, D. M., Blood group polymorphisms and production and type traits in dairy cattle: after forty years of research, in *Proc. 5th World Congress on Genetics Applied to Livestock Production*, Vol. 19, Univ. of Guelph, Ontario, Canada, 1994, 299.
61. Famula, T. R. and Medrano, J. F., Estimation of genotype effects for milk proteins with animal and sire transmitting ability models, *J. Dairy Sci.*, 77, 3153, 1994.
62. Knott, S. A., Elsen, J.-M., and Haley, C. S., Multiple marker mapping of quantitative trait loci in half-sib populations, in *Proc. 5th World Congress on Genetics Applied to Livestock Production*, Vol. 21, Univ. of Guelph, Ontario, Canada, 1994, 33.
63. Haley, C. S. and Knott, S. A., Interval mapping, in *Proc. 5th World Congress on Genetics Applied to Livestock Production*, Vol. 21, Univ. of Guelph, Ontario, Canada, 1994, 25.
64. Lebowitz, R. J., Soller, M., and Beckmann, J. S., Trait-based analyses for the detection of linkage between marker loci and quantitative trait loci in crosses between inbred lines, *Theor. Appl. Genet.*, 73, 556, 1987.
65. Darvasi, A. and Soller, M., Selective genotyping for determination of linkage between a marker locus and a quantitative trait locus, *Theor. Appl. Genet.*, 85, 353, 1992.
66. Plotsky, Y., Cahaner, A., Haberfeld, A., Lavi, U., and Hillel, J., Analysis of genetic association between DNA fingerprint bands and quantitative traits using DNA mixes, in *Proc. 4th World Congress on Genetics Applied to Livestock Production,* Vol. 13, Hill, W. G., Thompson, R., and Woolliams J. A., Eds., University of Edinburgh, 1990, 133.
67. Shalom, A., Darvasi, A., Barendse, W., Cheng H., and Soller, M., Single-parent segregant pools for allocation of markers to a specified chromosomal region in outcrossing species, *Anim. Genet.*, 27, 9, 1996.
68. Arnheim, N., Strange, C., and Erlich, H., Use of pooled DNA samples to detect linkage disequilibrium of polymorphic restriction fragments and human disease: studies on the HLA class II loci, *Proc. Natl. Acad. Sci. U.S.A.*, 82, 6970, 1985.
69. Michelmore, R. W., Paran, I., and Kesseli, R. V., Identification of markers linked to disease-resistance genes by bulked segregant analysis: a rapid method to detect markers in specified genomic regions by using segregating populations, *Proc. Natl. Acad. Sci. U.S.A.*, 88, 9828, 1991.
70. Darvasi, A., Khatib, H., and Soller, M., Selective genotyping with DNA pooling, *Anim. Genet.*, 23 (Suppl. 1), 108, 1992.
71. Hillel, J., Kalay, D., Gal, O., Plotsky, Y., Weisberger, P., and Haberfeld, A., Application of multilocus molecular markers in cattle breeding. 2. Use of blood mixes, *J. Dairy Sci.*, 76, 653, 1993.
72. Haley, C. S., Use of DNA fingerprints for the detection of major genes for quantitative traits in domestic species, *Anim. Genet.*, 22, 259, 1991.
73. Gotz, K. U. and Ollivier, L., Theoretical aspects of applying sib-pair linkage tests to livestock species, *Genet. Sel. Evol.*, 24, 29, 1992.
74. Blackwelder, W. C. and Elston, R. C., A comparison of sib-pair linkage tests for disease susceptibility loci, *Genet. Epidem.*, 2, 85, 1985.
75. Collins, A. and Morton, N. E., Nonparametric tests for linkage with dependent sib pairs, *Hum. Hered.*, 45, 311, 1995.
76. Hill, W. G. and Robertson, A., Linkage disequilibrium in finite populations, *Theor. Appl. Genet.*, 38, 226, 1968.
77. Kennedy, B. W., Quinton, M., and Van Arendonk, J. A., Estimation of effects of single genes on quantitative traits, *J. Anim. Sci.*, 70, 2000, 1992.

78. Briscoe, D., Stephens, J. C., and O'Brien, S. J., Linkage disequilibrium in admixed populations: applications in gene mapping, *J. Hered.*, 85, 59, 1994.
79. Henderson, C. R., *Applications of Linear Models in Animal Breeding,* Univ. of Guelph Press, Ontario, Canada, 1984.
80. Rocha, J. L., Taylor, J. F., Sanders, J. O., Openshaw, S. J., and Fincher, R., Genetic markers to manipulate QTL: the additive illusion, in *Proc. Annu. National Breeders Roundtable,* Poultry Breeders of America and Southeastern Poultry & Egg Association, St. Louis, Missouri, 1995, 12.
81. Bovenhuis, H., Van Arendonk, J. A., and Korver, S., Associations between milk protein polymorphisms and milk production traits, *J. Dairy Sci.,* 75, 2549, 1992.
82. Rothschild, M. F., Vaske, D. A., Tuggle, C. K., McLaren, D. G., Short, T. H., Eckardt, G. R., Mileham, A. J., Plastow, G. S., Southwood, O. I., and Van der Steen, H. A., Discovery of a major gene associated with litter size in the pig, in *Proc. Annu. National Breeders Roundtable,* Poultry Breeders of America and Southeastern Poultry & Egg Association, St. Louis, Missouri, 1995, 52.
83. Taylor, J. F., *Dairy Production Class Notes,* Texas A&M University, College Station, TX, 1990.
84. Misztal, I., Comparison of software packages in animal breeding, in *Proc. 5th World Congress on Genetics Applied to Livestock Production,* Vol. 22, Univ. of Guelph, Ontario, Canada, 1994, 3.
85. Fernando, R. L. and Grossman, M., Marker assisted selection using best linear unbiased prediction, *Genet. Sel. Evol.,* 21, 467, 1989.
86. Goddard, M. E., A mixed model for analyses of data on multiple genetic markers, *Theor. Appl. Genet.,* 83, 878, 1992.
87. Gibson, J. P., Kennedy, B. W., Schaeffer, L. R., and Southwood, O. I., Gametic models for estimation of autosomally inherited effects that are expressed only when received from either male or female parent, *J. Dairy Sci.,* 71 (Suppl. 1), 143, 1988.
88. Schaeffer, L. R., Kennedy, B. W., and Gibson, J. P., The inverse of the gametic relationship matrix, *J. Dairy Sci.,* 72, 1266, 1989.
89. Smith, S. P. and Maki-Tanilla, A., Genotypic covariance matrices and their inverses for models allowing for dominance and inbreeding, *Genet. Sel. Evol.,* 22, 65, 1990.
90. Tier, B. and Solkner, J., Analysing gametic variation with an animal model, *Theor. Appl. Genet.*, 85, 868, 1993.
91. Bink, M. C. and Van Arendonk, J. A., Marker-assisted prediction of breeding values in dairy cattle populations, in *Proc. 5th World Congress on Genetics Applied to Livestock Production,* Vol. 21, Univ. of Guelph, Ontario, Canada, 1994, 233.
92. Hoeschele, I., Bayesian QTL mapping via the Gibbs sampler, in *Proc. 5th World Congress on Genetics Applied to Livestock Production,* Vol. 21, Univ. of Guelph, Ontario, Canada, 1994, 241.
93. Hoeschele, I., Markov Chain Monte Carlo in genetic analysis, course notes, Dept. of Animal Breeding, Wageningen Agricultural University, The Netherlands, 1994.
94. Janss, L. L., Thompson, R., and Van Arendonk, J. A., Application of Gibbs sampling for inference in a mixed major gene-polygenic inheritance model in animal populations, *Theor. Appl. Genet.*, 91, 1137, 1995.
95. Graser, H.-U., Smith, S. P., and Tier, B., A derivative-free approach for estimating variance components in animal models by restricted maximum likelihood, *J. Anim. Sci.*, 64, 1362, 1987.
96. Wang, T., Fernando, R. L., Van der Beek, S., Grossman, M., and Van Arendonk, J. A., Covariance between relatives for a marked quantitative trait locus, *Genet. Sel. Evol.*, 27, 251, 1995.
97. Grignola, F. E., Hoeschele, I., and Meyer, K., Empirical best linear unbiased prediction to map QTL, in *Proc. 5th World Congress on Genetics Applied to Livestock Production,* Vol. 21, Univ. of Guelph, Ontario, Canada, 1994, 245.
98. Hoeschele, I., Elimination of quantitative trait loci equations in an animal model incorporating genetic marker data, *J. Dairy Sci.*, 76, 1693, 1993.
99. Hoeschele, I. and VanRaden, P. M., Bayesian analysis of linkage between genetic markers and quantitative trait loci. II. Combining prior knowledge with experimental evidence, *Theor. Appl. Genet.*, 85, 946, 1993.
100. Hoeschele, I. and VanRaden, P. M., Bayesian analysis of linkage between genetic markers and quantitative trait loci. I. Prior knowledge, *Theor. Appl. Genet.*, 85, 953, 1993.
101. Bovenhuis, H. and Weller, J. I., Mapping and analysis of dairy cattle quantitative trait loci by maximum likelihood methodology using milk protein genes as genetic markers, *Genetics*, 137, 267, 1994.

102. Georges, M., Nielsen, D., Mackinnon, M., Mishra, A., Okimoto, R., Pasquino, A. T., Sargeant, L. S., Sorensen, A., Steele, M. R., Zhao, X., Womack, J. E., and Hoeschele, I., Mapping quantitative trait loci controlling milk production in dairy cattle by exploiting progeny testing, *Genetics*, 139, 907, 1995.
103. Knott, S. A. and Haley, C. S., Maximum likelihood mapping of quantitative trait loci using full-sib families, *Genetics*, 132, 1211, 1992.
104. Weller, J. I., Experimental designs for mapping quantitative trait loci in segregating populations, in *Proc. 4th World Congress on Genetics Applied to Livestock Production,* Vol. 13, Hill, W. G., Thompson, R., and Woolliams, J. A., Eds., Edinburgh, 1990, 113.
105. Mackinnon, M. J. and Weller, J. I., Estimation of QTL parameters in a half-sib design using maximum likelihood methods, in *Proc. 3rd Australasian Gene Mapping Workshop*, Univ. of Queensland, Brisbane, Australia, 1992, 74.
106. Muir, W. M., Poultry improvement: integration of present and new genetic approaches for layers, in *Proc. 5th World Congress on Genetics Applied to Livestock Production*, Vol. 20, Univ. of Guelph, Ontario, Canada, 1994, 5.

# 8 Molecular Analysis of Epistasis Affecting Complex Traits

*Zhikang Li*

## CONTENTS



## 8.1 INTRODUCTION

Epistasis is a term originally used by Bateson in 1909 to describe genes which mask or cover the effects of other genes. This term has since acquired a more general meaning which is synonymous with nonlinear interactions between alleles at different loci. Epistasis is an important genetic basis underlying complex phenotypes, Wright's theory of evolution,[1,2] and founder effect models of speciation.[3] While the existence of gene interactions has been well established at physiological and molecular levels, detection and characterization of epistasis affecting complex quantitative traits have been challenging and unsolved problems. To date, numerous classical quantitative genetic studies using biometrical methods do not reveal pronounced epistasis affecting quantitative traits,[4] but these results are less convincing because the methodology has some unrealistic assumptions and is unable to dissect individual gene effects. On the other hand, two lines of indirect evidence from numerous evolutionary and population studies strongly suggest that epistasis may have played an important role in complex trait variation such as fitness and its components.[5-9] First, hybrid breakdown (reduced fertility and viability) resulted from incompatibility (unfavorable interactions) between genes of different species or subspecies. It has been invariably observed to be associated with the recombinant progenies from interspecific or intersubspecific hybrids in both animals and plants.[7,10,11] This observation suggests that epistasis is an important basis to maintain the genetic integrity of a species or subspecies. Second, differential phenotypic effects of genes or chromosomes

0-8493-7686-6/98/$0.00+$.50

in different genetic backgrounds have been reported in numerous cases in both animals and plants.[12-15]

Recent quantitative trait locus (QTL) mapping experiments using DNA markers (see the concepts and methodology described in the previous chapters) yield controversial results concerning the importance of epistasis affecting complex traits.[16] Most QTL mapping experiments could map only a limited number of QTLs for each trait studied, which collectively explain only a portion of the total trait variation. There is little evidence for the presence of epistasis between QTLs regardless of the amount of genetic variation in mapping populations and the genome coverage by DNA markers. However, results from a marker-assisted selection (MAS) experiment[14] and several recent mapping studies[15,17-21] have provided strong evidence suggesting that epistasis may be an important genetic basis underlying complex traits.

By transferring alleles at two QTLs controlling the inflorescence architecture of maize (*Zea mays* L. ssp. *mays*) and its progenitor teosinte (*Z. mays* ssp. *parviglumis*) into respective genetic backgrounds, Doebley et al.[15] have demonstrated that the phenotypic effects of QTLs may change dramatically depending on the genetic background. The epistatic effect between the two QTLs exceeded the total main effects of the two individual QTLs. Lark et al.[17] have also found that several QTLs affecting plant height interacted strongly with certain background loci in a large recombinant inbred soybean population. Using a unique experimental design and statistical tests, Cockerham and Zeng[18] were able to detect strong epistasis among linked loci which may behave as single 'overdominant' QTLs affecting many quantitative traits in maize. By genotyping an F2 population and phenotyping the derived F4 progeny from an intersubspecific rice cross, Li et al.[19,20] found that interactions between complementary loci from the same parents were largely responsible for hybrid breakdown (reduced fertility and grain yield components) in rice. These complementary loci do not appear to have main effects on quantitative traits when tested alone in a segregating population.[19] Although these studies have not provided a definite answer to the problem, they have demonstrated the usefulness of DNA markers in studying epistasis. In the following sections, the author intends to briefly exploit several aspects of detecting epistasis affecting complex traits using DNA markers.

## 8.2 DETECTION OF EPISTASIS AFFECTING COMPLEX TRAITS USING DNA MARKERS

### 8.2.1 Classification of Epistasis

Although several types of epistasis between major genes can be classified based on distinct phenotypes of different ratios in a segregating population,[22] epistasis between a pair of loci affecting a complex quantitative trait can only be detected by deviations in trait values from that expected based on the main effects of the two loci. Nevertheless, results from both classic evolutionary studies and recent QTL mapping experiments suggest possible presence of three types of epistasis affecting complex traits: (1) interactions between QTLs, (2) interactions between QTLs and 'background' (modifying) loci, and (3) interactions between 'complementary' loci. By using suitable experimental designs, these interactions can be further classified based on gene actions. For example, digenic interactions may include $a \times a$, $a \times d$, $d \times a$, and $d \times d$ components. Accordingly, high order interactions enjoy more possible categories.

The first type of epistasis has been well described in classic quantitative genetics theory,[4] in which polygenes (QTLs) having main (additive and dominance) effects on a quantitative trait are involved in epistasis affecting the same trait. Although there is little evidence from most QTL mapping studies in support of the importance of this type of epistasis,[16] it has been suggested that lack of epistasis between QTLs may be primarily due to the experimental designs and the statistical methods used in these studies.[18-20]

The second type of epistasis — interactions between QTLs and 'background' (or modifying) loci — has recently been shown to be a common type of epistasis affecting quantitative traits.[14,15,17,20]

The third type of epistasis, interactions between 'complementary' genes, is perhaps the most important one, but has received the least attention. This type of epistasis is suggested from evolutionary studies that alleles at interacting loci from the same gene pool interact to produce a balanced, intermediate phenotype with high fitness in the environment(s) it evolved.[5-8] Thus, the term complementary has two implications. First, interacting alleles from the same species, subspecies, or population are complementary or compatible while those from different species, subspecies, or populations are uncomplementary or incompatible. Second, in a population derived from crosses between distantly related parents, complementary loci will not be apparent as QTLs since the alleles at different loci have reciprocal effects on phenotype. This type of interaction between uncomplementary (incompatible) alleles has been shown to be an important genetic basis underlying complex fitness traits such as hybrid sterility and breakdown in the progenies from an *indica-japonica* cross of rice.[20] More importantly, various degrees of hybrid breakdown or hybrid weakness are also commonly observed by most plant and animal breeders in the progeny of crosses between well-adapted and closely related parents, suggesting that epistasis between complementary genes is an important basis for complex traits such as fitness and yield.

## 8.2.2 Quantitative Genetic Models for Epistasis

### 8.2.2.1 F2 Populations

There are many ways by which alleles at different loci may interact with one another. In a given type of mapping population, the phenotypic deviation of an individual arising from interactions between alleles of different loci can be described by inclusion of corresponding parameters in the quantitative genetic model. Consider the case of digenic epistasis between two QTLs (type 1 epistasis) with two alleles Aa and Bb at each locus: there are five possible types of genetic parameters associated with the nine genotypes in an F2 population (Table 8.1). These are two additive effects ($\alpha_B$ and $\alpha_A$) due to the allelic substitution at the two loci, two dominance effects ($h_A$ and $h_B$) associated with the heterozygotes at the two loci, four additive × additive effects ($\tau_{ij}$) arising from the interactions between homozygotes at the two loci, four additive × dominance effects ($\gamma_{ij}$) due to the interactions between the homozygotes and the heterozygotes, and one dominance × dominance effect ($\phi_{AB}$) attributable to the interaction between the two heterozygotes.

The assignment of genetic parameters in Table 8.1 is different from the classic genetic model in which only one additive digenic parameter, $i_{ab}$ ($i_{ab} = \tau_{AB} = \tau_{ab}$ and $-i_{ab} = \tau_{Ab} = \tau_{aB}$), and two additive × dominance parameters, $j_{ab}$ and $j_{ba}$ ($j_{ab} = \gamma_{AB} = -\gamma_{Ab}$ and $j_{ba} = \gamma_{BA} = -\gamma_{Ba}$), and one dominance × dominance parameter $l_{ab}$, are specified.[4] The approach by which the digenic parameters in Table 8.1 are defined is necessary and has important implications. First, in the epistasis model of Table 8.1, the estimates of the main effects of both loci are biased and confounded with both additive × additive and additive × dominance effects. For instance, the marginal effects of locus A is $(2\alpha_A + G(\tau_{AB} - \tau_{Ab} - \tau_{aB} + \tau_{ab}) + H(\gamma_{BA} + \gamma_{Ba})$. The estimate of the dominance effect, $[h_A + H\varphi_{AB} - J(\tau_{AB} + \tau_{Ab} + \tau_{aB} + \tau_{ab}) + G(\gamma_{AB} + \gamma_{Ab} - \gamma_{BA} - \gamma_{Ba})]$, is also confounded with all three types of digenic parameters.

The model of Table 8.1 is complicated and can be simplified under certain assumptions. For example, when the phenotypic effects from interactions between different allelic pairs are strictly additive, i.e., $\gamma_{AB} = H(\tau_{AB} + \tau_{aB})$, $\gamma_{BA} = H(\tau_{AB} + \tau_{Ab})$, $\gamma_{Ba} = H(\tau_{aB} + \tau_{ab})$, $\gamma_{Ab} = H(\tau_{Ab} + \tau_{ab})$, and $\varphi_{AB} = G(\tau_{AB} + \tau_{Ab} + \tau_{aB} + \tau_{ab})$, the model of Table 8.1 will become the additive epistatic model (Table 8.2). Under the complete dominance at both loci, i.e., $\gamma_{AB} = \gamma_{BA} = \varphi_{AB} = \tau_{AB}$, $\gamma_{Ab} = \tau_{Ab}$, and $\gamma_{Ba} = \tau_{aB}$, then, the model of Table 8.1 will become the dominance model in Table 8.3. Under the situation of one additive locus (A) interacting with a complete dominance locus (B), the model of Table 8.1 will become the mixed model in Table 8.4. In all these cases, the original nine digenic parameters in model 1 are replaced with only four additive digenic parameters. Fit of the different models to the real data may provide information about the relative importance of different types of gene action in epistasis.

**TABLE 8.1**
**The Genetic Model of Digenic Epistasis in an F2 Population**

| | AA | Aa | aa | Mean |
|---|---|---|---|---|
| BB | $\alpha_A + \alpha_B + \tau_{AB}$ | $\alpha_B + h_A + \gamma_{AB}$ | $-\alpha_A + \alpha_B + \tau_{aB}$ | $\frac{1}{2}h_A + \alpha_B + \frac{1}{4}(\tau_{AB} + \tau_{aB}) + \frac{1}{4}\gamma_{AB}$ |
| Bb | $\alpha_A + h_B + \gamma_{BA}$ | $h_A + h_B + \varphi_{AB}$ | $-\alpha_A + h_B + \gamma_{Ba}$ | $\frac{1}{2}h_A + h_B + \frac{1}{4}(\gamma_{BA} + \gamma_{Ba}) + \frac{1}{2}\varphi_{AB}$ |
| bb | $\alpha_A - \alpha_B + \tau_{Ab}$ | $h_A - \alpha_B + \gamma_{Ab}$ | $-\alpha_A - \alpha_B + \tau_{ab}$ | $\frac{1}{2}h_A - \alpha_B + \frac{1}{4}(\tau_{Ab} + \tau_{ab}) + \frac{1}{2}\gamma_{Ab}$ |
| Mean | $\frac{1}{2}h_B + \alpha_A + \frac{1}{4}(\tau_{AB} + \tau_{Ab}) + \frac{1}{2}\gamma_{BA}$ | $h_A + \frac{1}{2}h_B + \frac{1}{4}(\gamma_{AB} + \gamma_{Ab}) + \frac{1}{2}\varphi_{AB}$ | $\frac{1}{2}h_B - \alpha_A + \frac{1}{4}(\tau_{aB} + \tau_{ab}) + \frac{1}{2}\gamma_{Ba}$ | $\frac{1}{2}h_A + \frac{1}{2}h_B + \frac{1}{16}(\tau_{AB} + \tau_{Ab} + \tau_{aB} + \tau_{ab}) + \frac{1}{8}(\gamma_{AB} + \gamma_{Ab} + \gamma_{BA} + \gamma_{Ba}) + \frac{1}{2}\varphi_{AB}$ |

**TABLE 8.2**
**The Additive Genetic Model for Digenic Epistasis in an F2 Population**

| | AA | Aa | aa | Mean |
|---|---|---|---|---|
| BB | $\alpha_A + \alpha_B + \tau_{AB}$ | $\alpha_B - h_A + \frac{1}{2}(\tau_{AB} + \tau_{aB})$ | $-\alpha_A + \alpha_B + \tau_{aB}$ | $\frac{1}{2}h_A + \alpha_B + \frac{1}{2}(\tau_{AB} + \tau_{aB})$ |
| Bb | $\alpha_A + h_B + \frac{1}{2}(\tau_{AB} + \tau_{Ab})$ | $h_A + h_B + \frac{1}{4}(\tau_{AB} + \tau_{Ab}) + (\tau_{aB} + \tau_{ab})$ | $-\alpha_A + h_B + \frac{1}{2}(\tau_{aB} + \tau_{ab})$ | $\frac{1}{2}h_A + h_B + \frac{1}{4}(\tau_{AB} + \tau_{Ab} + \tau_{aB} + \tau_{ab})$ |
| bb | $\alpha_A - \alpha_B + \tau_{Ab}$ | $h_A - \alpha_B + \frac{1}{2}(\tau_{Ab} + \tau_{ab})$ | $-\alpha_A - \alpha_B + \tau_{ab}$ | $\frac{1}{2}h_A - \alpha_B + \frac{1}{2}(\tau_{Ab} + \frac{1}{4}\tau_{ab})$ |
| Mean | $\frac{1}{2}h_B + \alpha_A + \frac{1}{2}(\tau_{AB} + \tau_{Ab})$ | $h_A + \frac{1}{2}h_B + \frac{1}{4}(\tau_{AB} + \tau_{Ab} + \tau_{aB} + \tau_{ab})$ | $\frac{1}{2}h_B - \alpha_A + \frac{1}{2}(\tau_{aB} + \tau_{ab})$ | $\frac{1}{2}h_A + \frac{1}{2}h_B + \frac{1}{4}(\tau_{AB} + \tau_{Ab} + \tau_{aB} + \tau_{ab})$ |

**TABLE 8.3**
**The Dominance Epistasis Model for Interactions between Two Complete Dominant Gene Pairs in an F2 Population**

| | AA | Aa | aa | Mean |
|---|---|---|---|---|
| BB | $\alpha_A + \alpha_B + \tau_{AB}$ | $\alpha_B + h_A + \tau_{AB}$ | $-\alpha_A + \alpha_B + \tau_{aB}$ | $\frac{1}{2}h_A + \alpha_B + \frac{3}{4}\tau_{AB} + \frac{1}{4}\tau_{aB}$ |
| Bb | $\alpha_A + h_B + \tau_{AB}$ | $h_A + h_B + \tau_{AB}$ | $-\alpha_A + h_B + \tau_{AB}$ | $\frac{1}{2}h_A + h_B + \frac{3}{4}\tau_{AB} + \frac{1}{4}\tau_{aB}$ |
| bb | $\alpha_A - \alpha_B + \tau_{AB}$ | $h_A - \alpha_B + \tau_{Ab}$ | $-\alpha_A - \alpha_B + \tau_{aB}$ | $\frac{1}{2}h_A + \alpha_B + \frac{3}{4}\tau_{AB} + \frac{1}{4}\tau_{aB}$ |
| Mean | $\frac{1}{2}h_B + \alpha_A + \frac{3}{4}\tau_{AB} + \frac{1}{4}\tau_{Ab}$ | $h_A + \frac{1}{2}h_B + \frac{3}{4}\tau_{AB} + \frac{1}{4}\tau_{Ab}$ | $\frac{1}{2}h_B - \alpha_A + \frac{3}{4}\tau_{aB} + \frac{1}{4}\tau_{ab}$ | $\frac{1}{2}h_A + \frac{1}{2}h_B + \frac{9}{16}\tau_{AB} + \frac{3}{16}(\tau_{Ab} + \tau_{aB}) + \frac{1}{16}\tau_{ab}$ |

**TABLE 8.4**
**The Mixed Epistasis Model for Interactions between an Additive Gene (A) and a Complete Dominant Gene (B) in an F2 Population**

| | AA | Aa | aa | Mean |
|---|---|---|---|---|
| BB | $\alpha_A + \alpha_B + \tau_{AB}$ | $\alpha_A + h_A + \frac{1}{2}(\tau_{AB} + \tau_{aB})$ | $-\alpha_A + \alpha_B + \tau_{aB}$ | $\frac{1}{2}h_A + \alpha_B + \frac{1}{2}(\tau_{AB} + \tau_{aB})$ |
| Bb | $\alpha_A + h_B + \tau_{AB}$ | $h_A + h_B + \frac{1}{2}(\tau_{AB} + \tau_{aB})$ | $-\alpha_A + h_B + \tau_{aB}$ | $\frac{1}{2}h_A + h_B + \frac{1}{2}(\tau_{AB} + \tau_{aB})$ |
| bb | $\alpha_A - \alpha_B + \tau_{Ab}$ | $h_A - \alpha_B + \frac{1}{2}(\tau_{Ab} + \tau_{ab})$ | $-\alpha_A - \alpha_B + \tau_{ab}$ | $\frac{1}{2}h_A - \alpha_B + \frac{1}{2}(\tau_{Ab} + \tau_{ab})$ |
| Mean | $\frac{1}{2}h_B + \alpha_A + \frac{3}{4}\tau_{AB} + \frac{1}{4}\tau_{Ab}$ | $h_A + \frac{1}{2}h_B + \frac{3}{8}(\tau_{AB} + \tau_{aB}) + \frac{1}{8}(\tau_{Ab} + \tau_{ab})$ | $\frac{1}{2}h_B - \alpha_A + \frac{3}{4}\tau_{aB} + \frac{1}{4}\tau_{ab}$ | $\frac{1}{2}(h_A + h_B) + \frac{3}{8}(\tau_{AB} + \tau_{aB}) + \frac{1}{8}(\tau_{AB} + \tau_{ab})$ |

**TABLE 8.5**
**The Genetic Model of Digenic Epistasis in a RI or DH Population**

| | AA | aa | Mean |
|---|---|---|---|
| BB | $\alpha_A + \alpha_B + \tau_{AB}$ | $-\alpha_A + \alpha_B + \tau_{ab}$ | $\alpha_B + ½(\tau_{AB} + \tau_{aB})$ |
| bb | $\alpha_A - \alpha_B + \tau_{Ab}$ | $-\alpha_A - \alpha_B + \tau_{ab}$ | $\alpha_A + ½(\tau_{Ab} + \tau_{ab})$ |
| Mean | $\alpha_A + ½(\tau_{AB} + \tau_{Ab})$ | $-\alpha_A + ½(\tau_{aB} + \tau_{ab})$ | $¼(\tau_{AB} + \tau_{Ab} + \tau_{aB} + \tau_{ab})$ |

#### 8.2.2.2 RI and DH Populations

The digenic epistasis in a doubled haploid (DH) or recombinant inbred (RI) population is much simpler than that in an F2 population. There are four possible genotypes — AABB, AAbb, aaBB, and aabb, which are associated with two main effects and four digenic parameters in a RI or DH population, as shown in Table 8.5.

Again, it is noted that the marginal effects of the genotypes at the two loci contain the main effects confounded with the epistatic effects for interactions between QTLs. In other words, locus A would not be detected as a QTL unless the epistatic effects are consistent in direction with the main effects, [i.e., both $\alpha_A$ and ($\tau_{AB}$ + $\tau_{Ab}$) are either positive or negative], which means that the alleles at the two loci are synergetic. Otherwise, the main effects of either loci may easily be cancelled out by the epistatic parameters of opposing effects.

The model in Table 8.5 can be easily extended to cover cases of three-locus interactions, which is much more complicated with four additional digenic parameters and eight trigenic parameters. It can be shown that the mean marginal effects of single loci are confounded with both digenic and trigenic effects, and the mean marginal effects of digenic genotypes are confounded with some of the trigenic effects. This is generally true for higher order interactions.

It is noted that when the main effects ($\alpha$ and h) at either or both interacting loci in the above models (Tables 8.1 through 8.5) are removed, the models will represent the cases of type 2 interactions between QTLs and background loci, and type 3 (interactions between complementary loci) epistasis.

### 8.2.3 Statistical Models

#### 8.2.3.1 Two-Way ANOVA

The concept to detect and quantify epistasis affecting complex traits using linked DNA markers is much the same as that in QTL mapping described in the previous sections. Although the quantitative genetic models for epistasis described above are complicated, the statistical models for detecting epistasis are straightforword. Regardless of various types of mapping populations, the most commonly used statistical method to detect digenic epistasis between QTLs is two-way analysis of variance (ANOVA),[19,20,23-28] with the following general linear model:

$$y_{ijm} = \mu + \alpha_i + \alpha_j + \psi_{ij} + \varepsilon_{ijm}, \qquad \text{for } m = 1,2,\ldots,n_{ij} \tag{8.1}$$

where $y_{ijm}$ is the trait value of the mth individual with the digenic genotype at marker loci i and j, $\alpha_i$ and $\alpha_j$ are the main effects (the additive and the dominance effects) associated with the loci i and j, respectively; $\psi_{ij}$ is the effect arising from interactions between the alleles at loci i and j, and $e_{ijm}$ is the residual effect including the genetic effect unexplained by the two loci in the model plus the experimental error, which is assumed to be an identical and independent random variable having a normal distribution with zero mean and a variance of $\sigma^2$.

In a two-way ANOVA using the model (8.1) three hypotheses are tested simultaneously (including the main effects $\alpha_i$ and $\alpha_j$, associated with two loci and the interactions between alleles at the two loci), assuming markers i and j locate right on the two QTLs. The detection of epistasis between two QTLs then is to test the null hypothesis $H_0$: $\Sigma\hat{\psi}_{ij}^2 = 0$. The genetic expectations of the interaction effects, $\hat{\psi}_{ij}(\hat{\mu}_{ij} - \hat{\mu}_{i.} - \hat{\mu}_{.j} + \mu)$ in the model (8.1) which is estimated by unweighted sample means can be known from the genetic models described above. For example, in RI or DH populations

$$(i, j = 1,2),\ E(\psi_{11}) = \hat{\mu}_{11} - \hat{\mu}_{1.} - \hat{\mu}_{.1} + \hat{\mu} = E(\psi_{22}) = \tfrac{1}{4}(\tau_{AB} + \tau_{ab} - \tau_{Ab} - \tau_{aB})$$

and

$$E(\psi_{12}) = \hat{\mu}_{12} - \hat{\mu}_{1.} - \hat{\mu}_{.2} + \hat{\mu} = E(\psi_{21}) = \tfrac{1}{4}(\tau_{Ab} + \tau_{aB} - \tau_{AB} - \tau_{ab})$$

which equal $\hat{i}_{ab}$ and $-\hat{i}_{ab}$ in the classic model.[4] Thus, rejection of the null hypothesis will certainly indicate the presence of additive epistasis.

In an F2 population (i,j = 1, 2, 3), the genetic expectations of $\hat{\psi}_{ij}(\hat{\mu}_{ij} - \hat{\mu}_{i.} - \hat{\mu}_{.j} + \hat{\mu})$ include all three types of parameters. For example, based on the model of Table 8.1

$$E(\psi_{11}) = \tfrac{13}{16}\tau_{AB} - \tfrac{3}{16}(\tau_{ab} + \tau_{Ab} + \tau_{aB}) - \tfrac{3}{8}(\gamma_{AB} + \gamma_{BA}) + \tfrac{1}{8}(\gamma_{Ab} + \gamma_{Ba}) + \tfrac{1}{4}(\varphi_{AB} + \gamma_{Ba})$$

or

$$= i_{ab} - \tfrac{1}{2}(j_{ab} + j_{ba}) + \tfrac{1}{4}l_{ab}$$

in the classical model, etc. Thus, under any genetic models, rejection of $H_0$: $\Sigma\hat{\psi}_{ij}^2 = 0$ in the statistical model (Equation 8.1) would indicate the presence of the digenic epistasis. However, failure to reject the null hypothesis may not be an indication for the absence of epistasis since $\hat{\psi}_{ij}$ is a composite effect consisting of several types of digenic parameters which may differ in both sign and magnitude.

Thus, one of the major drawbacks of the model (Equation 8.1) is its inability to dissect individual digenic parameters without imposing certain assumptions. For example, for an F2 population in the case of additive model (Table 8.2)

$$\gamma_{AB} = \tfrac{1}{2}(\tau_{AB} + \tau_{aB}), \gamma_{BA} = \tfrac{1}{2}(\tau_{AB} + \tau_{Ab})$$

and

$$\varphi_{AB} = \tfrac{1}{4}(\tau_{AB} + \tau_{Ab} + \tau_{aB} + \tau_{ab})$$

then

$$E(\psi_{11}) = E(\psi_{22}) = \tfrac{1}{4}(\tau_{AB} + \tau_{ab} - \tau_{Ab} - \tau_{aB})$$

and

$$E(\psi_{12}) = E(\psi_{21}) = \tfrac{1}{4}(\tau_{Ab} + \tau_{aB} - \tau_{AB} - \tau_{ab})$$

which is the same as that in RI or DH populations. All nonadditive interaction effects ($\psi_{13}$, $\psi_{31}$, and $\psi_{33}$) are expected to be 0. In cases of the complete dominance (Table 8.3)

$$\gamma_{AB} = \gamma_{BA} = \varphi_{AB} = \tau_{AB}$$

and

$$\gamma_{Ab} = \tau_{Ab}, \gamma_{aB} = \tau_{aB}$$

then

$$E(\psi_{11}) = E(\psi_{31}) = E(\psi_{13}) = E(\psi_{33}) = \tfrac{1}{16}(\tau_{AB} + \tau_{aB} + \tau_{Ab} + \tau_{ab})$$

$$\hat{\psi}_{22} = \tfrac{9}{16}(\tau_{AB} + \tau_{ab} - \tau_{Ab} - \tau_{aB}), E(\psi_{11}) = E(\psi_{13}) = E(\psi_{12}) = E(\psi_{32}) = \tfrac{3}{16}(\tau_{Ab} + \tau_{aB} - \tau_{AB} - \tau_{ab})$$

In the case of one additive locus (A) and the other (B) being completely dominant (Table 8.4)

$$E(\psi_{11}) = E(\psi_{13}) = \tfrac{1}{8}(\tau_{AB} + \tau_{ab} - \tau_{Ab} - \tau_{aB})$$

$$E(\psi_{21}) = E(\psi_{23}) = \tfrac{1}{8}(\tau_{Ab} + \tau_{aB} - \tau_{AB} - \tau_{ab})$$

$$E(\psi_{12}) = \tfrac{3}{8}(\tau_{Ab} + \tau_{aB} - \tau_{AB} - \tau_{ab})$$

$$E(\psi_{22}) = \tfrac{3}{8}(\tau_{AB} + \tau_{ab} - \tau_{Ab} - \tau_{aB})$$

and

$$E(\psi_{31}) = E(\psi_{33}) = E(\psi_{32}) = 0$$

Again, individual digenic parameters are inestimable using sample means of digenic genotypes in any of these cases since there are fewer independent equations than the unknown parameters. However, fit of the observed $\hat{\psi}_{ij}$ to the genetic expectations from different models of Tables 8.2–8.4, may provide information about the relative importance of different gene actions in the observed epistasis. Another problem of the two-way ANOVA using the model (8.1) is that it may have very high probability of false positive interactions arising from the background genetic effects,[19] which will be discussed in the later section.

Other than two-way ANOVA, a likelihood ratio test was used to detect epistasis in a soybean RI population by Lark et al.,[17] in which the null hypothesis $H_0$: $\delta = (\mu_{AB} + \mu_{ab}) - (\mu_{Ab} + \mu_{aB}) = 0$ was tested. This test is virtually equivalent to the ANOVA method in that

$$E(\delta) = (\hat{\mu}_{AB} + \hat{\mu}_{ab}) - (\hat{\mu}_{Ab} + \hat{\mu}_{aB}) = (\tau_{AB} + \tau_{ab} + \tau_{Ab} + \tau_{aB})$$

in the model of Table 8.5, or equals $4i_{ab}$ in the classic genetic model.[4]

#### 8.2.3.2 Multiple Regression Model — Control of 'Background Genetic' Effects

Results from most recent QTL mapping experiments indicate that it is generally true that for a quantitative trait, a significant proportion of trait variation is attributable to a limited number of

QTLs with relatively large phenotypic effects.[16] In such a situation, false positive interactions between two random markers may arise as a result of the background genetic effects from the nonrandom sampling of segregating QTLs. This can be a very serious problem in the detection of epistasis, particularly with small mapping populations, and/or segregating QTLs having very large phenotypic effects.[19] There are at least two ways by which the background genetic effects can be controlled or minimized.

The first and the most efficient way to control the background genetic effects is to use multiple regression analyses. The theoretical properties of multiple regression analysis in QTL mapping and control of background genetic effects have been fully demonstrated by Zeng.[29,30] Assuming that other than the epistatic loci i and j, there are k independent QTLs segregating in a population, the linear model of the multiple regression to detect the interaction between loci i and j, is shown as follows

$$y_{ijm} = b_0 + \sum_k b_k x_{mk} + b_i x_{mi} + b_j x_{mj} + b_{mij} x_{ij} + \varepsilon_{ijm} \quad \text{for } k = 1,2,\ldots k,\ k \neq i,j,\ m = 1,2,\ldots,n_{ij} \quad (8.2)$$

where $y_{ijm}$ is the trait value of the individuals with the same digenic genotype at marker loci i and j (i, j, = 1, 2, 3 in an F2 population, or 1, 2 in a RI or DH population), $b_0$ is the mean of the model, $b_k$ is the partial regression coefficient of the phenotype on the kth QTL, $\alpha_k$ is the main effect associated with the kth QTL, $b_i$, $b_j$, and $b_{ij}$ (equivalent to $\alpha_i$, $\alpha_j$, and $\psi_{ij}$ in Equation 8.1) are partial coefficients (the main effects and the interaction effects) of phenotype y on the ith and jth markers conditional on all k QTLs, and $\varepsilon_{ijm}$ is the residual genetic effects plus the error, which is assumed to be identically and independently distributed variable with zero mean and a variance of $\sigma^2$. Since the interaction effects $b_{ij}$ is tested conditional to all segregating QTLs, not only the background genetic effects from nonrandom sampling of the QTLs can be effectively controlled, but the power to detect epistasis is greatly improved,[29,30] provided that the QTLs are independent from one another. This method was successfully utilized by Li et al.[19] to control background genetic effects in the detection of epistasis affecting three grain yield components of rice. In their experiment, Li et al.[19] found that 40 to 70% of statistically significant ($p < 0.001$) interactions using Equation 8.1 could be attributable to background genetic effects of segregating QTLs depending on different traits studied. They propose that detection of epistasis requires that the null hypothesis $H_0$: $\Sigma\hat{\psi}_{ij}^2 = 0$ be rejected (by at least $p < 0.001$) in both Equations 8.1 and 8.2 in order to avoid serious false positive problems. It is also important to point out that scales, or the ways by which complex quantitative traits are measured or recorded, can have a significant impact on the detection of epistasis since certain types of nonallelic interactions may be removed by appropriate data transformation.[4]

The second way to control the background genetic effects is to construct specific genetic materials such as near isogenic lines (NILs) or introgression lines by introducing specific interacting gene pairs into the same genetic backgrounds and evaluating these materials in comparable environments.[15,21] For example, when a pair of unlinked interacting genes (Aa and Bb) is identified in a diploid plant species using DNA markers, four pure near isogenic lines with respective digenic genotypes AABB, AAbb, aaBB and aabb, can be easily generated by a marker-assisted backcrossing procedure. Heterozygous genotypes can be generated by making crosses between these lines. Phenotypic evaluation of the nine genotypes in various environments will provide accurate information of all types of gene actions involved in the interacting genes.

It is important to point out that in the above statistical models, the loci i and j are genetic markers linked to the presumed interacting genes with genetic distances $r_i$ and $r_j$. Thus, in the statistical model(s), all the effects in the model including the main effects and the interaction effects should be adjusted accordingly, as shown in the interval mapping of QTLs described in the previous chapters. The theory and the methodology of interval mapping of QTLs[31] should be applicable to the mapping of epistatic loci. For example, to identify digenic epistasis affecting a complex trait using a complete genetic map, one may scan two genome regions for all possible two-way interactions between any two points flanked by four markers. A LOD peak over a predetermined threshold in a three dimensional surface with X and Y axes representing two unlinked genomic regions each flanked by two markers, and Z axis representing the LOD, would suggest possible presence of epistatic genes in the respective regions.

### 8.2.4 Other Important Factors

Results from large numbers of theoretical studies on QTL mapping suggest that several other factors are also important in detection of epistasis.

#### 8.2.4.1 Experimental Design

Different types of experimental designs influence detection of epistasis. For instance, an F2 population has the advantage to give a complete description of all three types of epistasis and allows an assessment of different gene actions in epistasis. However, it suffers two major disadvantages. The first one is the fact that all phenotypic data of unreplicated F2 individuals are single measurement and subject to large environmental noises unless progeny testing is used. Second, a very large population is required to obtain statistically reliable results because of the maximum number of genotypes in an F2 population. The RI or DH populations are powerful for detecting additive epistasis but suffer the inability to detect nonadditive component of epistasis. Design III and its modified forms (BCnF1 lines generated by mating random progenies to their parents) are particularly powerful experimental designs which allow varied aspects of epistasis to be quantified using different statistical methods.[18]

#### 8.2.4.2 Population Size

Reliable detection and quantification of epistasis requires larger population size than normal QTL mapping since the number of genotypes increase geometrically as the number of loci increases regardless of the types of mapping populations. For instance, in an F2 mapping population with codominant markers and no segregation distortion, to detect digenic interactions in an F2 population, only $^1/_{16}$ of the total individuals contribute to each of the four additive × additive parameters, J to each of the four additive × dominance effects, and G to the dominance × dominance effects. Such a reduction in the effective population size will certainly be associated with large errors from trait measurements and sampling of segregating QTLs.

## 8.3 SUMMARY

In summary, epistasis is an important genetic basis for complex traits. Several lines of evidence in recent QTL mapping studies indicate that epistasis is commonly detected between QTLs (with main effects) and background loci, and between complementary loci which do not appear to have significant main (additive and/or dominance) effects. Lack of interactions between QTLs in common experimental designs and statistical methods used in most QTL mapping studies appears due to lack of power, and modified designs clearly show that epistasis is a common feature for most loci influencing complex traits. Although most commonly used experimental designs and statistical methods do allow detection of epistasis, accurate estimation of epistatic parameters between specific gene pairs remains a challenging problem and may require specifically constructed materials and modified experimental designs. Thus, classification of epistasis and development of epistatic genetic models have important implications in development of statistical methods in detecting and quantifying epistasis influencing complex traits. With DNA markers and development of statistical methodology, it is anticipated that more complete understanding of the role of epistasis in genetic variation of complex traits can be achieved.

## REFERENCES


1. Wright, S., The roles of mutation, inbreeding, crossbreeding and selection in evolution, *Proc. VI. Intern. Congr. Genet.*, 1, 356–366, 1932.
2. Wright, S., The genetic structure of populations, *Ann. Eugenics,* 15, 323–354, 1951.
3. Templeton, A. R., The theory of speciation via the founder principle, *Genetics,* 94, 1011–1038, 1980.
4. Mather, K. and Jinks, J. L., *Biometrical Genetics*, 3rd ed., Chapman and Hall, London, 1982, chap. 5.
5. Dobzhansky, T., Studies on hybrid sterility. II. Localization of sterility factors in *Drosophila pseudoobscura* hybrids, *Genetics,* 21, 113–135, 1936.
6. Muller, H. J. and Pontecorvo, G., Recombinants between *Drosophila* species, the F1 hybrids of which are sterile, *Nature* (London), 146, 199, 1940.
7. Stebbins, G. L., The inviability, weakness, and sterility of interspecific hybrids, *Adv. Genet.,* 9, 147–215, 1958.
8. Oka, H. I., Function and genetic bases of reproductive barriers, in *Origin of Cultivated Rice*, Jpn. Scientific Society Press, Elsevier, New York, 1988.
9. Allard, R. W., Genetic basis of the evolution of adaptedness in plants, *Euphytica,* 92(1–2), 1–11, 1996.
10. Wu, C. I. and Davis, A. W., Evolution of postmating reproductive isolation: the composite nature of Haldaneís rule and its genetic bases, *Amer. Nat.,* 142, 187–212, 1993.
11. Wu, C. I. and Palopoli, M. F., Postmating reproductive isolation in animals, *Annu. Rev. Genet.,* 28, 283–308, 1994.
12. Spassky, B., Dobzhansky, T., and Anderson, W. W., Genetics of natural populations. XXXVI, Epistatic interactions of the components of the genetic load in *Drosophila pseudoobscura, Genetics,* 52, 653–664, 1965.
13. Kinoshita, T. and Shinbashi, N., Identification of dwarf genes and their character expression in the isogenic background, *Japan. J. Breed,* 32, 219–231, 1982.
14. Tanksley, S. D. and Hewitt, J. D., Use of molecular markers in breeding for soluble solids in tomato — a re-examination, *Theor. Appl. Genet.,* 75, 811–823, 1988.
15. Doebley, J., Stec, A., and Gustus, C., *Teosinte branchedl* and the origin of maize: evidence for epistasis and the evolution of dominance, *Genetics,* 141, 333–346, 1995.
16. Paterson, A. H., Molecular dissection of quantitative traits: progress and prospects, *Genome Res.,* 5, 321–333, 1995.
17. Lark, K. G., Chase, K., Adler, F., Mansur, L. M., and Orf, J. H., Interactions between quantitative trait loci in soybean in which trait variation at one locus is conditional upon a specific allele at another, *Proc. Natl. Acad. Sci. U.S.A.,* 92, 4656–4660, 1995.
18. Cockerham, C.C. and Zeng, Z. B., Design III with marker loci, *Genetics,* 143, 1437–1456, 1996.
19. Li, Z. K., Pinson, S. R. M., Park, W. D., Paterson, A. H., and Stansel, J. W., Epistasis for three grain yield components in rice (*Oryza sativa* L.), *Genetics,* 145, 453–465, 1997.
20. Li, Z. K., Pinson, S. R. M., Park, W. D., Paterson, A. H., and Stansel, J. W., Genetics of hybrid sterility and hybrid breakdown in rice (*Oryza sativa* L.), *Genetics,* 147, (April), 1997.
21. Eshed, Y. and Zamir, D., Less-than-additive interactions of QTL in tomato, *Genetics,* 143(4), 1807–1817, 1996.
22. Suzuki, D. T., Griffiths, A. J. F., and Lewontin, R. C., *An Introduction to Genetic Analysis,* Third ed., W. H. Freeman and Company, New York, 1987.
23. Edwards, M. D., Stuber, C. W., and Wendel, J. F., Molecular-marker-facilitated investigations of quantitative-trait loci in maize. I. Numbers, genomic distribution and types of gene action, *Genetics,* 116, 113–125, 1987.
24. Stuber, C. W., Edwards, M. D., and Wendel, J. F., Molecular marker-facilitated investigations of quantitative-trait loci in maize. II. Factors influencing yield and its components' traits, *Crop Sci.,* 27, 239–248, 1987.
25. Stuber, C. W., Lincoln, S. E., Wolff, D. W., Helentjaris, T., and Lander, E. S., Identification of genetic factors contributing to heterosis in a hybrid from two elite maize inbred lines using molecular markers, *Genetics,* 132, 823–839, 1993.
26. Paterson, A. H., Lander, S. E., Hewitt, J. D., Peterson, S., Lincoln, H. D. et al., Resolution of quantitative traits into Mendelian factors by using a complete linkage map of restriction fragment length polymorphisms, *Nature,* 335, 721–726, 1988.

27. deVicente, M. C. and Tanksley, S. D., QTL analysis of transgressive segregation in an interspecific tomato cross, *Genetics,* 134, 585–596, 1993.
28. Xiao, J. H., Li, J., Yuan, L. P., and Tanksley, S. D., Dominance is the major genetic basis of heterosis in rice as revealed by QTL analysis using molecular markers, *Genetics,* 140, 745–754, 1995.
29. Zeng, Z.B., Theoretical basis of separation of multiple linked gene effects on mapping quantitative trait loci, *Proc. Natl. Acad. Sci. U.S.A.*, 90, 10,972–10,976, 1993.
30. Zeng, Z.B., The precision mapping of quantitative trait loci, *Genetics,* 136, 1457–1468, 1994.
31. Lander, E. S. and Botstein, D., Mapping Mendelian factors underlying quantitative traits using RFLP linkage maps, *Genetics,* 121, 185–199, 1989.

# 9 QTL Mapping in DNA Marker-Assisted Plant and Animal Improvement

*Andrew H. Paterson*

**CONTENTS**



## 9.1 INTRODUCTION

Scientific breeding of plants and animals has long been a cornerstone in the productivity of modern agriculture, and will remain so for the forseeable future. Intrinsic genetic solutions to the challenges that face plant/animal productivity and quality are usually of moderate cost, have negligible environmental impact, are readily delivered to the producer or consumer, accrue cumulative benefits over many years, and provide a stepping stone to still higher levels of performance.

During the initial domestication of productive crop plants from their wild ancestors, discrete loss-of-function mutations played a major role (see Chapter 13, this volume). Changes such as reduced seed dispersal, enhanced "strength" of the inflorescence as a carbohydrate sink, altered timing of flowering and synchrony of reproduction to optimize yield in temperate environments, and development of compact plants amenable to mechanized harvest, were among the key phenotypic changes.

However, as these basic features that distinguish crops from their ancestors became fixed in elite gene pools, different genes with more subtle phenotypic effects were exposed as the primary determinants of phenotypic variation. In many crop plants, the differences between world-class cultivars and obsolete breeding lines are so subtle as to be detectable only by large-scale replicated testing, often over both locations and years. A growing body of molecular data support phenotypic

0-8493-7686-6/98/$0.00+$.50

and pedigree information in suggesting that elite gene pools for many of the world's crops have a very small "effective population size," often tracing back to fewer than 10 genotypes.

As early as the 1920s, plant geneticists were pioneering the concept that individual genes responsible for very subtle phenotypic differences might be "mapped," in a manner similar to that which was, by then, well understood for discrete loci. Sax[1] associated differences in bean size with discrete variations in seedcoat pigmentation, and outlined most of the basic tenets of "QTL mapping." Many investigators, working in a wide range of plant and animal taxa, and using morphological and later protein markers, applied Sax's concepts over the next 60 years. However, the paucity of genetic markers available for most plants was a persistent constraint to QTL mapping.

In the 1980s, the advent of DNA markers made it possible to develop comprehensive genetic maps in virtually any plant (or animal), and apply these maps to gaining better understanding of variation in many gene pools. This new capability resulted in a veritable explosion of activity in "genome mapping", encompassing a large number of plant and animal species, and a wide range of traits. Some examples are listed in Table 9.1.

Experiments in the late 1980s began to bear out the prospect first raised by Sax, that complex traits might be dissected into individual "quantitative trait loci" (QTLs).[2] Using closely linked DNA probes, QTLs might be readily manipulated in breeding programs, accelerating progress toward objectives that would otherwise be cumbersome.

Although "DNA marker-assisted breeding" has developed more slowly than anticipated, it seems clear that genome analysis will be an enduring addition to the toolbox of modern plant breeding. The basis for this assertion is that genome analysis enables us to extract more information from breeding populations. Even today, genome analysis is considered by some to be the latest in a series of "fads" that have transiently influenced the thinking of agricultural researchers over the years. The ability to design more precise experiments and make more efficient progress, using molecular tools, is likely to influence how plant breeding is done. Moreover, genetic mapping establishes conduits that enable breeders to take advantage of information from many new sources.

### 9.1.1 Interface Between Genetics and Breeding

QTL mapping research often represents an interface between "genetics" and "breeding." The literature of QTL mapping is most closely allied with the experimental methods and statistical tests used in genetics — however, applications of QTL mapping frequently address traits of importance to breeders. In the following sections, I will discuss how QTL mapping has influenced our understanding of basic transmission genetics, and suggest some ways in that this new information affects plant and animal breeding.

## 9.2 GENE NUMBERS AND EFFECTS

Geneticists have long debated the degree of complexity of quantitative traits.[3] A continuum of theories, ranging from "virtually infinite numbers of genes with tiny effects", to "few genes with large effects" have been proposed, championed, questioned, revised, rejected, and reincarnated. Geneticists have long realized that some assumptions used to simplify quantitative models, such as equality of gene effects and additivity of gene action, were unlikely to precisely describe individual QTLs. It was no particular surprise that QTL mapping showed such assumptions to be incorrect (see below). However, it has remained controversial whether the results of QTL mapping experiments reflect the true complexity of quantitative inheritance, or simply detect only a subset of (relatively large) gene effects.

The classical assumption of equal phenotypic effects for different genes controlling a quantitative trait was the first casualty. Figure 9.1 represents a model for phenotypic effects of individual QTLs, that has emerged both from theoretical considerations, and from genetic mapping studies of recent years. A relatively small number of genes account for very large portions of phenotypic variance, with increasing numbers of genes accounting for progressively smaller portions of

**TABLE 9.1**
**Representative Examples of Phenotypes Which Have Been Analyzed by QTL Mapping**

**Animals**

***Complex behavioral characteristics such as***
Avoidance, exploration[42]
Substance abuse[43,44]
Reading disability (see Chapter 19, this volume).

***Medically important phenotypes***
High blood pressure[45]
Hypertension[46]
Obesity[47,48]
Lactation[49]
Muscular development[50]
Weaver disease[51]

***Interactions between organisms***
Effectiveness of a human disease agent (malaria) at parasitizing its vector (mosquito)[52]

**Plants**

***Parameters of vegetative development***
Height[10,53,54]
Flowering time[10,53,55]
Rhizomatousness and tillering[56]
Size and shape of organs[57]

***Yield components***
Size, number, and harvestability of seed[11,56,58-63]
Biomass and/or growth rates[27,64]

***Quality parameters***
Composition of fruit or seed[6,12,14,27,36,65,66]
Shape of tubers[67]
Specific gravity of wood[68]
Cotton fiber quality[68a]

***Impact of adversities***
Diseases[69-73]
Insects[74-75]
Water use efficiency[76]
Nutrient use efficiency[77]

***Evolutionary novelties***
The maize ear[78]
Floral characteristics which influence pollinator preference[79]

variance.[4,5] First-generation QTL mapping experiments usually detect only genes with relatively large effects, and may not even detect all of these (see Chapter 10, this volume). Further, if genes explaining large portions of phenotypic variance are rendered homozygous (Reference 6 and Chapter 15 this volume), additional genes explaining smaller portions of phenotypic variance may be exposed.

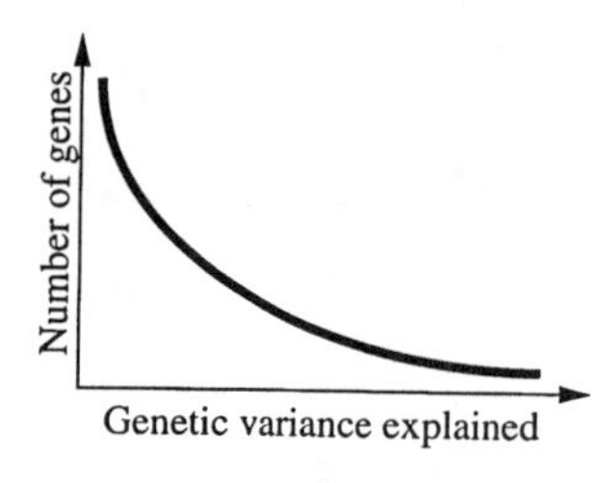


**FIGURE 9.1** Conceptual model for inheritance of complex traits. Recent data from QTL mapping suggest that relatively few genes may account for a large portion of variance in many traits, with a much larger number of genes accounting for smaller portions of variance.

### 9.2.1 Gene Action and Interaction

Following the assumption of "equal effects," the assumption of additive gene action for individual QTLs was the second casualty. QTLs have been found to exhibit the entire range of conceivable dose-responses, including additivity, dominance/recessiveness, and over/underdominance, with all gradations in between (see Reference 7).

The independence of QTL action has remained controversial. Intuitively, no gene can function completely independently of all other genes in the genome. However, until recently, QTL mapping experiments have shown very little evidence in support of the importance of epistasis, with nonlinear interactions among DNA marker loci reaching statistical significance at approximately the frequency that would be expected to occur by chance (see Reference 4).

Classical evidence has strongly suggested the importance of epistasis, or nonlinear interactions between unlinked genetic loci, in quantitative inheritance.[8-12] Hints of epistasis among QTLs have derived from the demonstration of "genetic background effects" on quantitative traits in *Drosophila,*[13] rice,[14,15] and tomato,[16] and from the discovery of occasional loci reported to show interaction with multiple unlinked sites in a genome.[17]

Modified experimental designs may reconcile QTL mapping results with the importance attributed to epistasis in classical studies. Doebley and colleagues[18] developed genetic stocks differing by two QTLs suspected to interact epistatically, but otherwise uniform in genetic background — and found strong evidence for epistasis between the loci. Lark and colleagues[19] utilized recombinant inbred lines to reduce the complexity of interactions, and replicate phenotypic measurements — and found evidence of epistasis between QTLs, in genetic control of several agronomic traits.

Li and colleagues (see Chapter 8, this volume) employed an unusually high level of replication, together with remarkably stringent statistical criteria, to show that epistasis between unlinked genetic loci occurred far more often than could be explained by chance. Moreover, favorable interactions tended to be between alleles from the same gene pool, while unfavorable interactions tended to be between loci from different gene pools. QTLs themselves were only rarely involved with interactions — however, traits for those few QTLs that could be mapped showed a greater preponderance of interactions.

All of these results suggest that the absence of epistasis in prior QTL mapping studies may have been due to minimal replication, and/or minimal statistical resolution to detect interactions, in the presence of many QTLs with large main effects. While the effects of some QTLs appear independent of interacting loci, in at least some cases it is becoming clear that "the whole" is, indeed, greater than the sum of the parts. Epistasis may account for a portion of the "genetic difference between parents" that was previously going unexplained by QTL mapping.

Epistasis may appear in forms that are not obvious. The occasional discovery of unexpected "transgressants" in breeding populations has been mirrored in recent years by the discovery of valuable genes from unexpected sources. In tomato,[20,21] rice,[22] sorghum,[23] and cotton,[23a] alleles from inferior parental stocks have been associated with improvements in agricultural productivity or quality.

### 9.2.2 Statistical Significance Thresholds

One of the most important considerations in analysis and interpretation of QTL data is the threshold employed for inferring statistical significance. Because QTL mapping involves analysis of many independent (unlinked) markers throughout a genome, there are many opportunities for false-positive results. Nominal significance criteria of 99.8% or more for any single QTL are usually necessary to assure an "experiment-wide" confidence level of 95% for all QTLs reported across a genome. Appropriate criteria are often described in detail accompanying development of an analytical approach.[24] Alternatively, means for empirical calculation of criteria appropriate to particular data sets have been described.[25,26] As opportunities for "comparative analysis" of previously published QTLs become more prevalent,[23,27] it becomes ever more important that the literature of QTL mapping be based upon stringent statistical criteria that minimize the likelihood of false-positive results.

While rigorous statistical criteria are important in published data to assure the usefulness of the published literature, plant and animal breeders may rightfully decide to consider less-stringent criteria in making breeding decisions. To assure that published literature is sound, most scientists consider it prudent to err on the side of (statistical) conservatism, attributing significance to only those results that are "beyond a reasonable doubt." In the context of QTL mapping, "beyond a reasonable doubt" usually means setting significance thresholds such that there is less than a 5% chance that even one of the many QTLs that might be found in a single experiment represents a false positive.

Relaxation of statistical criteria in applied plant and animal breeding does not by itself encumber the scientific literature, since most data contributing to breeding decisions are never published. The vast majority of breeding decisions are made with almost no attention to formal statistics and with the *a priori* knowledge that both false-positive and false-negative error rates will be high. The ultimate test of breeding decisions comes from productivity of resulting germplasm. A successful breeder is often one who quickly makes a vast number of decisions with reasonable accuracy, rather than one who becomes entangled in making a small number of decisions perfectly. Both theoretical[5] and empirical[7] data suggest that DNA marker data can improve this success rate, especially in selection for traits of low heritability.

An example is shown in Figure 9.2 — in this study of a backcross population, a region of tomato chromosome 1 was loosely associated with the concentration of soluble solids in the fruit, however, the effect was too small to reach significance. By contrast, a QTL reducing fruit size was mapped near the opposite end of the chromosome.[17] By DNA marker-assisted breeding, a genetic stock was developed that retained the part of the chromosome tentatively associated with soluble solids, but was free of the part associated with reduced fruit size. The increase in soluble solids persisted. Finally, the chromosomal region was tested in near-isogenic stocks, and the heterozygote showed a significant increase in "solids yield" (the product of soluble solids concentration and fruit yield is a measure of how much economic product is harvested). Had the region simply been dismissed as "not statistically significant," this prospective gain would have been overlooked.

While subthreshold associations may often prove to be false, they may also less frequently represent "macromutations" associated with domestication or gross developmental differences, and therefore be more useful in breeding programs. Productivity of resulting germplasm may ultimately provide a definitive test of such subthreshold associations that can then take their rightful place in the scientific literature.

As a second-order phenomenon (or higher), analysis of epistasis is even more subject to the problem of false-positive results than analysis of individual QTLs (as discussed above). It is especially important to use stringent statistical criteria for inferring statistical significance of interactions between genetic loci, to control "experiment-wise" error rates. Li and colleagues (see Chapter 8, this volume) provide a good example of such criteria.

## 9.3 CHOICE OF GENE POOLS

A frequent criticism of QTL mapping has been that the populations studied were not representative of the elite gene pools relevant to mainstream improvement of many crops. Indeed, interspecific crosses between crop cultivars and their wild relatives remain common targets of genome analysis

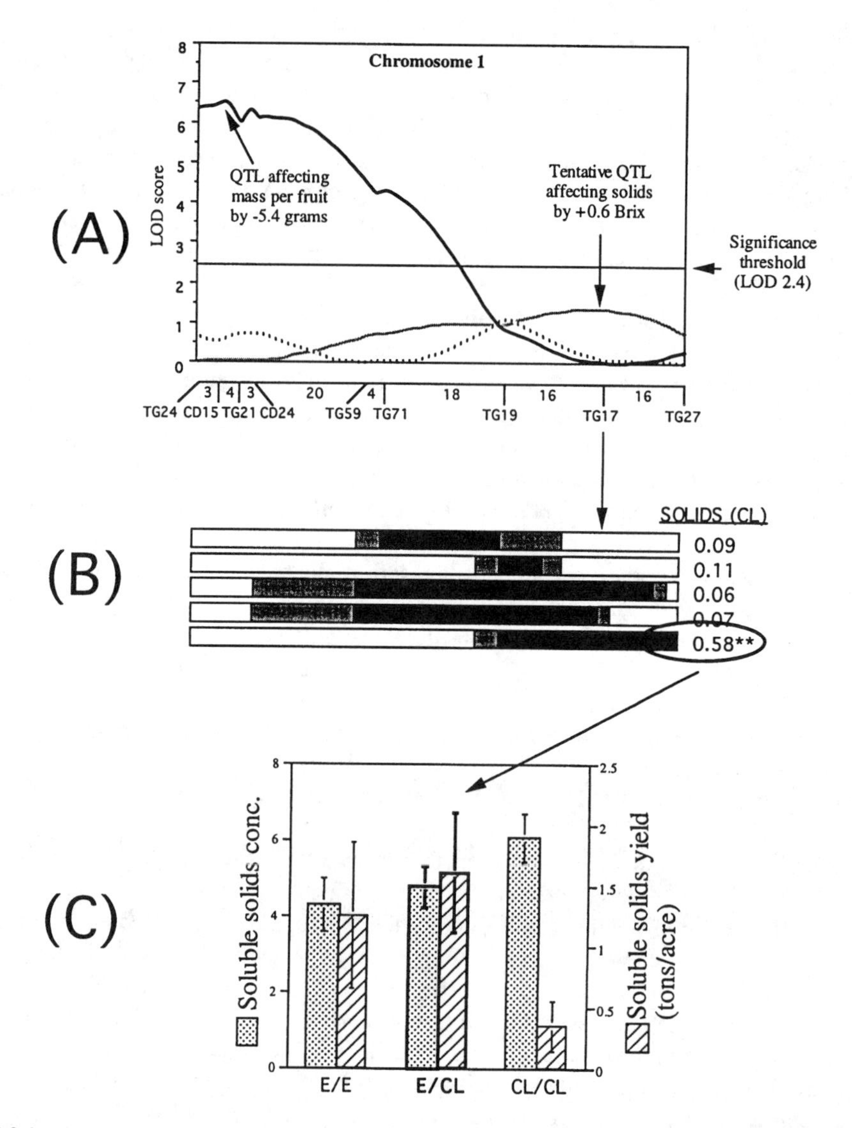


**FIGURE 9.2** QTLs with small phenotypic effects may be important in crop improvement, as illustrated by a 3-generation experiment in DNA marker-assisted introgression. (A) In the BC1 progeny of a tomato cultivar crossed to its wild relative *Lycopersicon chmielewskii*, a region of chromosome 1 from the wild parent was loosely associated with the concentration of soluble solids in the fruit, however the effect was too small to reach significance (see LOD threshold shown). By contrast, a QTL reducing fruit size was mapped near the opposite end of the chromosome.[17] (B) By DNA marker-assisted breeding, several genetic stocks were developed that retained the part of the chromosome tentatively associated with soluble solids, but were free of the part associated with reduced fruit size. The increase in soluble solids persisted, and appeared to require the terminal portion of the chromosome to harbor the wild (*L. chmielewskii*) allele (shown in black). (C) Finally, the chromosomal region was tested in near-isogenic stocks. The heterozygote showed higher soluble solids concentration than the cultivated parent, and higher fruit yield than either parent, for a significant increase in "solids yield" (the product of soluble solids concentration and fruit yield, a measure of how much economic product is harvested). Had the region simply been dismissed as not important, based on its low LOD score in the BC1 study, this prospective gain would have been overlooked.

even today. Molecular analysis of elite crop gene pools has lagged behind other areas of genome analysis, especially in self-pollinated crops. In cross-pollinated crops such as maize and brassica, a larger reservoir of genetic variation persists in the elite gene pool.

Molecular mapping of elite germplasm in many crops has faced two challenges:

1. Levels of DNA polymorphism are low, and it is difficult to find DNA marker loci at which genotypes carry different alleles. New technology (see Chapter 2, this volume) is gradually overcoming this limitation.
2. Apparent phenotypic variation is small, as elite genotypes have often been selected for a common set of criteria.

These "challenges" are symptomatic of the need for expediting use of exotic germplasm to broaden the genetic base of major crops. While genome mapping tools can be used to make analysis of elite crop cultivars more tractable, they do not introduce new variation into the narrow and vulnerable gene pool of cultivated cotton. This objective is addressed in more detail below.

However, clearly it is also important to gain a better understanding of the composition of elite crop gene pools, the structure of genetic variation therein, the distribution and phenotypic effects of genes that still segregate in these gene pools, and the prospects of making further improvements by selecting directly in elite germplasm. Long-term selection experiments in many species show that even within a limited founder population enduring progress toward novel phenotypes is possible.

### 9.3.1 Associating Patterns of Genome Composition with Important Traits

It is well-established that the gene pools of many crops are largely derived from a small number of recent ancestors.[28] Thus, modern cultivars of many crops can be thought of as mosaics of chromosome segments from these ancestors. Given a sufficient number of DNA polymorphisms, leading cultivars might be described in terms of their repertoire of ancestral chromosome segments. A simple example would be the "fingerprinting" of genomic regions introgressed into near-isogenic lines (NIL) in association with selection for either simple or complex traits (see Ref. 16).

The "founder effect," together with isolation of particular breeding populations, create conditions amenable to mapping of specific genes using genealogical information. In human populations, mutations that have been introduced into a population within the past 30 to 40 generations have been mapped based on the discovery of common DNA marker genotypes along small chromosomal regions (ca. 2 cM) in affected individuals.[29] Such analyses require densely populated maps of highly polymorphic markers such as microsatellites (see Chapter 2, this volume), but are increasingly within the reach of plant and animal genetics. Moreover, by increasing density of markers along a map, one may have the opportunity to reach back to identify genes based on more ancient "introgression" events.

Early examples of the use of genealogical information to track QTLs may derive from interspecific introgression events in cotton. Recurring patterns of genome composition have been revealed over more than a century of breeding progress in *Gossypium barbadense*, in independent breeding programs in the Caribbean, Egypt, and U.S.[30] At least five specific chromosome segments derived from *G. hirsutum* have persisted through many generations of recombination and selection in diverse environments. Recombination within these chromosomal regions reveals specific locations in the genome of *G. barbadense* at which the *G. hirsutum* allele is retained. One hypothesis to account for such a result would suggest that particular alleles or allele combinations have been of long-standing importance in *G. barbadense* cotton improvement. Verification of such predictions can employ routine QTL mapping procedures, and if corroborated, provide both basic information and DNA markers for accelerated improvement of future elite types.

Such an "historical" approach may efficiently capture an enormous body of information lying latent in the gene pools of major crops. Identification of DNA markers diagnostic of genes/genomic regions important to quality and/or productivity, would help the breeder to eliminate quickly those genotypes that are destined to failure and focus efforts on achieving new gains rather than reconstituting prior progress.

### 9.3.2 Population Improvement and Broadening of the Genetic Base Using Exotic Germplasm

About 130 to 200 million years of plant evolution has led to the existence of a remarkable diversity of flora found on our planet. However, only a tiny fraction of this diversity is represented in modern crop cultivars, due to the fact that very few plant taxa have been "domesticated" for use as crops, and further that even the gene pools of these select few have been subjected to "genetic bottlenecks" during domestication.[28]

At least three general problems can be directly attributed to the use of exotic germplasm in breeding programs. First, most temperate crop plants have been domesticated from taxa of tropical or subtropical origin. Therefore, exotic germplasm remains adapted to its tropical climate and confers traits such as short-day (photoperiodic) flowering that are adaptive in the native environment, but are not suitable for temperate cultivation.[23,27] Second, crop gene pools have been selected intensively for alterations of harvest index, partitioning a maximum of photosynthate to specific economic organs such as seeds, in a single growing season. Exotic germplasm is often perennial, subject to selection criteria that involve a balance between vegetative and reproductive growth (see Chapter 2, this volume), and therefore tends to transmit reduced yield and other undesirable traits. Third, because a large number of genes frequently differ between exotic and cultivated types, there exists a high likelihood that any one desirable gene will be genetically linked to other undesirable genes; "linkage drag" will therefore reduce the gains that might otherwise be realized if a single valuable gene could be transmitted from its exotic source to a recipient cultivar.

To motivate use of exotic germplasm, the value of a specific trait from an exotic source must substantially outweigh the difficulties associated with use of exotic germplasm. This tradeoff has sometimes been sufficiently favorable to motivate "introgression" of major genes with large effects on important traits such as disease resistance.(See Reference 31.) However, the greater complications associated with introgression of multiple genes conferring a valuable, complex trait have usually discouraged such efforts, although a few have enjoyed some success.[32] Largely for these reasons, the potential contribution of wild and feral germplasm to mainstream plant breeding has not been realized. "Prebreeding" programs, designed to reduce problems associated with use of exotics, have had significant impact.[33,34] However, such programs require a major investment of resources and often progress only slowly.

QTL mapping is reducing the obstacles to use of exotic germplasm, in two specific ways. First, many major genes that have interfered with the use of exotic germplasm, such as short-day flowering, have been mapped,[cf. 10] and can now be quickly eliminated from breeding or "prebreeding" programs using DNA markers. Second, methods such as "advanced-backcross-QTL[21]"-based breeding, in which selection against undesirable major genes is exercised during early generations, and QTL mapping applied after two to three backcrosses, are revealing genes valuable for improvement of complex traits, even from sources with inferior phenotypes.[20,22,35]

### 9.3.3 Comparative Crop Genome Analysis — A Conduit for Flow of Genetic Information

Vavilov's "law of homologous series in variation"[36] was perhaps the earliest recognition of fundamental similarity between different cultivated species. Most plant breeders now recognize that similarities between their crop(s) and related taxa transcend diversity in breeding objectives. However, except for long-term high-cost efforts to clone individual genetic loci, it has previously been difficult to identify corresponding genes in taxa that could not be intermated.

"Comparative mapping," the study of similarities and differences in gene order along the chromosomes of taxa that cannot be hybridized, has recently been used to demonstrate that only a modest number of chromosomal rearrangements (inversions and/or translocations) distinguish many major crops and model systems. Moreover, comparative maps provide a conduit for communication — permitting information gathered during study of one species to be quickly and efficiently applied to related species. Detailed comparative maps of many species within common taxonomic families have been established.[38] The recent discovery of small chromosomal regions retaining similar gene order in one monocot (sorghum) and two dicots (*Arabidopsis* and cotton), suggest that comparative mapping may ultimately reach across much greater "evolutionary distances" than have been spanned to date.[38]

Comparative genetic mapping can help to provide a more comprehensive catalog of genes that potentially influence a trait. Most measures of quality and/or productivity of crop plants can potentially be influenced by allelic variation at a large number of genetic loci. It is virtually inconceivable to identify a single pedigree that segregates for allelic variants at all genetic loci influencing a trait (although one advantage of studying very wide crosses has been the possibility to find a maximal number of allelic variants per population studied). Even if one pedigree could be identified that segregated for allelic variants at all genetic loci influencing a trait, statistical considerations would delimit the number of QTLs that could be mapped with confidence (see Chapter 10, this volume). By aligning genetic maps of different populations, one can begin to assemble such a "catalog" of genes that can potentially affect a trait.

Comparative maps aligning genes or QTLs mapped in many different crosses find many applications. In breeding programs, one might predict the locations of genes conferring resistance to new races of pests, based on prior analyses in other species or populations. Through the collective efforts of a large number of investigators, such comparative maps are gradually becoming a reality. For example, a recent study drew inferences based upon 185 QTLs or discrete mutants affecting height and/or flowering time in maize, sorghum, rice, wheat, and barley.[23] Increasingly, electronic databases are providing useful summaries of the repertoires of genes/QTLs known to affect a particular phenotype (see Chapter 12, this volume).

## 9.4 REDUCING BARRIERS TO MORE WIDESPREAD USE OF MOLECULAR TOOLS

Many techniques are now available for visualization of DNA markers (see Chapter 2, this volume), however infrastructure and cost remain constraints to the widespread use of DNA markers in crop improvement.

The restricted fragment length polymorphism (RFLP) technique remains the single most widely used DNA marker assay in crop plants. Mapped DNA probes are available in many plant species, and the technique is readily transferrable between different labs. For the purpose of genetic research — mapping genes or QTLs, and relating these data to results from other populations or species — RFLPs remain a valuable tool that affords economies of scale to the astute user.

Well-known limitations of the RFLP technique have motivated development of several alternative technologies. In particular, these limitations are the quantity of DNA required (about 50 to 200 μg per individual, to generate a DNA fingerprint of the entire genome), and the allelic richness of elite germplasm for RFLP alleles. Polymerase chain reaction (PCR)-based assays reduce the demand for genomic DNA by a factor of 10- to 100-fold, and are very efficient if only one or a few genotypes per individual are needed. Although the availability of DNA sequence information was once a factor limiting application of PCR and impelled development of "arbitrary-primer" techniques,[39-41] ready availability of DNA sequence has overcome this limitation. Ultimately, efficient low-cost robotics are likely to make PCR-based assays the method of choice even for generating detailed maps of small populations — however at present, the possibility to use a single Southern blot 10 to 20 times (RFLP technique) retains considerable appeal over the need to run 10 to 20 separate gels to obtain a similar quantity of data (PCR).

On-site implementation of DNA marker analysis in breeding programs, empowering modern plant breeders with the best available tools, will require further technological simplification. "Coarse-resolution" studies such as introgression of exotic germplasm, mapping of major genes or QTLs with large effects, or comparative analyses of different taxa are rightfully done in well-equipped genetics labs (although preferably with the benefit of collaboration with enthusiastic plant breeders). However, the "fine-tuning" that distinguishes elite cultivars from also-ran breeding lines remains the art of the plant breeder, and will remain so indefinitely. An increasing emphasis on training of today's plant breeding students in basic molecular genetics is providing a generation of scientists prepared to accomplish this integration. However, continuing simplification of molecular marker technology remains necessary to bring the cost and infrastructural demands into reach of most breeding programs. The solution is NOT increased instrumentation, or more efficient robotics — although these things may help make progress toward a solution. Society needs the plant breeder to exercise his/her creativity and practice his/her art rather than to become a DNA robotics specialist. The solution may be a fundamentally different sort of assay, preferably derived from existing DNA probes so as to benefit from the wealth of genome-related information accumulated over the past decade but requiring nominal time and nominal investment in equipment.

## REFERENCES


1. Sax, K., The association of size differences with seedcoat pattern and pigmentation in Phaseolus vulgaris, *Genetics*, 8, 552, 1923.
2. Geldermann, H., Investigations on inheritance of quantitative characters in animals by gene markers. I. Methods, *Theor. Appl. Genet.*, 46, 319, 1975.
3. Dove, W. F., The gene, the polygene, and the genome, *Genetics,* 134, 999, 1993.
4. Paterson, A. H., Molecular Dissection of Quantitative Traits: Progress and Prospects, *Genome Res.*, 5, 321, 1996.
5. Lande, R. and Thompson, R., Efficiency of marker-assisted selection in the improvement of quantitative traits, *Genetics,* 124, 743, 1990.
6. Paterson, A. H., Deverna, J. W., Lanini, B., and Tanksley, S. D., Fine mapping of quantitative trait loci using selected overlapping recombinant chromosomes in an interspecies cross of tomato, *Genetics*, 124, 735, 1990.
7. Paterson, A. H., Damon, S., Hewitt, J. D., Zamir, D., Rabinowitch, H. D., Lincoln, S. E., Lander, E. S., and Tanksley, S. D., Mendelian factors underlying quantitative traits in tomato: comparison across species, generations, and environments, *Genetics*, 127, 181, 1991.
8. Falconer, D. S., *Introduction to Quantitative Genetics,* 2nd ed., Longman Press, London, 1981.
9. Mather, K. P. and Jinks, J. L., *Biometrical Genetics,* 3rd ed., Chapman and Hall, London, 1982.
10. Pooni, H. S., Coombs, D. J., and Jinks, P. S., Detection of epistasis and linkage of interacting genes in the presence of reciprocal differences, *Heredity,* 58, 257, 1987.
11. Spickett, S. G. and Thoday, J. M., Regular response to selection. 3. Interaction between located polygenes, *Genet. Res.*, 7, 96, 1966.
12. Allard, R. W., Genetic changes associated with the evolution of adaptedness in cultivated plants and their wild progenitors, *J. Hered.*, 79, 225, 1988.
13. Spassky, B., Dobzhansky, T., and Anderson, W. W., Genetics of natural populations. XXXVI. Epistatic interactions of the components of the genetic load in *Drosophila pseudoobscura, Genetics,* 52, 653, 1965
14. Kinoshita, T. and Shinbashi, N., Identification of dwarf genes and their character expression in the isogenic background, *Jpn. J. Breed.*, 32, 219, 1982.
15. Sato, S. and Sakamoto, I., Inheritance of heading time in isogenic line rice cultivar, Taichung 65 carrying earliness genes from a reciprocal translocation homozygote, T3-7, *Jpn. J. Breed.*, 33, 118, 1983.
16. Tanksley, S. D. and Hewitt, J. D., Use of molecular markers in breeding for soluble solids in tomato: a re-examination, *Theor. Appl. Genet.*, 75, 811–823, 1988.
17. Paterson, A. H., Lander, E. S., Hewitt, J. D., Peterson, S., Lincoln, S. E., and Tanksley, S. D., Resolution of quantitative traits into Mendelian factors by using a complete map of restriction fragment length polymorphisms, *Nature*, 335, 721, 1988.
18. Doebley, J., Stec, A., and Gustus, C., *Teosinte branched 1* and the origin of maize: evidence for epistasis and the evolution of dominance, *Genetics,* 141, 333, 1995.

19. Lark K. G., Chase, K., Adler, F., Mansur, L. M., and Orf, J. H., Interactions between quantitative trait loci in soybean in which trait variation at one locus is conditional upon a specific allele at another, *Proc. Natl. Acad. Sci. U.S.A.*, 92, 4656, 1995.
20. DeVicente, M. C. and Tanksley, S. D., QTL analysis of transgressive segregation in an intraspecific tomato cross, *Genetics,* 134, 585, 1993.
21. Tanksley, S. D. and Nelson, J. C., Advanced backcross QTL analysis: a method for the simultaneous discovery and transfer of valuable QTLs from unadapted germplasm into elite breeding lines, *Theor. Appl. Genet.*, 92, 191, 1996.
22. Xiao, J., Li, J., Grandillo, S., Ahn, S. N., McCouch, S. R., Tanksley, S. D., and Yuan, L., A wild species contains genes that may significantly increase the yield of rice, *Nature* (in press), 1996.
23. Lin, Y. R., Schertz, K. F., and Paterson, A. H., Comparative mapping of QTLs affecting plant height and flowering time in the Gramineae, in reference to an interspecific *Sorghum* population, *Genetics,* 141, 391, 1995.
23a. Paterson, A. H., Unpublished data, 1997.
24. Lander, E. S. and Botstein, D., Mapping Mendelian factors underlying quantitative traits using RFLP linkage maps, *Genetics*, 121, 185, 1989; and Corrigendum, *Genetics,* 136, 705, 1994.
25. Churchill, G. A. and Doerge, R. W., Empirical threshold values for quantitative trait mapping, *Genetics,* 138, 963, 1994.
26. Rebai, A., Goffinet, B., and Mangin, B., Comparing power of different methods for QTL detection, *Biometrics,* 51, 87, 1995.
27. Paterson, A. H., Lin, Y. R., Li, Z., Schertz, K. F., Doebley, J. F., Pinson, S. R. M., Liu, S. C., Stansel, J. W., and Irvine, J. E., Convergent domestication of cereal crops by independent mutations at corresponding genetic loci, *Science,* 269, 1714, 1995.
28. National Academy of Sciences, *Genetic vulnerability of major crops,* Cotton, Washington, D.C., 1972, 269, chap. 15.
29. Varilo, T., Nikali, K., Suomalainen, A., Lonnqvist, T., and Peltonen, L., Tracing an ancestral mutation: Genealogical and haplotype analysis of the infantile onset spinocerebellar ataxia locus, *Genome Res.,* 6, 870, 1996.
30. Wang, G., Dong, J., and Paterson, A. H., Genome composition of cultivated *Gossypium barbadense* reveals both historical and recent introgressions from *G. hirsutum, Theor. Appl. Genet.*, 91, 1153, 1995.
31. Meredith, W. R., Jr., Contributions of introduced germplasm to cotton cultivar development, *Agron. Abstr.*, 91, 1989.
32. Rick, C. M., High soluble-solids content in large-fruited tomato lines derived from a wild green-fruited species, *Hilgardia,* 42, 493, 1974.
33. Stephens, J. C., Miller, F. R., and Rosenow, D. T., Conversion of alien sorghums to early combine genotypes, *Crop Sci.*, 7, 396, 1967.
34. McCarty, J. C. and Jenkins, J. N., Cotton germplasm: characteristics of 79 day-neutral primitive race accessions, *Miss. Ag. For. Expt. Stn. Tech. Bull.*, 184, 1992.
35. Tanksley, S. D., Grandillo, S., Fulton, T. M., Zamir, D., Eshed, Y., Petiard, V., Lopez, J., and Beck-Bunn, T., Advanced backcross QTL analysis in a cross between an elite processing line of tomato and its wild relative *L. pimpinellifolium, Theor. Appl. Genet.*, 92, 213, 1996.
36. Vavilov, N. I., The law of homologous series in variation, *J. Genet.,* 12, 1922.
37. Paterson, A. H., Comparative gene mapping in crop improvement. Chapter 2 in *Plant Breeding Rev.*, 1997.
38. Paterson, A. H., Lan, T. H., Reischmann, K. P., Chang, C., Lin, Y. R., Liu, S. C., Burow, M. D., Kowalski, S. P., Katsar, C. S., DelMonte, T. A., Feldmann, K. A., Schertz, K. F., and Wendel, J. F., Toward a unified map of higher plant chromosomes, transcending the monocot-dicot divergence, *Nat. Genet.*, in press.
39. Welsh, J. and McClelland, M., Fingerprinting genomes using PCR with arbitrary primers, *Nucleic Acids Res.*, 18, 7213, 1990.
40. Williams, J. G. K., Kubelik, A. R., Livak, K. J., Rafalski, J. A., and Tingey, S. V., Oligonucleotide primers of arbitrary sequence amplify DNA polymorphisms which are useful as genetic markers, *Nucleic Acids Res.,* 18, 6531, 1990.
41. Baum, T. J., Gresshoff, P. M., Lewis, S. A., and Dean, R. A., DNA amplification fingerprinting (DAF) of isolates of four common Meloidogyne species, and their host races, *Phytopathology,* 82, 1095, 1992.

42. Neiderheiser, J. M., Plomin, R., and McClearn, G. E., The use of CXB recombinant inbred mice to detect quantitative trait loci in behavior, *Physiol. Behav.*, 52, 429, 1992.
43. Crabbe, J. C., Belknap, J. K., and Buck, K. J., Genetic animal models of alcohol and drug abuse, *Science*, 264, 1715, 1994.
44. Quock R. M., Mueller, J. L., Vaughn, L. K., and Bellnap, J. K., Nitrous oxide (N-2O) antinociception in BXD recombinant inbred (RI) mouse strains and identification of quantitative trait loci (QTL), *FASEB J.*, 8, A628, 1994.
45. Rapp, J. P., Wang, S., and Dene, H., A genetic polymorphism in the renin gene of Dahl rats cosegregates with blood pressure, *Science*, 243, 542, 1989.
46. Jacob, H. J., Lindpainter, K., Lincoln, S. E., Kusumi, K., Bunker, R. K., Mao, Y.-P., Ganten, D., Dzau, V. J., and Lander, E. S., Genetic mapping of a gene causing hypertension in the stroke-prone spontaneously hypertensive rat, *Cell*, 67, 213, 1991.
47. Andersson, L., Haley, C. S., Ellegren, H., Knott, S. A., Johansson, M., Andersson, K., Andersson-Eklund, L., Edfors-Lilja, I., Fredholm, M., Hansson, I., Hakansson, J., and Lundstrom, K., Genetic mapping of quantitative trait loci for growth and fatness in pigs, *Science*, 263, 1771, 1994.
48. Pelleymounter, M. A., Cullen, M. J., Baker, M. B., Hecht, R., Winters, D., Boone, T., and Collins, F., Effects of the obese gene product on body weight regulation in ob/ob mice, *Science*, 269, 540, 1995.
49. Georges, M., Nielsen, D., MacKinnon, M., Mishra, A., Okimoto, R., Pasquino, A. T., Sargeant, L. S., Sorensen, A., Steele, M. R., Zhao, X., Womack, J. E., and Hoeschele, I., Mapping quantitative trait loci controlling milk production in dairy cattle by exploiting progeny testing, *Genetics*, 139, 907, 1995.
50. Cockett, N. E., Jackson, S. P., Shay, T. L., Nielsen, D., Moore, S. S., Steele, M. R., Barendse, W., Green, R. D., and Georges, M., Chromosomal localization of the callipyge gene in sheep (Ovis aries) using bovine DNA markers, *Proc. Natl. Acad. Sci. U.S.A.*, 91, 3019, 1994.
51. Georges M., Dietz, A. B., Mishra, A., Nielsen, D., Sargeant, L. S., Sorensen, A., Steele, M. R., Zhao, X., Leipold, H., Womack, J. E., and Lathrop, M., Microsatellite mapping of the gene causing Weaver disease in cattle will allow the study of an associated quantitative trait locus, *Proc. Natl. Acad. Sci. U.S.A.*, 90, 1058, 1994.
52. Severson, D. W., Thathy, V., Mori, A., Zhang, Y., and Christensen, B. M., Restriction fragment length polymorphism mapping of quantitative trait loci for malaria parasite susceptibility in the mosquito *Aedes aegypti*, *Genetics*, 139, 1711, 1995.
53. Koester, R. P., Sisco, P. H., and Stuber, C. W., Identification of quantitative trait loci controlling days to flowering and plant height in two near isogenic lines of maize, *Crop Sci.*, 33, 1209, 1993.
54. Pereira, M. G., Lee, M., and Rayapati, P. J., Comparative RFLP and QTL mapping in sorghum and maize, Poster 169 in the Second Internal Conference on the Plant Genome, Scherago Internal, Inc., New York, 1994.
55. Kowalski, S. D., Lan, T.-H., Feldmann, K. A., and Paterson, A. H., Comparative mapping of *Arabidopsis thaliana* and *Brassica oleracea* chromosomes reveal islands of conserved gene order, *Genetics*, 138, 499, 1994.
56. Paterson, A. H., Schertz, K. F., Lin, Y.-R., Liu, S.-C., and Chang, Y.-L., The weediness of wild plants: molecular analysis of genes influencing dispersal and persistence of johnsongrass, *Sorghum halepense* (L.) Pers., *Proc. Natl. Acad. Sci. U.S.A.*, 92, 6127, 1995.
57. Kennard, W. C., Slocum, M. K., Figdore, S. S., and Osborn, T. C., Genetic analysis of morphological variation in Brassica oleracea using molecular markers, *Theor. Appl. Genet.*, 87, 721, 1994.
58. Stuber, C.W., Edwards, M. D., and Wendel, J. F., Molecular-marker-facilitated investigations of quantitative-trait loci in maize. II. Factors influencing yield and its component traits, *Crop Sci.*, 27, 639, 1987.
59. Stuber, C. W., Lincoln, S. E., Wolff, D. W., Helentjaris, T., and Lander, E. S., Identification of genetic factors contributing to heterosis in a hybrid from two elite inbred lines using molecular markers, *Genetics*, 132, 823, 1992.
60. Abler, B. S. B., Edwards, M. D., and Stuber, C. W., Isozymatic identification of quantitative trait loci in crosses of elite maize inbreds, *Crop Sci.*, 31, 267, 1991.
61. Fatokun, C. A., Menacio-Hautea, D. I., Danesh, D., and Young, N. D., Evidence for orthologous seed weight genes in cowpea and mungbean, based upon RFLP mapping, *Genetics*, 132, 841, 1992.
62. Doebley J., Bacigalupo, A., and Stec, A., Inheritance of kernel weight in two maize-teosinte hybrid populations: Implications for crop evolution, *J. Heredity*, 85, 191, 1994.

63. Schon, C. C., Melchinger, A. E., Boppenmaier, J., Brunklaus-Jung, E., Herrmann, R. G. et al., RFLP mapping in maize: quantitative trait loci affecting testcross performance of elite European flint lines, *Crop Sci.,* 34, 378, 1994.
64. Bradshaw, H. D. and Stettler, R. F., Molecular genetics of growth and development in Populus. IV. Mapping QTLs with large effects on growth, form, and phenology of traits, *Genetics,* 139, 963–973, 1995.
65. Weller, J. I., Maximum likelihood techniques for the mapping and analysis of quantitative trait loci with the aid of genetic markers, *Biometrics,* 42, 627, 1986.
66. Teutonico, R. A. and Osborn, T. C., Mapping of RFLP and qualitative trait loci in *Brassica rapa* and comparison to the linkage maps of *B. napus, B. oleracea,* and *Arabidopsis thaliana, Theor. Appl. Genet.,* 89, 885, 1994.
67. Van Eck, H. J., Jacobs, J. M. E., Stam, P., Ton, J., Stiekema, W. J., and Jacobsen, E., Multiple alleles for tuber shape in diploid potato detected by qualitative and quantitative genetic analysis using RFLPs, *Genetics,* 137, 303, 1994.
68. Groover, A., Devey, M., Fiddler, T., Lee, J., Megraw, R., Mitchell-Olds, T., Sherman, B., Vujcic, S., Williams, C., and Neale, D., Identification of quantitative trait loci influencing wood specific gravity in an outbred pedigree of loblolly pine, *Genetics,* 138, 1293, 1994.
68a. Paterson, A. H. et al., in preparation.
69. Bubeck, D. M., Goodman, M. M., Beavis, W. D., and Grant, D., Quantitative trait loci controlling resistance to gray leaf spot in maize, *Crop Sci.,* 33, 838, 1993.
70. Leonards-Schippers, C., Gieffers, W., Schaefer-Pregl, R., Ritter, E., Knapp, S. J., Salamini, F., and Gebhardt, C., Quantitative resistance to *Phytophthora infestans* in potato: a case study for QTL mapping in an allogamous plant species, *Genetics,* 137, 68, 1994.
71. Wang, G., MacKill, D. J., Bonman, J. M., McCouch, S. R., Champoux, M. C., and Nelson, R. J., RFLP mapping of genes conferring complete and partial resistance to blast in a durably resistant rice cultivar, *Genetics,* 136, 1421, 1994.
72. Li, Z., Pinson, S. R. M., Marchetti, M. A., Stansel, J. W., and Park, W. D., Characterization of quantitative trait loci in cultivated rice contributing to field resistance to sheath blight *(Rhizoctonia solani), Theor. Appl. Genet.,* 91, 374, 1995.
73. Jung, M., Weldekidan, T., Schaff, D., Paterson, A., Tingey, S., and Hawk, J., Generation means analysis and genetic mapping of anthracnose stalk rot resistance in maize, *Theor. Appl. Genet.,* in press.
74. Nienhuis, J., Helentjaris, T., Slocum, M., Ruggero, B., and Schaefer, A., Restriction fragment length polymorphism analysis of loci associated with insect resistance in tomato, *Crop Sci.,* 27, 797, 1987.
75. Bonierbale, M. W., Plaisted, R. L., Pineda, O., and Tanksley, S. D., QTL analysis of trichome-mediated insect resistance in potato, *Theor. Appl. Genet.,* 87, 973, 1994.
76. Martin, B., Nienhuis, J., King, G., and Schaefer, A., Restriction fragment length polymorphisms associated with water use efficiency in tomato, *Science,* 243, 1725, 1989.
77. Reiter, R. S., Coors, J. G., Sussman, M. R., and Gabelman, W. H., Genetic analysis of tolerance to low-phosphorus stress in maize using restriction fragment length polymorphisms, *Theor. Appl. Genet.,* 82, 561, 1991.
78. Doebley, J., Stec, A., Wendel, J., and Edwards, M., Genetic and morphological analysis of a maize-teosinte F2 population: implications for the origin of maize, *Proc. Natl. Acad. Sci. U.S.A.,* 87, 9888, 1990.
79. Bradshaw, H. D., Wilbert, S. M., Otto, K. G., and Schemske, D. W., Genetic mapping of floral traits associated with reproductive isolation in monkeyflowers (Mimulus), *Nature* (London), 376, 762, 1995.

# 10 QTL Analyses: Power, Precision, and Accuracy

*William D. Beavis*

## CONTENTS



## 10.1 INTRODUCTION

Historically the term quantitative trait has been used to describe variability in expression of a trait that shows continuous variability and is the net result of multiple genetic loci possibly interacting with each other or with the environment. Recently, the term complex trait has been used to describe any trait that does not exhibit classic Mendelian inheritance attributable to a single genetic locus.[1] The distinction between the terms is subtle and for purposes of this chapter the two terms can be used synonymously.

It has been estimated that 98% of human genetic diseases are complex traits,[2] and it is likely that a similar percentage could be ascribed to economically important quantitative traits in domesticated plants and animals. Quantitative traits tend to be classified as oligogenic or polygenic. Such a classification scheme is based on the perceived numbers and magnitudes of segregating genetic factors, i.e., quantitative trait loci, affecting the variability in expression of the trait. Unfortunately, perception is seldom based on carefully obtained empirical evidence. Biometric techniques designed to estimate the number of underlying quantitative trait loci (QTL) responsible for the variability of a quantitative trait require large samples of segregating progeny from population structures that are seldom available outside model species.[3]

The development of ubiquitous polymorphic genetic markers that span the genome have made it possible for quantitative and molecular geneticists to investigate what Edwards et al.[4] referred to

0-8493-7686-6/98/$0.00+$.50

as the numbers, magnitudes, and distributions of QTL. Since 1987, there have been at least 250 manuscripts published on mapping and analysis of QTL. Most of these have been conducted using plant species and have been based on an experimental paradigm in which segregating progeny derived from a single cross of two inbred lines are genotyped at multiple marker loci and evaluated for one to several quantitative traits. QTL are identified as significant statistical associations between genotypic values and phenotypic variability among the segregating progeny. This is an ideal experimental paradigm because the F1 parents all have the same linkage phase, all segregating progeny are informative and linkage disequilibrium is maximized.

The inbred line cross experimental paradigm has been used extensively with a wide array of plant species from *Arabidopsis* to maize where it is possible to obtain inbred lines. Some of these studies were initiated with crosses from divergent germplasm, e.g., interspecific inbred lines, with the goal of identifying QTL associated with obvious and easy-to-classify morphological differences between the species. Other studies were initiated with crosses from convergent germplasm, e.g., intraspecific breeding lines, with the goal of identifying QTL associated with agronomically important traits that exhibit continuous variability in the segregating progeny. From these studies QTL have been identified for both easy-to-classify morphological traits, e.g., inflorescence architecture, and for traits that exhibit continuous variability, e.g., grain yield. Although a study of this extensive literature can provide numerous valuable lessons, it is the intent of this chapter to highlight common features of results from QTL studies based on progeny from inbred line crosses and to propose explanations for these features based on the statistical issues of power, precision, and accuracy that are inherent to QTL analyses. In Section 10.2 of this chapter results from selected studies will be compared to illustrate both the consistent and inconsistent aspects of QTL studies. In Section 10.3, the empirical results reported in Section 10.2 are explained as functions of factors that affect power, precision, and accuracy of QTL analyses. In Section 10.4 the implications of the empirical results and statistical issues on marker assisted selection (MAS), in plant breeding are discussed.

## 10.2 LESSONS FROM EXPERIMENTAL RESULTS

### 10.2.1 RESULTS BASED ON PROGENY FROM INTERSPECIFIC CROSSES

The estimated numbers and magnitudes of genetic effects from several studies based on progeny from interspecific crosses are summarized in Table 10.1.[5-10] Quantitative trait loci were identified and analyzed in these studies for a variety of morphological traits that distinguished the species involved in the cross. These traits describe the plant architecture (e.g., number of lateral branches), inflorescence architecture (e.g., arrangement of cupules and glume morphology), and fruit architecture (e.g., disarticulation of seeds, and seed or fruit size). The types of progeny included backcross, F2, and replicated F2-derived lines. The number of progeny ranged from 60 to 370. The number of QTL identified for any given trait ranged from 1 to 7 and the estimated magnitude of genetic effects, as expressed by the amount of phenotypic variability among the progeny explained by any one of the QTL, ranged from about 5 to as much as 86%, although for most the maximum was in the range of 40 to 50%. Features of these studies that are not shown include the distribution of the estimated genetic effects and genomic locations of the QTL. In all of the studies except one[9] the distribution was characterized by one or two QTL with large estimated genetic effects and several additional QTL that explained a relatively small amount of the phenotypic variability.

In some of these examples, a comparison of specific genomic locations and estimated QTL effects can be made between studies. Paterson et al.[6] showed that half of the QTL identified for three quantitative traits using backcross and F2 progeny from the two interspecific tomato crosses mapped to the same genomic regions. Doebley and Stec[8] found that the largest QTL mapped to the same genomic site in two maize × teosinte populations for six of nine morphological traits. Because cereal species are largely syntenic,[11-13] Paterson et al.[14] were able to compare QTL associated with seed size and seed dispersal from the maize × teosinte crosses, with those identified using an interspecific Sorghum cross,[10] and a cross between two divergent subspecies of rice.

**TABLE 10.1**
**Estimated Numbers and Magnitudes QTL for Morphological Traits in Segregating Populations Derived from Divergent Germplasm**

| Population[a] | Number of progeny | Number of QTL | Magnitude of effects[b] | |
|---|---|---|---|---|
| | | | Minimum | Maximum |
| *Lycopersicon esculentum* × *L. chmielewski* (BC)[5] | 237 | 4–6 | 4 | 24 |
| *Lycopersicon esculentum* × *L. cheesmanii* (F2)[6] | 350 | 4–7 | 5 | 42 |
| *Maize* × *Teosinte* (F2)[7] | 260 | 2–6 | 4 | 42 |
| *Maize* × *Teosinte* (F2)[8] | 290 | 4–5 | 4 | 42 |
| *Glycime max* × *G. soja* (F2:3,[9] | 60 | 1–3 | 16 | 25 |
| *Sorghum bicolor* × *S. propinquum* (F2)[10] | 370 | 3–6 | 4 | 86 |

[a] Populations are referenced and indicated by the cross and type of progeny.
[b] Magnitude of QTL effects are reported as the minimum and maximum percent of phenotypic variability explained by the significant QTL.

Genomic locations of the QTL with estimated large effects mapped to syntenic regions across all three genera.

Thus, when compared across independent studies, QTL identified for complex but easily classified morphological traits using segregating progeny derived from crosses of divergent lines show a number of consistent characteristics: there are a relatively small number of QTL responsible for morphological divergence between domestic crop species and their wild progenitors,[8,14] and most of the phenotypic variability can be accounted for by one or two QTL with large estimated effects that map to similar regions across comparable studies. Investigation of such traits using divergent lines has thus been useful for drawing inferences about the genomic sites of genetic mutations responsible for the origin of domestic species, but are the inferences accurate or even useful for understanding quantitative trait variability exhibited within plant breeding populations?

### 10.2.2 Results Based on Progeny from Intraspecific Crosses

Identification of QTL for agronomically important traits has been pursued through progeny derived from intraspecific crosses of adapted inbred lines. Often these lines exhibit only slight morphological differences, but their progeny can exhibit considerable genetic variability for the traits of interest. Variability exhibited for quantitative traits of interest to plant breeders is assumed to be either oligogenic or polygenic and due to more QTL than the morphological traits that distinguish divergent germplasm. For example, two quantitative traits that have been routinely studied and reported in maize QTL experiments include plant height and grain yield.[15-21] The numbers of QTL thought to be associated with the variability of plant height from inbred line crosses of maize breeding germplasm are generally regarded as being five to ten, whereas the numbers of QTL affecting variability in grain yield (Mg/ha) are considered by maize breeders to be at least 20. Although variability in grain yield is assumed to be due to many more genes with smaller effects than those responsible for variability in plant height, there is little rigorous experimental evidence to support the assumption.

The estimated numbers and magnitudes of genetic effects from several QTL studies on plant height QTL in maize are summarized in Table 10.2.[15,17,19-21] The numbers of progeny evaluated in these experiments ranged from about 100 to 400 and the types of progeny included F2-derived lines evaluated per se backcrossed to the inbred parents, and topcrossed to unrelated inbred testers. The estimated numbers of QTL identified ranged from three to seven and the estimated magnitudes of the genetic effects, as expressed by the amount of phenotypic variability among the progeny explained by any one of the QTL, ranged from about 5 to 40%, although the maximum for most of the studies was about 25%.

**TABLE 10.2**
**Estimated Numbers and Magnitudes for Plant Height QTL in Maize**

| Population[a] | Number of progeny | Number of QTL | Magnitude of effects[d] | |
|---|---|---|---|---|
| | | | Minimum | Maximum |
| B73 × Mo17(F2:4)[15] | 112 | 6 | 12 | 23 |
| B73 × G35(F2:4)[15] | 112 | 4 | 11 | 24 |
| K05 × W65(F2:3)[15] | 144 | 3 | 12 | 23 |
| J90 × V94(F2:3)[15] | 144 | 3 | 17 | 25 |
| C0159 × Tx303(F2)[17] | 187 | 4 | 12 | 27 |
| B73 × Mo17(F2:3)[b] | 100 | 5 | 13 | 30 |
| B73 × Mo17(F2:3) × B73[c] | 264 | 4 | 4 | 21 |
| B73 × Mo17(F2:3) × Mo17[c] | 264 | 6 | 6 | 12 |
| Mo17 × H99(F2:3)[19] | 150 | 5 | 6 | 40 |
| B73 × Mo17(F2:3) × V78[20] | 112 | 5 | 6 | 14 |
| KW1265 × D146(F2:3) × KW4115[21] | 380 | 7 | 4 | 17 |
| KW1265 × D146(F2:3) × KW5361[21] | 380 | 4 | 5 | 32 |

[a] Populations are referenced and indicated by the parents and types of progeny.
[b] Beavis, W. D., Hallauer, A. R., and Lee, M., unpublished, 1991.
[c] Stuber, L. S. et al., personal communication, 1991.
[d] Magnitude of QTL effects are reported as the minimum and maximum percent of phenotypic variability explained by the significant QTL.

Data from Beavis, W. D., *49th Annual Corn and Sorghum Research Conference,* American Seed Trade Assoc., Washington, D.C.

The estimated numbers and magnitudes of genetic effects for yield QTL are summarized in Table 10.3.[16-18,20] The numbers of progeny used to evaluate grain yield were in the range of 100 to 250 and the types of progeny included F2-derived lines evaluated per se backcrossed to the inbred parents or topcrossed to an inbred tester. The estimated numbers of QTL identified ranged from two to eight and the estimated magnitudes of the genetic effects for any single QTL explained from 5 to 25% of the phenotypic variability.

Although not shown in the tables, the distribution of the estimated genetic effects for both plant height QTL and yield QTL was characterized by one or two loci with large estimated effects and several additional QTL that explained relatively smaller amounts of the phenotypic variability. Thus, the estimated numbers, magnitudes, and distribution of genetic effects were similar for both traits and were similar to the results reported for morphological traits in progeny derived from divergent germplasm. At first glance, these similarities challenge the assumption that variability for morphological traits that distinguish species is due to fewer QTL than the number of QTL responsible for variability in plant height and grain yield. However, closer inspection of the results suggest other explanations.

Beavis et al.[15] compared the genomic sites for plant height QTL from four studies that were based on about 100 to 150 F2-derived lines per se and found that no QTL mapped to the same genomic sites across all four sampled families. The QTL did show congruency with many mapped mutants known to have major effects on plant height in maize. They proposed that the most likely explanation for the lack of congruency among studies was that different sets of polymorphic alleles were segregating in the different genetic backgrounds.

In order to remove the confounding aspect of genetic background, a comparison to consider is one where QTL were identified in the same genetic background. Beavis et al.,[20] reported that yield QTL identified using F2:4 progeny, from the maize cross B73 × Mo17, herein referred to as the PHI progeny, did not map to the same genomic sites as the yield QTL identified by Stuber et al.,[20]

**TABLE 10.3**
**Estimated Numbers and Magnitudes of Grain Yield QTL in Maize**

| Population[a] | Number of progeny | Number of QTL | Magnitude of effects[c] Minimum | Maximum |
|---|---|---|---|---|
| Oh43 × Tx303(F2:3) × B73[16] | 216 | 6 | NR[d] | NR |
| Oh43 × Tx303(F2:3) × Mo17[16] | 216 | 6 | NR | NR |
| C0159 × Tx303(F2)[17] | 187 | 3 | 6 | 17 |
| B73 × Mo17(F2:3) × B73[18] | 264 | 6 | 6 | 18 |
| B73 × Mo17(F2:3) × Mo17[18] | 264 | 8 | 6 | 14 |
| B73 × Mo17(F2:3) × V78[20] | 112 | 2 | 9 | 13 |
| B73 × Mo17(F2:4)[20] | 112 | 5 | 8 | 23 |
| B73 × Mo17(F2:3)[b] | 100 | 5 | 8 | 21 |

[a] Populations are referenced and described by the inbred parents and types of progeny.
[b] Beavis, W. D., Hallauer, A. R., and Lee, M., unpublished, 1991.
[c] Magnitude of QTL effects are reported as the minimum and maximum percent of phenotypic variability explained by significant QTL.
[d] NR = not reported.

Data from Beavis, W. D., *49th Annual Corn and Sorghum Research Conference*, American Seed Trade Assoc., Washington, D.C.

in an independent set of progeny derived from the same cross, herein referred to as the NCS progeny. Although both studies used the same data analysis techniques on progeny from the same genetic background, there were still a number of confounding aspects with the comparison. (1) Different sets of genetic marker loci were used in each of the studies. Thus, relative placement of QTL within linkage groups from the two studies was tentative, although most of the differences were among linkage groups, rather than placement within linkage groups. (2) The sources of the parental lines used to generate the populations were not the same, so there might have been different sets of QTL with segregating alleles. (3) Progeny from each study were evaluated in different sets of environments. (4) Different samples of progeny were evaluated as either backcross or F2:4 progeny. So, although the same sets of QTL alleles were segregating in the progeny, the progeny were not evaluated at the same level of inbreeding. It is possible that epistatic and/or epigenetic factors influenced the expression of the QTL.[22] Finally, sampling may play a role in which different sets of QTL are identified in any given experiment.

In order to remove some of the confounding aspects of the comparison, a third independent set of 100 F2:3 lines derived from B73 × Mo17,[23] referred to herein as the ISU progeny, were investigated for plant height and yield QTL. Each of the ISU progeny were restriction fragment length polymorphism (RFLP)-typed using the same 96 RFLP markers used to genotype the PHI lines and QTL data analyses were the same as those applied to the PHI and NCS progeny. Although the parental sources used to generate the ISU progeny were different from the PHI sources, the RFLP patterns at all 96 markers were the same (data not shown). The estimated numbers, magnitudes, and distribution of genetic effects based on the ISU progeny were similar to those identified with the NCS and PHI progeny (Tables 10.2 and 10.3).[23a] However, the estimated genomic sites for plant height and yield QTL were not the same as those identified with the PHI progeny (Table 10.4), nor were they the same as those identified with the NCS progeny.

Although the comparison between the PHI and ISU QTL is based on similar numbers and types of progeny from the same genetic background with the same RFLP patterns, the comparison still has confounded factors. First, the progeny were not evaluated in the same sets of environments, although there was little evidence for environmental influence on the identification of QTL within any of the studies.[18,24] Second, the progeny from the studies were not evaluated at the same level

**TABLE 10.4**
**Estimated Genomic Position and Amount of Phenotypic Variability Described by Plant Height and Yield QTL Identified in Two Independent Sets of F2 Derived Lines (Denoted ISU and PHI) from the Maize Cross B73 × Mo17**

| Estimated genomic position | | Estimated amount of phenotypic variability (%) explained by the QTL[a] | |
|---|---|---|---|
| Chromosome | Flanking Markers | ISU Progeny | PHI Progeny |
| | **Plant Height (cm) QTL** | | |
| 1 | php1122/bnl7.21 | 17 | 7 |
| 1 | bnl8.10/php20518 | 16 | — |
| 2 | umc131/php20005 | — | 8 |
| 3 | bnl8.35/umc10 | — | 10 |
| 3 | umc60/bnl6.16 | 9 | — |
| 4 | umc42/umc19 | — | 5 |
| 6 | umc62/php20599 | 4 | — |
| 8 | bnl12.30/bnl10.24 | 7 | — |
| 9 | wx1/css1 | — | 10 |
| 10 | php15013/php10033 | — | 12 |
| | **Yield (Mg/ha) QTL** | | |
| 1 | umc13/php1122 | 14 | — |
| 1 | bnl8.10/php20518 | — | 8 |
| 2 | umc34/php10012 | 26 | — |
| 2 | umc36/php20622 | — | 10 |
| 3 | bnl6.16/umc63 | 7 | — |
| 4 | umc31/bnl5.46 | — | 7 |
| 5 | bnl6.10/php60012 | — | 9 |
| 6 | umc62/php20599 | 6 | — |
| 8 | bn19.11/bnl10.39 | 13 | — |
| 9 | php10005/bz1 | — | 23 |

[a] RFLP genotyping, development and evaluation of plant height and yield, and QTL analyses are described elsewhere.[20,23]

Data from Beavis, W. D., *49th Annual Corn and Sorghum Research Conference,* American Seed Trade Assoc., Washington, D.C.

of inbreeding, but before invoking epistasis, it is important to consider that the lack of congruency may have been an artifact of sampling. That is, if there are a large number of small-effect QTL, and if the number of progeny used in the experimental design provides little power, then only a few QTL will be identified in any given experiment. Also, it is unlikely that QTL identified with one sample of progeny will be identified with a second independent sample of progeny.

## 10.3 LESSONS ON POWER, PRECISION, AND ACCURACY

### 10.3.1 Definitions and Background

There is a tendency to make the inferential leap from QTL to physiology of gene effects at genetic loci, but QTL are, by definition, merely significant statistical associations. As statistical constructs the results of QTL analyses can be characterized by type I errors, power, precision, and accuracy. Type I errors occur when QTL are claimed to exist in regions of the genome where no actual QTL exist. Statistical significance of a QTL is determined by the frequency ($\alpha$) of false positive associations that the scientist is willing to accept. Power is the probability of identifying a QTL of known magnitude, given the predetermined frequency of false positive associations, i.e., $\alpha$. Precision is a measure of the dispersion of repeated independent estimates of genomic positions or genetic effects of the alleles at the QTL and is often reported by inverse measures such as standard errors or confidence intervals. Accuracy is a measure of how close the estimates are to the true values. In practice, accuracy is very difficult to evaluate for experimental results because the true values are unknown.

The choice of $\alpha$ depends on the goals of the experiment. For reasons that are unclear, QTL researchers have tended to use $\alpha = 0.05$ as an acceptable error rate, but for exploratory QTL experiments $\alpha = 0.25$ could be acceptable. Although choice of $\alpha$ is not a technical issue, determining the appropriate threshold for the test statistic to assure $\alpha$ is. Unfortunately, many QTL studies of plant species have reported an incorrect value of $\alpha$ by reporting the values provided by commercial software packages. With the development of ubiquitous polymorphic markers that span the genome it became apparent that the usual inferences about significance of calculated test statistics could not be applied because hundreds of test statistics may be calculated within each QTL experiment. Furthermore, these tests are not independent because the markers are genetically linked. The threshold can be determined analytically for a genome with an infinite number of markers and estimated through simulations for less saturated genomes.[25] Although values based on simulated genomes are an improvement relative to the values provided by commercial software packages, many authors, including myself, have inappropriately applied results from simulations to actual experimental genomes. A significant improvement and one that is intuitively more appealing is to obtain empirical estimates suitable for each experiment through permutation tests.[26,27]

After the threshold for which a chosen $\alpha$ has been determined, the power, precision, and accuracy can be evaluated for different statistical analysis methods or experimental designs. Numerous data analysis techniques have been proposed and developed for the identification and mapping of QTL in inbred line cross experiments. These can be classified as marker-trait (MT) methods,[28-32] interval mapping (IM) methods,[25,33] and multiple QTL model (MQM) methods.[34-40] MT methods utilize t-tests and F-tests to detect significant statistical associations between segregating marker genotypes and quantitative trait variation.[28,29] Interval mapping, as originally proposed, maximizes the likelihood function and utilizes genetic information from flanking markers to find the most likely position and genetic effects of a single QTL.[25] Multiple QTL models are based on the integration of multiple regression methods with IM and were proposed to increase the probability of including significant QTL in the model.[34] Within the context of the inbred line cross the experimental design is determined by the type of progeny [e.g., backcross, F2, F2-derived lines, recombinant inbreds, doubled haploids (DH)], the number of progeny, type of genetic markers (i.e., dominant or co-dominant), number of genetic markers, and precision of phenotypic measurement. Thus, power, precision, and accuracy of experimental results are affected by the reproductive biology of the species, availability of data analysis methods and experimental resources.

### 10.3.2 Methods for Evaluating Power, Precision, and Accuracy

Typically power and precision of a test statistic are evaluated based on the asymptotic distribution of the test statistic. The development of MT methods have been based on t and F statistics and have relied on the theoretical properties of these distributions to assess power and precision of contrasting marker genotypic classes.[29,32,41] Evaluation of power and precision is straightforward if the asymptotic distribution of the test statistic is known, but test statistics generated by QTL analyses, such as IM and MQM, do not always follow known distributions.[25] If the asymptotic distribution of the test statistic is unknown, it is still possible to evaluate the power, precision, and accuracy through Monte Carlo simulations. In the context of QTL experiments, the idea is to simulate a set of QTL with known genetic locations and effects in a segregating population and then evaluate power, precision, and accuracy of the known simulated QTL.[42,43] Although Monte Carlo simulations can be very informative, they are computationally intensive and time consuming. Recently, deterministic sampling was proposed as a more efficient means of obtaining information on power and precision of QTL analyses.[44]

### 10.3.3 Evaluation of Data Analysis Methods

Based on Monte Carlo simulations, differences in power, precision, and accuracy have been shown among QTL data analysis methods. IM provides improved precision and accuracy relative to MT analyses, but improvements to power are negligible for genomes with markers distributed at densities greater than one per 20 cM.[45,46] Although IM represented a significant conceptual contribution to QTL analyses, it is based on the null hypothesis of no QTL; an incorrect assumption for quantitative traits. Unless multiple QTL are included in the genetic model and the effects are estimated simultaneously, the estimates of genetic effects as well as inferences about statistical significance will be biased. Jansen[34] first proposed that multiple regression methods be integrated with IM to increase the probability of including all significant QTL in the model. Such MQM methods were developed simultaneously and independently by Jansen,[36] Rudolphe and LeFort,[38] and Zeng.[39] Monte Carlo simulations of large effect QTL showed that MQM methods produced more accurate and precise estimates than IM, but inclusion of too many cofactors reduced the power to identify QTL relative to IM.[37,40,47] Thus, the challenge with MQM is to develop decision rules for including or excluding markers as cofactors.[37] Following the theme of improving the accuracy of the genetic and statistical models that underlie the data analyses, recent developments in data analysis methods for the inbred line cross experimental paradigm have focused on including parameters for unequal variances[48] and epistasis.[49]

### 10.3.4 Evaluation of Experimental Design Parameters

In addition to data analyses, experimental factors such as number of progeny, type of progeny, or precision of phenotypic measurement influence power, precision, and accuracy of QTL results. Based on asymptotic theory, it has been shown that the type of progeny developed in the experiment will affect the power to identify QTL using MT methods.[29,32,41] DH are most powerful for estimating additive effects, while backcross progeny from the North Carolina Design III are least powerful.[32] F2-derived progeny are about two times as powerful as backcross progeny.[29] The use of replicated progeny in MT evaluations also have been shown through asymptotic theory to increase the power of QTL detection and precision of estimated genetic effects.[50,51] Because resources are limited and it is often cheaper to evaluate the phenotypes of progeny in field plots than to genotype the progeny, these results[50,51] made it tempting to evaluate a small sample of progeny in a large number of replicated field trials. However, based on asymptotic theory and ignoring QTL by environment interaction effects, Knapp and Bridges[52] showed that it is more efficient to evaluate a single replication of a large sample of progeny.

As previously mentioned, it is also possible to evaluate the influence of factors such as number of progeny, type of progeny, or precision of phenotypic measurement on power, precision, and accuracy of IM or MQM through the application of Monte Carlo simulations. For example, suppose that we would like to investigate the potential effects of sample size and heritability on the power, precision, and accuracy of QTL experiments similar to those reported in Section 10.2.

#### 10.3.4.1 Simulation Design

To illustrate, ideal experiments were simulated in which all markers provided accurate, complete codominant genotypic information in F2 populations with independently segregating QTL. Given the genetic size and distribution of genetic information in the maize genome, it is possible to have up to 40 independently segregating QTL, so polygenic traits based on 10 or 40 QTL were simulated. Each QTL was randomly assigned to the middle of an independent linkage group. The genome consisted of 75 independent linkage groups, each consisting of 20 recombinants with no interference per 100 gametes produced by the F1 between two inbred parents. Marker loci were assigned to the ends of each linkage group. Thus, the genome of each F2 population consisted of about 1600 cM and was completely and uniformly covered with genetic markers. Two hundred simulated F2 populations were generated for each of the 18 sets of experimental conditions. The 18 sets of experimental conditions consisted of 10 or 40 QTL that explained 30, 63, or 95% of the phenotypic variability (heritability) in 100, 500, or 1000 F2 progeny. All of the simulated genotypic variability was due to equal additive effects with no dominance at the QTL. All of the positive effects came from one of the parents. The phenotypic value that was assigned to each F2 individual was calculated by adding random error, which was normally distributed with mean 0 and variance determined by the heritability, to the sum of the additive effects. The magnitudes of the genetic effects of each QTL can be represented as the percentage of the phenotypic variability that each contributes. For example, if there are 10 segregating QTL that are responsible for a trait that is 30% heritable, then each QTL contributes 3% to the phenotypic variability. Although idealistic, the intent of these simulations was to evaluate the potential power, precision, and accuracy.

#### 10.3.4.2 Data Analyses

For most of the experiments reported in Tables 10.2 to 10.4, likelihood-based interval mapping[53] was used to analyze each of the 3600 simulated data sets. This is because, until recently,[54,55] it was the only statistical method implemented in a publicly available computer package. Since most QTL experiments have exploratory research goals where the impact of missing real QTL is costly, $\alpha = 0.25$ was chosen. The threshold associated with $\alpha = 0.25$ for declaring the presence of a QTL was determined by evaluating 200 data sets with no simulated QTL and choosing a maximum test statistic found in no more than 25% of these data sets (LOD = 2.5).

#### 10.3.4.3 Results

I have previously reported some of the following results in a nonrefereed publication.[56] Consider first the power to identify simulated QTL given $\alpha = 0.25$, (Table 10.5). For the case where there were ten simulated QTL that accounted for 63% of the variability among 100 F2 progeny the power to identify QTL was ~0.33. In other words, of the 2000 QTL that were generated in the 200 simulations with ten segregating QTL that explained 63% of the phenotypic variability among 100 F2 progeny, 653 were correctly identified to be on one of the 20-cM linkage groups with a simulated QTL. It was possible to consistently identify virtually all ten independently segregating QTL, but only if they were responsible for at least 63% of the phenotypic variability and 1000 F2 progeny were used to evaluate the trait. If 40 independent QTL were segregating in the population, then it was not possible to consistently identify all of the QTL even if the heritability among 1000 progeny was 95%.

Consider next the precision, or standard error, of estimated genetic effects and genomic positions, (Table 10.5). The estimated standard errors of each decreased with increasing heritability and number of progeny. The distribution of the estimated genetic effects of each correctly identified QTL from the simulated data sets where the heritability was 65% among 100 progeny indicates that the estimates were not symmetrically distributed (Figure 10.1). Notice that the estimates consist of a few QTL with large estimated effects and many QTL with relatively small estimated effects. This represents the same pattern of estimated genetic effects observed in experimental QTL studies. The distribution of the estimated genomic positions was symmetric about the mean, but trimodal with an unusually large frequency of estimated QTL being placed at the molecular markers (Figure 10.2).

Finally, consider the accuracy of the estimated effects and genomic positions (Table 10.5). The averaged estimated magnitudes of genetic effects associated with correctly identified QTL were greatly overestimated if only 100 progeny were evaluated, slightly overestimated if 500 progeny were evaluated and fairly close to the actual magnitude when 1000 progeny were evaluated. The bias at small sample sizes was due to overestimates of both additive and dominance effects. Recall dominance effects were simulated to be zero. Of the 654 correctly identified QTL in the 200 simulations with only 10 independently segregating QTL in 100 F2 progeny, the most frequent estimates were fairly close to the actual value of about 6%, although a few were estimated to

**TABLE 10.5**
**Effects of Heritability and Sample Size on the Power, Precision, and Accuracy of QTL Identified in F2 Progeny with Either 10 or 40 Simulated QTL**

| | | Magnitude of Genetic Effects[b] | | | | | |
|---|---|---|---|---|---|---|---|
| | | Variance Explained | | Additive Effects | | | |
| Simulated conditions[a] | Power | Simulated | Estimated | Simulated | Estimated | Estimated dominance | Estimated genomic site[c] |
| 10–30–100 | 9 | 3.00 | 16.76 ± 0.40 | 2.45 | 4.96 ± 0.10 | 3.28 ± 0.18 | 1.30 ± 0.55 |
| 10–30–500 | 57 | 3.00 | 4.33 ± 0.05 | 2.45 | 2.89 ± 0.02 | 1.01 ± 0.02 | 0.53 ± 0.17 |
| 10–30–1000 | 85 | 3.00 | 3.02 ± 0.03 | 2.45 | 2.56 ± 0.01 | 0.68 ± 0.01 | 0.80 ± 0.12 |
| 10–63–100 | 33 | 6.25 | 12.65 ± 0.20 | 3.55 | 4.68 ± 0.04 | 1.80 ± 0.06 | 0.51 ± 0.26 |
| 10–63–500 | 86 | 6.25 | 7.08 ± 0.06 | 3.55 | 3.73 ± 0.02 | 0.94 ± 0.02 | 0.96 ± 0.11 |
| 10–63–1000 | 98 | 6.25 | 6.34 ± 0.04 | 3.55 | 3.60 ± 0.01 | 0.01 ± 0.02 | 1.04 ± 0.08 |
| 10–95–100 | 39 | 9.50 | 18.68 ± 0.18 | 4.36 | 5.85 ± 0.04 | 2.33 ± 0.06 | 0.58 ± 0.26 |
| 10–95–500 | 94 | 9.50 | 10.10 ± 0.07 | 4.36 | 4.49 ± 0.02 | 0.88 ± 0.02 | 1.08 ± 0.10 |
| 10–95–1000 | 100 | 9.50 | 9.67 ± 0.05 | 4.36 | 4.44 ± 0.01 | 0.01 ± 0.02 | 1.19 ± 0.08 |
| 40–30–100 | 3 | 0.75 | 15.78 ± 0.41 | 1.22 | 4.40 ± 0.14 | 3.69 ± 0.24 | 0.83 ± 0.88 |
| 40–30–500 | 11 | 0.75 | 3.17 ± 0.05 | 1.22 | 2.35 ± 0.02 | 1.36 ± 0.04 | 0.17 ± 0.35 |
| 40–30–1000 | 25 | 0.75 | 1.46 ± 0.02 | 1.22 | 1.85 ± 0.01 | 0.82 ± 0.02 | 0.17 ± 0.22 |
| 40–63–100 | 4 | 1.56 | 16.31 ± 0.35 | 1.77 | 4.71 ± 0.10 | 3.59 ± 0.20 | 0.45 ± 0.65 |
| 40–63–500 | 29 | 1.56 | 3.54 ± 0.03 | 1.77 | 2.59 ± 0.01 | 1.13 ± 0.02 | 0.13 ± 0.21 |
| 40–63–1000 | 59 | 1.56 | 1.96 ± 0.02 | 1.77 | 2.09 ± 0.01 | 0.74 ± 0.01 | 0.37 ± 0.12 |
| 40–95–100 | 6 | 2.40 | 16.55 ± 0.29 | 2.18 | 5.02 ± 0.09 | 3.27 ± 0.15 | 0.45 ± 0.51 |
| 40–95–500 | 46 | 2.40 | 3.97 ± 0.03 | 2.18 | 2.79 ± 0.01 | 1.06 ± 0.02 | 0.12 ± 0.15 |
| 40–95–1000 | 77 | 2.40 | 2.58 ± 0.02 | 2.18 | 2.36 ± 0.01 | 0.70 ± 0.01 | 0.29 ± 0.09 |

[a] Numeric values denote number of QTL–heritability–number of progeny.

[b] The simulated genetic effects were additive and equal at all QTL for each set of conditions. There were no simulated dominance effects and all positive alleles came from one of the parents. Phenotypic values for each F2 were calculated as the sum of the additive effects from the QTL and random error which was distributed with mean 0 and variance determined by the heritability. Estimated effects are given as the averaged value for all correctly identified QTL ± standard error.

[c] Each simulated QTL was located equidistant (–10 cM) from two genetically linked markers. The estimated genomic site, based on IM, is given as the averaged deviation (cM) ± the standard error of the estimated QTL from the simulated QTL site.

Some data previously reported in Beavis, W. D., *49th Annual Corn and Sorghum Research Conference*, American Seed Trade Assoc., Washington, D.C., and in Smith, S. and Beavis, W. D., *The Impact of Plant Molecular Genetics*, Sobral, B. W. S., Ed., Birkhäuser, Boston, 1966.

explain as much as 35% of the phenotypic variability (Figure 10.1). On the other hand, when 40 independently segregating QTL were responsible for variability of the trait in 100 F2 progeny, the magnitude of the estimated genetic effects were severely biased for all 352 correctly identified QTL. The average estimated genomic position of the QTL showed little bias under any set of experimental conditions.

#### 10.3.4.4 Discussion

Previous reports of simulation studies have investigated the power of IM to identify one to six independently segregating QTL,[57,58] while those reported herein focused on polygenic inheritance. The results of all these studies have shown that there is very little power to identify small-effect QTL with a small number of progeny (<500). Also, estimates of genetic effects and genomic locations are imprecise, and estimates of genetic effects are biased. The analyses used in all of these studies were based on IM prior to the release of software based on MQM methods. To date,

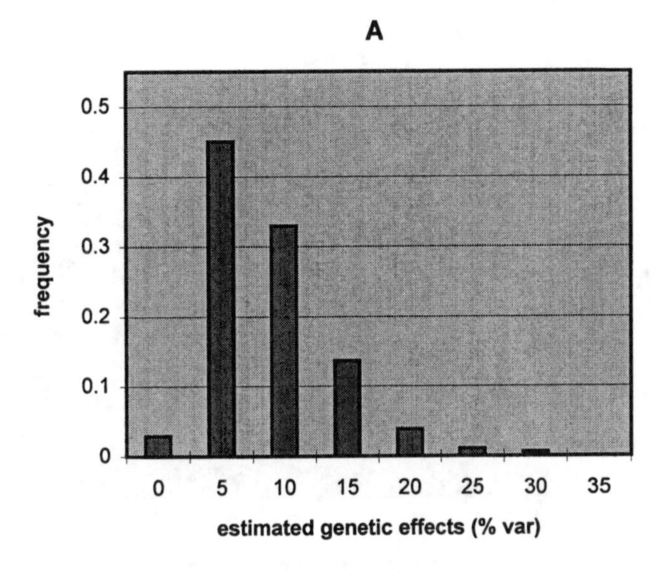


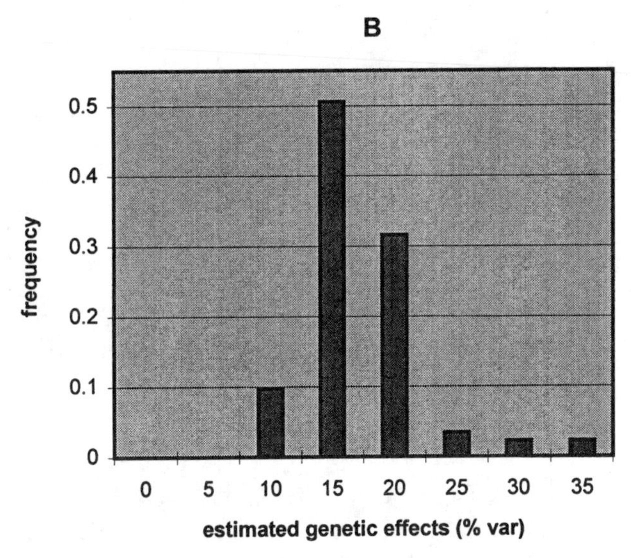


**FIGURE 10.1** Frequency distribution of the estimated genetic effects, expressed as the percentage of the phenotypic variability explained by QTL that were identified on one of the linkage groups with a simulated QTL. (A) The estimated QTL were identified in 200 samples of 100 F2 progeny with 10 simulated QTL that explained 63% of the phenotypic variability, i.e., the simulated additive effects of each QTL accounted for 6.25% of the total phenotypic variability. (B) The estimated QTL were identified in 200 samples of 100 F2 progeny with 40 simulated QTL that explained 63% of the phenotypic variability, i.e., the simulated additive effects of each QTL accounted for 1.56% of the total phenotypic variability.

evaluation of MQM methods has been based on simulations of large-effect QTL.[37,40,47] Evaluation of power, precision, and accuracy of MQM methods for small-effect QTL such as those reported herein still need to be investigated. However, a reanalysis of the PHI and ISU progeny using MQM revealed little change in the inferences (data not shown). The analyses provided more precise estimates of genomic position and slightly changed the estimated genetic effects, but the results were not sufficient to change the lack of congruency of QTL between the studies. This suggests that data analyses based on more accurate genetic and statistical models have improved the precision and accuracy of large effect QTL, but the primary limitation on information about QTL for

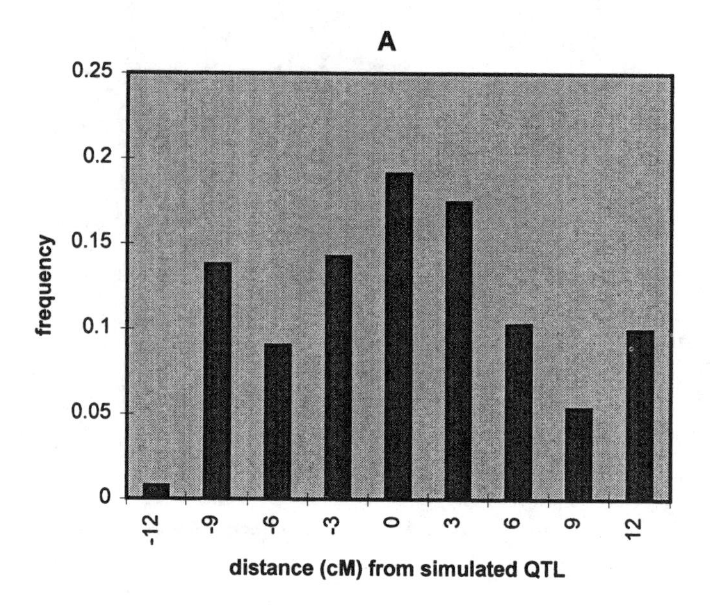


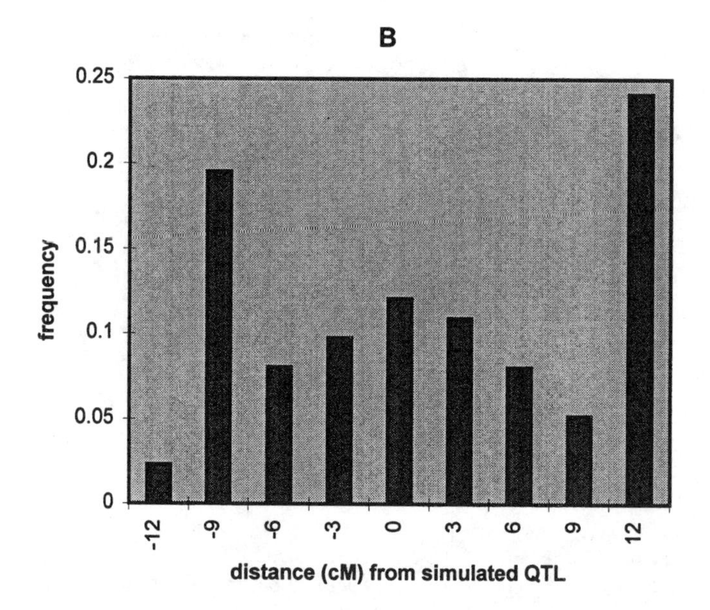


**FIGURE 10.2** Frequency distribution of the genomic sites of estimated QTL that were identified on one of the linkage groups with a simulated QTL. Estimated genomic sites are denoted as the deviation (cM) from the genomic site of the simulated QTL. (A) The estimated QTL were identified in 200 samples of 100 F2 progeny with 10 simulated QTL that explained 63% of the phenotypic variability, i.e., the simulated additive effects of each QTL accounted for 6.25% of the total phenotypic variability. (B) The estimated QTL were identified in 200 samples of 100 F2 progeny with 40 simulated QTL that explained 63% of the phenotypic variability, i.e., the simulated additive effects of each QTL accounted for 1.56% of the total phenotypic variability.

oligogenic and polygenic traits is due to constraints imposed by the experimental design and the true underlying genetic complexity.

Most QTL experiments reported to date (Section 10.2) have been conducted with 100 to 250 progeny in replicated tests where the estimated broad sense heritability has been in the range of 60 to 90%. Interpolating from the results in the simulations, the expectation is to identify 3 to 7 QTL with unequal effects when there are actually 10 equal-effect QTL segregating in 100 to 250 F2 progeny. On the other hand, if there are 40 independently segregating QTL in 100 to 250 F2 progeny responsible for the expression of a trait that exhibits 65 to 95% heritability, then the expectation is to identify 2 to 8 of the QTL and to estimate their effects unequally. Obviously, inferences about the number and magnitude of QTL for complex traits cannot be drawn from inbred line cross experiments conducted to date.

Monte Carlo simulations of QTL can provide insight into the results of QTL experiments, but inferences are limited. The simulated QTL had equal additive effects with all positive alleles from one of the inbred parents. They were placed in the middle of independently segregating linkage groups that were 20 cM long and flanked with genetic marker loci. There were no missing genotypic data, nor were there any errors in the genotypic data. Obviously, the simulations were designed to optimize the conditions for identification of small effect QTL by IM. Thus, results from IM of experimental data will be less powerful, less precise, and less accurate than observed in these simulations. Also, application of MQM methods to simulated polygenic traits still needs to be investigated so inferences about the potential improvements through improved data analyses can be made.

A question to consider is whether the traits of interest to plant breeders are polygenic with many QTL of relatively small effects or effectively oligogenic with a few QTL responsible for most of the variability. There has been a tendency to cite studies such as those reported in Section 10.2 as indicative of the latter. However, based on results from the simulations, unless sufficiently large numbers of progeny have been evaluated, the estimated genetic effects from experimental studies are suspect. To date, there are no genome-wide QTL studies in plant species that have used "large" numbers of progeny, but in an inbred line cross experiment with mice based on a moderate sample (n = 535) of F2 progeny, 16 growth-rate QTL were identified and the estimated genetic effects accounted for 2 to 10% of the phenotypic variability.[59] To emphasize, with a moderate number of progeny there were a large number of QTL identified and none accounted for a large amount of the phenotypic variability. As larger samples of progeny are evaluated in inbred cross experiments, will we see greater evidence for the infinitesimal model upon which historical quantitative genetics is based?

## 10.4 LESSONS FOR PLANT BREEDING

Our focus has been on QTL analyses for quantitative or complex traits in plant species with reproductive capacities that are amenable to an experimental paradigm based on evaluation of selfed progeny derived from a single cross of inbred lines. A primary motivation for development of molecular markers in crop species has been the potential for increased efficiency through marker assisted selection (MAS). To date, molecular markers have been used to assist backcross breeding of single genes from transgenic and unadapted germplasm, but have not been shown to be more efficient or effective than existing breeding methods for quantitative traits. Stuber,[60] has claimed success using MAS, but his selected lines were not contrasted with phenotypically selected lines derived from the same germplasm; i.e., could the same results have been realized through routine breeding methods? Because most agronomically important traits in crops are quantitative, it seems fair to say that molecular markers have had little impact on crop improvement despite hundreds of published QTL experiments during the last 10 years. Explanations for this lackluster application of molecular markers include inadequate experiments and failure to integrate methods into existing breeding methods.

Results from the Monte Carlo simulations suggest that experimental QTL studies in plant species have been inadequate for drawing inferences about numbers, magnitudes, and distribution of QTL for most quantitative traits. If there are numerous independently segregating QTL with equal or near equal small effects, all QTL will have a similar small probability of being identified in any given experiment. Thus, there is little reason to expect the same set of QTL to be identified among independent experiments. The converse also can be inferred, i.e., if independent QTL experiments identify noncongruent sets of QTL, then there are likely a large number of QTL responsible for the variability in the trait. Furthermore, the results from the simulations also suggest that the estimates of genomic locations and genetic effects from most experimental studies have been imprecise and biased.

Because inbred line cross experiments have been inadequate to accurately estimate the numbers, magnitudes, and distribution of QTL, it should not be surprising if MAS does not produce exceptional results. It has been shown that the greatest efficiency from MAS will be for those traits that exhibit low heritability, assuming that the QTL are linked to selectable markers.[61-64] Unfortunately there is little power to identify markers linked to QTL or accurately estimate their effects in traits that exhibit low heritability, unless a large number of progeny are evaluated (Table 10.5). Thus, unless large numbers of progeny are evaluated for QTL, MAS will have minimal impact on plant breeding.[65] These results also suggest that new breeding strategies based on evaluation of large numbers of progeny will be necessary to realize the potential of MAS.

Tanksley and Nelson[66] proposed that another reason for the lack of success with MAS for quantitative traits is that QTL discovery and varietal development have been treated as separate processes. Indeed the experimental paradigm used in most QTL studies bears little resemblance to applied plant breeding. The populations used for QTL identification consist of 100 to 400 progeny from a cross of two inbred lines, whereas plant breeding populations consist of 500 to 10,000 progeny from multiple crosses of partially inbred lines. Based on 50 years of experience, successful plant breeders understand that useful genetic variability for most agronomic traits is best evaluated using progeny from hundreds of crosses among adapted lines.[67] Thus, if QTL analyses could be integrated into existing breeding strategies the power, precision, and accuracy associated with large numbers of progeny could be realized. Another advantage of identifying QTL in breeding populations is that QTL effects will be predictive across breeding germplasm because they will have been evaluated in multiple genetic backgrounds.

There has been some effort to adapt the analysis methods developed for biparental experimental populations to idealized breeding populations.[68-70] However, the assumptions that underlie analysis methods for inbred line cross experimental populations are not adequate for breeding populations because the genetic structures are not the same. Selection of the inbred lines used in the experimental paradigm is often based on maximizing phenotypic and molecular marker differences between the lines. As a consequence, the F1 is fully informative, linkage disequilibrium is maximized, linkage phase is known, there are only two QTL alleles and, except for backcross progeny, the frequency of each QTL allele is 0.5. In contrast, selection of lines for breeding is based on maximizing useful genetic variability for agronomic performance. As a consequence, the crosses are not necessarily informative at marker loci and QTL, linkage disequilibrium exists among progeny within families, but not necessarily across the breeding population; linkage phase may not be consistent across the breeding population, multiple QTL alleles can exist and the frequency of each will be between 0 and 1. These genetic structures also exist in human and animal breeding populations,[71-74] although there are distinctions. For example, outbred populations are not derived from partially inbred grandparents and do not consist of partially inbred progeny. Nonetheless, I anticipate that methods based on mixed effects models[75] and random effects models[76] will be useful for plant breeding populations that arise from multiple related crosses.

## ACKNOWLEDGMENTS

I am grateful to my colleagues, Drs. Oscar Smith, Steven Openshaw, and David Grant for numerous philosophical and technical discussions on the genetics of quantitative traits, and I'd like to gratefully acknowledge Dr. Heike Kross for carefully reading this manuscript and for her numerous helpful suggestions.

## REFERENCES

1. Lander, E. S. and Schork, N. J., Genetic dissection of complex traits, *Science,* 265, 2037, 1994.
2. Strohman, R. C., Ancient genomes, wise bodies, unhealthy people: limits of a genetic paradigm in biology and medicine, *Perspect. Biol. Med.,* 37, 112, 1993.
3. Zeng, Z.-B., Correcting the bias of Wright's estimates of the number of genes affecting a quantitative character: a further improved method, *Genetics,* 131, 987, 1992.
4. Edwards, M. D., Stuber, C. W., and Wendel, J. F., Molecular-marker-facilitated investigation of quantitative-trait loci in maize. I. Numbers, genomic distribution and types of gene action, *Genetics,* 115, 113, 1987.
5. Paterson, A. H., Lander, E. S., Hewitt, J. D., Peterson, S., Lincoln, S. E., and Tanksley, S. D., Resolution of quantitative traits into Mendelian factors by using a complete linkage map of restriction fragment length polymorphisms, *Nature,* 335, 721, 1988.
6. Paterson, A. H., Damon, S., Hewitt, J. D., Zamir, D., Rabinowitch, H. D., Lincoln, S. E., Lander, E. S., and Tanksley, S. D., Mendelian factors underlying quantitative traits in tomato: comparison across species, generations, and environments, *Genetics,* 127, 181,1991.
7. Doebley, J., Stec, A., Wendel, J., and Edwards, M., Genetic and morphological analysis of a maize-teosinte F2 populations: implications for the origin of maize, *Proc. Natl. Acad. Sci. U.S.A.,* 87, 9888, 1990.
8. Doebley, J. and Stec, A., Inheritance of morphological differences between maize and teosinte: comparison of results for two F2 populations, *Genetics,* 134, 559, 1993.
9. Keim, P., Diers, B. W., Olson, T. C., and Shoemaker, R. C., RFLP mapping in soybean: association between marker loci and variation in quantitative traits, *Genetics,* 126, 735, 1990.
10. Lin, Y.-R., Schertz, K. F., and Paterson, A. H., Comparative analysis of QTLs affecting plant height and maturity across the Poaceae, in reference to an interspecific sorghum population, *Genetics,* 141, 391, 1995.
11. Hulbert, S. H., Richter, T. E., Axtell, J. D., and Bennetzen, J. L., Genetic mapping and characterization of sorghum and related crops by means of maize DNA probes, *Proc. Natl. Acad. Sci. U.S.A.,* 87, 4251, 1990.
12. Ahn, S. and Tanksley, S. D., Comparative linkage maps of the rice and maize genomes, *Proc. Natl. Acad. Sci. U.S.A.,* 90, 7980, 1993.
13. Moore, G., Gale, M. D., Kurata, N., and Flavell, R. B., Molecular analysis of small grain cereal genomes: current status and prospects, *Bio/tech,* 11, 584, 1993.
14. Paterson, A. H., Lin, Y.-R., Li, Z., Schertz, K. F., Doebley, J. F., Pinson, S. R. M., Liu, S.-C., Stansel, J. W., and Irvine, J. E., Convergent domestication of cereal crops by independent mutations at corresponding genetic loci, *Science,* 269, 1714, 1995.
15. Beavis, W. D., Grant, D., Albertsen, M., and Fincher, R., Quantitative trait loci for plant height in four maize populations and their associations with quantitative genetic loci, *Theor. Appl. Genet.,* 83, 141, 1991.
16. Stuber, C. W. and Sisco, P. H., Marker-facilitated transfer of QTL alleles between elite inbred lines and responses in hybrids, in *Proc. Forty-sixth Annu. Corn and Sorghum Industry Research Conf.,* ASTA, Washington, D.C., 1991, 104–133.
17. Edwards, M. D., Helentjaris, T., Wright, S., and Stuber, C. W., Molecular-marker facilitated investigations of quantitative trait loci in maize, *Theor. Appl. Genet.,* 83, 765, 1992.
18. Stuber, C. W., Lincoln, S. E., Wolff, D. W., Helentjaris, T., and Lander, E. S., Identification of genetic factors contributing to heterosis in a hybrid from two elite maize inbred lines using molecular markers, *Genetics,* 132, 823, 1992.
19. Veldboom, L. R., Lee, M., and Woodman, W. L., Molecular marker-facilitated studies in an elite maize population. I. Linkage analysis and determination of QTL for morphological traits, *Theor. Appl. Genet.,* 88, 7, 1994.
20. Beavis W. D., Smith, O. S., Grant, D., and Fincher, R., Identification of quantitative trait loci using a small sample of topcrossed and F4 progeny from maize, *Crop Sci.,* 34, 882, 1994.
21. Schon, C. C., Melchinger, A. E., Boppenmaier, J., Brunklaus-Jung, E., Herrmann, R. G., and Seitzer, J. F., RFLP mapping in maize: quantitative trait loci affecting testcross performance of elite European flint lines, *Crop Sci.,* 34, 378, 1994.

22. Frankel, W. N., Taking stock of complex trait genetics in mice, *Trends Genet.,* 12, 471, 1995.
23. Covarrubius-Prieto, J., Hallauer, A. R., and Lamkey, K. R., Intermating F2 populations of maize, *Genetika,* 21, 111, 1989.
23a. Beavis, W. D., Hallauer, and Lee, Unpublished data.
24. Beavis, W. D. and Keim, P., Identification of QTL that are affected by environment. In *Genotype by Environment Interactions: New Perspectives.* Kang, M. S. and Gauch, H., Eds., CRC Press, Boca Raton, FL, pp. 123, 1995.
25. Lander, E. S. and Botstein, D., Mapping Mendelian factors underlying quantitative traits using RFLP linkage maps, *Genetics,* 121, 185, 1989.
26. Churchill, G. A. and Doerge, R. W., Permutation tests for multiple loci affecting a quantitative character, *Genetics,* 138, 963, 1994.
27. Doerge, R. W. and Churchill, G. A., Empirical threshold values for quantitative trait mapping, *Genetics,* 142, 285, 1996.
28. Thoday, J. M., Location of polygenes, *Nature,* 191, 368, 1961.
29. Soller, M., Brody, T., and Genizi, A., On the power of experimental design for the detection of linkage between marker loci and quantitative loci in crosses between inbred lines, *Theor. Appl. Genet.,* 47, 35, 1976.
30. Cowen, N. M., Multiple linear regression analysis of RFLP data sets used in mapping QTLs, in *Development and Application of Molecular Markers to Problems in Plant Genetics.* Helentjaris, T. and Burr, B., Eds., Cold Spring Harbor Laboratory, Cold Spring Harbor, NY, 1989, 113.
31. Romero-Severson, J., Lotzer, J., Brown, C., and Murray, M., The use of RFLPs for the analysis of quantitative trait loci in maize. In *Development and Application of Molecular Markers to Problems in Plant Genetics.* Helentjaris, T. and Burr, B., Eds., Cold Spring Harbor Laboratory, Cold Spring Harbor, NY, 1989, 97.
32. Moreno-Gonzales, J., Genetic models to estimate additive and non-additive effects of marker-associated QTL using multiple regression techniques, *Theor. Appl. Genet.,* 85, 435, 1992.
33. Haley, C. S. and Knott, S. A., A simple method for mapping quantitative trait loci in line crosses using flanking markers, *Heredity,* 69, 315, 1992.
34. Jansen, R. C., A general mixture model for mapping quantitative trait loci using molecular markers, *Theor. Appl. Genet.,* 85, 252, 1992.
35. Knapp, S. J., Bridges, W. C., and Liu, B. H., Mapping quantitative trait loci using nonsimultaneous and simultaneous estimators and hypothesis tests, in *Plant Genomes: Methods for Genetic and Physical Mapping,* Beckman, J. S. and Osborn, T. S., Eds., Kluwer, Dordrecht, The Netherlands, 1992, 209.
36. Jansen, R. C., Interval mapping of multiple quantitative trait loci, *Genetics,* 135, 205, 1993.
37. Jansen, R. C., Controlling the type I and type II errors in mapping quantitative trait loci, *Genetics,* 138, 871, 1994.
38. Rodolphe, F. and Lefort, M., A multi-marker model for detecting chromosomal segments displaying QTL activity, *Genetics,* 134, 1277, 1993.
39. Zeng, Z.-B., Theoretical basis of precision mapping of quantitative trait loci, *Proc. Natl. Acad. Sci. U.S.A.,* 90, 10972, 1993.
40. Zeng, Z.-B., Precision mapping of quantitative trait loci, *Genetics,* 136, 1457, 1994.
41. McMillan, I. and Robertson, A., The power of methods for the detection of major genes affecting quantitative characters, *Heredity,* 32, 349, 1974.
42. Fraser, A. and Burnell, D., *Computer Models in Genetics,* McGraw-Hill, NY, 1970.
43. Tinker, N. A. and Mather, D. E., GREGOR: software for genetic simulation, *Heredity,* 84, 237, 1992.
44. Mackinnon, M. J., van der Beek, S., and Kinghorn, B. P., Use of deterministic sampling for exploring likelihoods in linkage analysis for quantitative traits, *Theor. Appl. Genet.,* 92, 130, 1996.
45. Darvasi, A., Weinreb, A., Minke, V., Weller, J. I., and Soller, M., Detecting marker-QTL linkage and estimating QTL gene effect and map location using a saturated genetic map, *Genetics,* 134, 943, 1993.
46. Rebai, A., Goffinet, B., and Mangin, B., Comparing power of different methods for QTL detection, *Biometrics,* 51, 87 1995.
47. Utz, H. F. and Melchinger, A. E., Comparison of different approaches to interval mapping of quantitative trait loci, in *9th Meeting Eucarpia Section on Biometrics in Plant Breeding,* Waginen, The Netherlands, 1994, 195.

48. Korol, A. B., Ronin, Y. I., Tadmore, Y., Bar-Zur, A., Kirzhner, V. M., and Nevo, E., Estimating variance effect of QTL: an important prospect to increase the resolution power of interval mapping, *Genet. Res. Camb.*, 67, 187, 1996.
49. Cockerham, C. and Zeng, Z.-B., Design III with marker loci, *Genetics,* 143, 1437–1456, 1996.
50. Cowen, N. M., The use of replicated progenies in marker-based mapping of QTLs, *Theor. Appl. Genet.*, 75, 857, 1988.
51. Soller, M. and Beckman, J. S., Marker-based mapping of quantitative trait loci using replicated progenies, *Theor. Appl. Genet.*, 80, 205, 1990.
52. Knapp, S. J. and Bridges, W. C., Using molecular markers to estimate quantitative trait locus parameters: power and genetic variances for unreplicated and replicated progeny, *Genetics,* 126, 769, 1990.
53. Lincoln S. E. and Lander, E. S., Mapping genes controlling quantitative traits using MAPMAKER/QTL, *Tech. Report of Whitehead Inst. for Biomed. Res.*, Cambridge, MA, 1990.
54. Utz, H. F. and Melchinger, A. E., PLABQTL: a computer program to map QTL. Institut für Pflanzenzuechtung. Saatgutforschung und Populationsgenetik, Universität Hohenheim, Stuttgart, Germany, 1995.
55. Basten, C. J., Weir, B. S., Zeng, Z.-B., QTL cartographer: a reference manual and tutuorial for QTL mapping, Program in Statistical Genetics, North Carolina State Univ., Raleigh, NC, 1996.
56. Beavis, W. D., The power and deceit of QTL experiments: lessons from comparative QTL studies, in *Proc. Forty-ninth Annu. Corn and Sorghum Industry Research Conf.,* ASTA, Washington, D.C., 1994, 250.
57. van Ooijen, J. W., Accuracy of mapping quantitative trait loci in autogamous species, *Theor. Appl. Genet.,* 84, 803, 1992.
58. Carbonell, E. A., Asins, M. J., Baselga, M., Balansard, E., and Gerig, T. M., Power studies in the estimation of genetic parameters and the localization of quantitative trait loci for backcross and doubled haploid populations, *Theor. Appl. Genet.,* 86, 411, 1993.
59. Cheverud, J. M., Routman, E. J., Duarte, F. A. M., van Swinderen, B., Cothran, K., and Perel, C., Quantitative trait loci for murine growth, *Genetics,* 142, 1305, 1996.
60. Stuber, C. W., Successes in the use of molecular markers for yield enhancement in corn, in *Proc. Forty-ninth Annu. Corn and Sorghum Industry Research Conf.,* ASTA, Washington, D.C., 1994, 232.
61. Lande, R. and Thompson, R., Efficiency of marker-assisted selection in the improvement of quantitative traits, *Genetics,* 124, 743, 1990.
62. Page, N., A computer simlulation evaluation of the utility of marker-assisted selection, unpublished Ph.D. Diss., Dept. of Statistics, Univ. Minnesota, St. Paul MN, 1991.
63. Zhang, W. and Smith, C., Simulation of marker assisted selection utilizing linkage disequilibrium: the effects of sevearl additional factors, *Theor. Appl. Genet.,* 86, 492, 1993.
64. de Koning, G. J. and Weller, J. I., Efficiency of direct selection on quantitative trait loci for a two-trait breeding objective, *Theor. Appl. Genet.,* 88, 669, 1994.
65. Gimelfarb, A. and Lande, R., Simulation of marker assisted selection in hybrid populations, *Genet. Res. Camb.,* 63, 39,1994.
66. Tanksley, S. D. and Nelson, J. C., Advanced backcross QTL analysis: a method for the simultaneous discovery and transfer of valuable QTLs from unadapted germplasm into elite breeding lines, *Theor. Appl. Genet.,* 92, 191, 1996.
67. Zehr, B. E., Dudley, J. W., Chojecki, J., Some practical considerations for using RFLP markers to aid in selection during inbreeding in maize, *Theor. Appl. Genet.,* 84, 704, 1992.
68. Hill, A. P., Quantitative linkage: a statistical procedure for its detection and estimation, *Ann. Hum. Genet. Lond.,* 38, 439, 1975.
69. Rebai, A. and Goffinet, B., Power of tests for QTL detection using replicated progenies derived from a diallel cross, *Theor. Appl. Genet.,* 86, 1014, 1993.
70. Fu, Y. B. and Ritland, K., On estimating the linkage of marker genes to viability genes controlling inbreeding depression, *Theor. Appl. Genet.,* 88, 925, 1994.
71. Haseman, J. K. and Elston, R. C., The investigation of linkage between a quantitative trait and a marker locus, *Beh. Genet.,* 2, 3, 1972.
72. Goldgar, D. E., Multipoint analysis of human quantitative genetic variation, *Am. J. Hum. Genet.*, 47, 957, 1990.
73. Fernando, R. L. and Grossman, M., Marker assisted selection using best linear unbiased prediction, *Gen. Sel. Evol.*, 21, 467, 1989.

74. Knott, S. A. and Haley, C. S., Maximum likelihood mapping of quantitative trait loci using full-sib families, *Genetics,* 132, 1211, 1992.
75. van Arendonk, J. A. M., Tier, B., and Kinghorn, B. P., Use of multiple genetic markers in prediction of breeding values, *Genetics,* 137, 319, 1994.
76. Xu, S. and Atchley, W. R., A random model approach to interval mapping of quantitative trait loci, *Genetics,* 141, 1189, 1995.

# 11 High-Resolution Mapping of QTLs

*Andrew H. Paterson*

## CONTENTS



## 11.1 INTRODUCTION

A central issue in genetic mapping is "precision" — defined eloquently by Beavis (Chapter 10) as "a measure of the dispersion of repeated independent estimates of genomic positions … often reported by inverse measures such as standard errors or confidence intervals."

In mapping of discrete traits, conferred by single genes for which phenotype is a perfect indicator of genotype, precision of genetic mapping is based solely on the number of recombinant gametes assayed. To achieve a predetermined level of precision, one assays a number of individuals (gametes) which is readily calculated by the binomial probability distribution function.

However, in QTL mapping, achievement of a high degree of precision is more complex. A QTL is a statistical description of the phenotypic effects of a genetic locus or closely linked group of loci in a defined (set of) population(s) and environment(s). The precision of this description is affected by many factors, including genetic variation at other loci in the study population(s), nonlinear interactions between multiple genetic loci, measurement error, and vagaries of the study environment(s).

0-8493-7686-6/98/$0.00+$.50

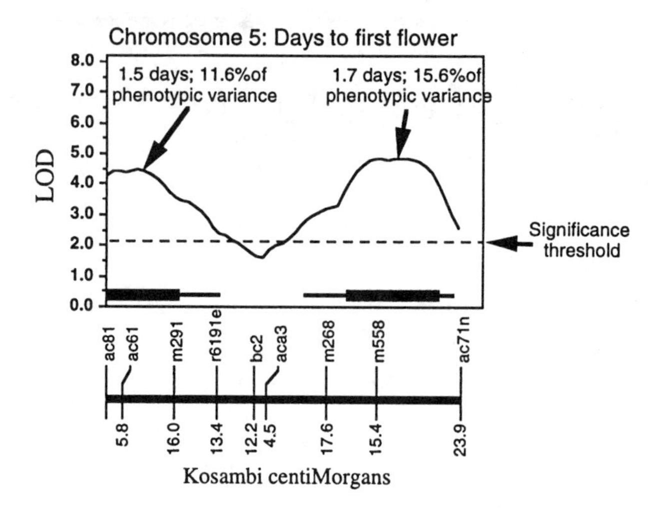


**FIGURE 11.1** Likelihood intervals for QTLs affecting flowering time, on *Arabidopsis* chromosome 5. These two QTLs were mapped in a cross between the early flowering *Arabidopsis* ecotype Wassilewskija (WS), and the late-flowering ecotype Hannover/Munden (HM), which flowered about 8 d apart in the study environment. "Likelihood intervals" for the locations of the QTLs are below the graph, represented as bars (90% likelihood) and whiskers (99% likelihood). At the bottom, the genetic map of chromosome 5 for this population is presented, with distances between markers in centiMorgans.

### 11.1.1 An Example of the Importance of Precision in QTL Mapping

To exemplify the dilemma, consider an example presented in Figure 11.1. Two different QTLs explaining 15.6 and 11.6% of the phenotypic variance for flowering date in a study population of 198 individuals, were mapped to the lower part of *Arabidopsis* chromosome 5 by interval analysis.[1] Each QTL can be asserted with a likelihood of 90% to fall somewhere within a 20 to 30 cM interval between marker loci indicated in the figure. In total, the *Arabidopsis* genome is estimated to span approximately 500 cM, so each interval represents about 5% of the genome. Further, the number of genes in the *Arabidopsis* genome is conservatively estimated as 25,000.[2] Within the 90% likelihood interval of a QTL, an average of 1250 other genes are likely to be present.

The relatively low precision at which QTLs are mapped has many consequences, including:

1. Difficulty at distinguishing pleiotropic effects of a single gene from independent effects of individual genes which happen to be genetically linked.
2. Reduced gains from DNA marker-assisted selection, as a result of undesirable effects associated with genes closely linked to the target gene.
3. Difficulties in comparative analysis of mutations conferring common phenotypes in divergent taxa (cf. comparative QTL mapping[3]), especially between taxa in which chromosomal rearrangement has reduced synteny to small chromosomal segments,[4] shorter than the typical 20 to 30 cM length of a likelihood interval.
4. Difficulty in molecular cloning of the gene(s) underlying specific QTLs, especially in model taxa such as *Arabidopsis* in which the density of genes along the chromosomes is very high.

### 11.1.2 General Approaches to Improving the Precision of QTL Mapping

Three basic approaches have been described for substantially improving the precision of QTL mapping:

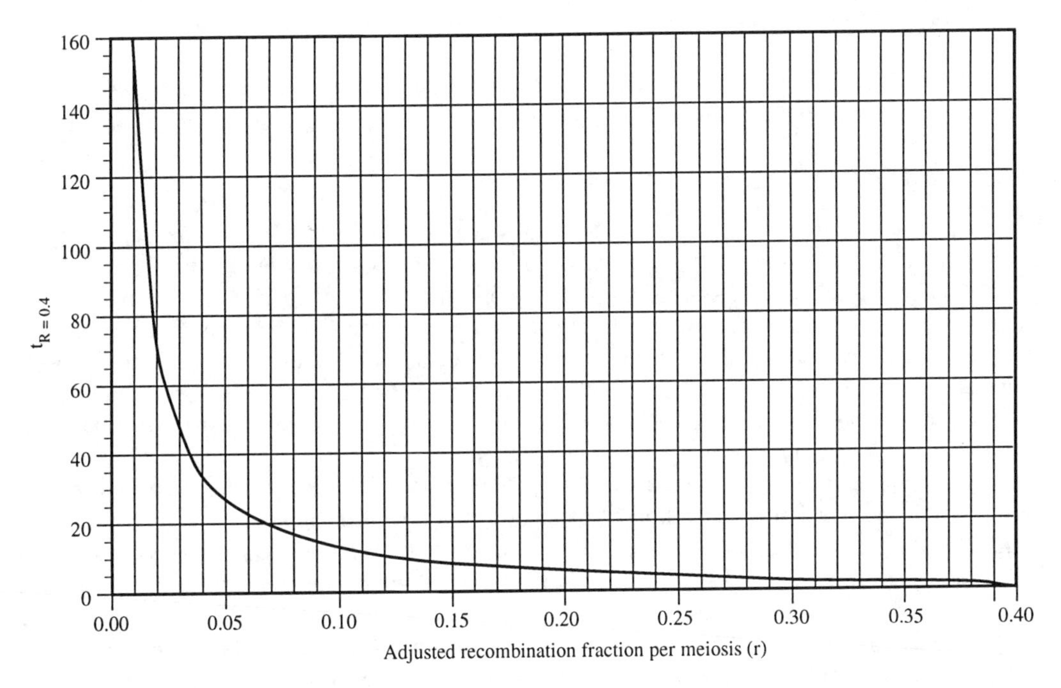


**FIGURE 11.2** Number of generations of random mating necessary to approach linkage equilibrium between loci starting from various genetic distances. In this example, derived from Liu et al.,[26] we start with an F2 population derived from a cross between homozygous lines, and calculate the number of generations of intermating necessary until genetic linkage can no longer be discerned in a reasonable population size. (For the purposes of the example, we assumed that a per-meiosis recombination fraction of 0.4 was the maximum discernible). For loci which show a recombination fraction per meiosis of 0.01, more than 150 generations of random mating are necessary to reach a recombination fraction of 0.4 (a distance at which linkage becomes difficult to discern from independent assortment, in populations of manageable size). Wright[13] has previously shown that for such loci to reach true random mating (recombination fraction of 0.5) may require as many as 1000 generations.

1. Mathematical approach — using models that attempt to simultaneously estimate locations and effects of multiple QTLs rather than independently estimate individual locations of single QTLs, might improve the precision of QTL mapping. This approach may enable one to extract more information from an existing data set at the cost of some computer time. However, this approach is not likely to be sufficient by itself to overcome any of the consequences listed above.
2. Recombinational approach — by reducing the level of linkage disequilibrium in the study population, one can change the relationship between the recombinational size of a QTL likelihood interval and its DNA content. This approach can be applied to many QTLs simultaneously, segregating in a common population. However, it requires additional generations of breeding (see Figure 11.2). While such a breeding approach may in some instances be adequate to address 1 through 3 of the above-raised consequences it will almost certainly fail to address consequence 4.
3. "Substitution mapping" — by screening large populations of progeny, one can determine the QTL genotype of individuals that contain recombination events near a QTL. This approach, essentially as described,[5] enables the investigator to delineate a QTL with a level of precision determined by the number of recombinants screened and the number of DNA markers available to distinguish between these recombinants.

## 11.2 IMPLEMENTATION OF HIGH-RESOLUTION QTL MAPPING

The following three sections will present a more detailed discussion of each of these two sources of information for high-resolution mapping of QTLs.

### 11.2.1 Mathematical Approaches

It is a well-established principle of statistics that residual variance is minimized, and sensitivity of a model thereby maximized, if all statistically significant variables are included in the model. Both classical inferences about QTLs based on "single-point" analysis and the early implementations of "interval mapping"[6] employed simplified statistical models, which tested the phenotypic variance attributable to a single locus against a combined "residual" term including both true "error" as well as genetic variance due to other loci in the genome. (Interval analysis differs from "single-point analysis" in that the single point under consideration is usually at a location between two markers, rather than at a location delineated by a single marker.)

In principle, early implementations of interval mapping acknowledged that additional QTLs might be resolved, or better resolution of existing QTLs obtained, by simultaneous mapping of multiple QTLs.[6] The first implementation of MAPMAKER-QTL included algorithms for simultaneous mapping of multiple QTLs, however, in the first test case, employment of these algorithms did not identify any additional QTLs.[7]

More recently, several authors have described alternative QTL mapping algorithms based on mapping QTLs by "multiple regression" approaches.[8-11] More details regarding several of these techniques are provided by Liu (Chapter 4).

It is not the objective of this chapter to recommend one QTL analysis method over another. Each of the proposed approaches are well described by leading investigators, are theoretically sound, and enjoy advocates. In many instances, adoption of a particular approach may be based on availability of nearby expertise, or historical reasons such as a prior record of publication using a particular approach. Perhaps the most important message to those seeking to extract additional data from existing or new data sets, is that use of one of the available multiple regression approaches may offer some advantages.

#### 11.2.1.1 Example of Information Gain Using the Mathematical Approach

An excellent example of the gains which might be realized by using the "mathematical approach", was found in an experiment to map genes associated with flowering time of sorghum.[12] In this experiment, a bimodal distribution of flowering times was found in a study population of 370 F2 individuals, and the presence of a single gene with a very large phenotypic effect was postulated. A "genome scan" using classical "interval analysis" revealed that a single genetic locus, *Ma-1*, accounted for about 86% of phenotypic variance in the study population. No other QTLs in the genome could be deemed significant.

A modified mathematical model, which removed the phenotypic effects of the *Ma-1* locus from the "error" term, revealed two additional QTLs that met stringent statistical thresholds. The importance of these QTLs was further verified by a completely independent experiment — specifically, by evaluating patterns of introgression in exotic genetic stocks which had been crossed with an elite, early flowering cultivar, then recurrently backcrossed to the exotic parent but with selection for early flowering. The selection experiment revealed that the chromosomal regions harboring each of the two new QTLs were, in fact, associated with introgression of chromatin from the early flowering parent.

While the two additional QTLs had relatively small effects, two independent approaches show them to be real and the "mathematical approach" enabled more information to be resolved from existing data.

### 11.2.2 The Recombinational Approach — Reduction of Linkage Disequilibrium between Markers and QTLs

Genetic mapping is based upon use of defined pedigrees, in which alleles at genetic marker loci can be used to predict the genotype of chromosomal regions between the marker loci. The size of a chromosomal segment for which DNA markers are an effective proxy is determined by the breeding history of the study population. Wright[13] describes the relationship between the number of generations of random mating and the length of chromosomal intervals over which alleles remain coupled, as illustrated in Figure 11.2.

Linkage disequilibrium, the strength of relationship between marker loci, and surrounding chromosomal regions, can be manipulated in experimental populations. Plant breeders have long recognized that "correlations between traits" can be reduced (when undesirable) by several generations of random intermating between individuals selected from a founder population segregating for particular attributes by maintaining heterozygosity (and therefore recombination).[14-21] In a similar manner, the practice of backcrossing to a cultivar in order to eliminate undesirable genes from exotic germplasm, reduces the average length of the introgressed chromosome segment by maintaining it in a heterozygous condition such that recombination continues to be effective.[22]

Curiously, the ease at which genetic mapping populations in plants can be made has imposed an inadvertant constraint on the map resolution achieved by such populations. Many plants are amenable to self-pollination, a breeding scheme by which homozygosity is achieved at a maximal rate, affording little opportunity for closely linked loci to be separated by recombination. The widespread use in plant genetics of F2 populations derived by selfing F1 hybrids, has the consequence that genetic mapping populations are in strong linkage disequilibrium. This is an advantage in the early stages of assembling a genetic map, when one seeks to quickly identify linkages between widely scattered marker loci. However, as a map becomes densely populated with markers, strong linkage disequilibrium limits the number of recombination events available to resolve the order of closely linked markers.

Plant genetic mapping can benefit from approaches used in mammalian genetics. Genetic mapping in the mouse has traditionally relied heavily upon development of homozygous (and therefore reproducible) populations, by recurrent brother-sister mating, a form of inbreeding which eliminates heterozygosity at exactly half the rate of selfing. Because heterozygosity persists, additional opportunities for recombination reduce the extent of disequilibrium between genetic markers and genes. Recombinant inbred populations derived by sib mating offer a maximum of fourfold expansion of the observed frequency of recombinants, R, when the per-meiosis recombination rate, r, approaches 0.[23,24]

There is some confusion in the plant-genetics literature about the extent to which plant genetic mapping has, in fact, achieved these potential benefits. The basis for confusion arises from the fact that recombinant inbred (RI) populations of plants are usually made by selfing rather than by sib mating. RI populations generated by recurrent self-pollination (i.e., single seed descent[25]) lose 50% of remaining heterozygosity each generation yielding a maximum of twofold expansion of R when r approaches 0.[26] RI-selfed individuals are thus derived from two putatively identical copies of a single gamete, which has been through the equivalent of two cycles of recombination. By contrast, F2 individuals are derived from two different gametes which have each been through only one cycle of recombination — providing equivalent information for resolving close linkages. The loss of heterozygosity during selfing is so rapid that gains of information from new recombination are exactly canceled out. By contrast, mammalian RI individuals are derived from two putatively-identical copies of a single gamete that has been through the equivalent of four cycles of recombination, with the slower loss of heterozygosity affording twice as much information as RI-selfed individuals for resolving close linkages. Consequently, suggestions that RI populations of plants "permit higher mapping resolution for short linkage distances" than F2 populations,[27-29] are often misleading.

Recently, plant populations have been developed which combine the advantages of plant and mammalian genetic mapping strategies. By starting with inbred strains, making a heterozygous F1, then selfing to produce an F2 population, one can manipulate linkage disequilibrium to a predetermined degree by randomly intermating among segregants for an appropriate number of generations. The first report of the method showed enhanced recombination along one chromosome in an intermated population of maize.[30] A later report provided a detailed comparison of genome-wide recombination in F2, RI, and intermated populations of *Arabidopsis*, all derived from a common F2 population.[26] By applying the constraint that selfing (which is expected at a low rate in a "random-mating" population) is not allowed, a reasonably large population (>100 individuals) can be intermated for many generations with virtually linear gains in recombinational information.[26,31] In many plant systems, genetic male sterility, or chemical emasculants, can be used to simplify crosses.

By applying several generations of random intermating, followed by selfing to homozygosity, one can derive an "intermated RI (IRI) population" which achieves the resolution of mammalian RI populations but in fewer generations. While the length of time needed to develop such populations is a constraint, the large potential improvement in "recombinational information per individual"[26,31] warrants development of IRI populations to serve as long-term resources for genetic mapping in most major crop plants.

In animal populations, while costs usually limit use of multigeneration breeding schemes to experimental systems, long-term accumulation of pedigree information sometimes offers a basis for fine-resolution associations between genetic markers and nearby QTLs. Such analysis is considered in detail by Taylor and Rocha (Chapter 7).

#### 11.2.2.1 Examples of Information Gained by the Recombinational Approach

The "recombinational approach" is relatively new, and to date, this author is not aware of specific examples in which this approach has been used to improve the resolution of QTLs. Primary mapping of many QTLs in the mouse has benefitted *a priori* from this approach. In plants, the long time needed to develop suitable populations is often invested in other approaches such as substitution mapping, however, as intermated plant populations come into existence it seems likely that examples of high-resolution mapping of QTLs will appear.

### 11.2.3 Substitution Mapping

While experimental manipulation of linkage disequilibrium offers some additional information regarding the precise location(s) of gene(s) responsible for QTLs, a fundamentally different experiment facilitates precision mapping of QTLs. This approach, deemed substitution mapping,[5] utilizes progeny testing to determine the QTL genotype of individual recombinants. By associating phenotypic variation with differences in the genomic composition of recombinants, one can map individual QTLs to a resolution which is equivalent to that of discrete genes. Substitution mapping is best applied to QTLs one at a time and has prerequisites of a high-resolution genetic map, a QTL likelihood interval established by prior mapping, and availability of closely spaced recombinants in the likelihood interval.

Substitution mapping works essentially as follows and as illustrated in Figure 11.3.

1. A QTL likelihood interval is delineated using techniques described above. Adjuncts such as multiple-QTL approaches to data analysis (see Chapters 10 and 4) and/or use of intermated populations (above), can provide some gains in resolution.
2. Identification of recombinants in the QTL likelihood interval, and QTL fine mapping. Delineation of the QTL to as small an interval as possible will facilitate gene isolation, by minimizing the number of candidate transcripts which must be evaluated. Further, a large number of closely spaced recombinants may shed light on the possibility, often

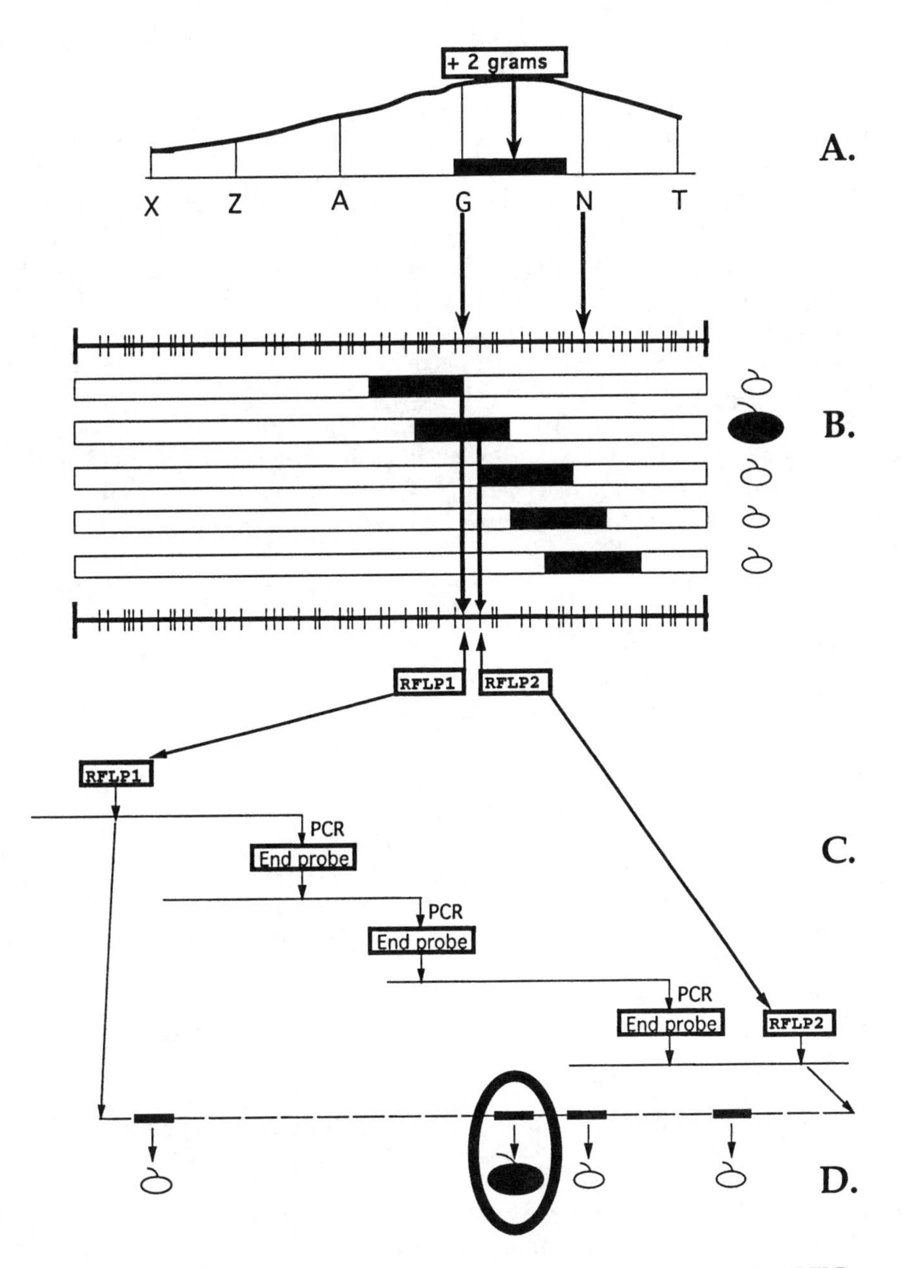


**FIGURE 11.3** A schematic for positional cloning of QTLs, using substitution mapping. RFLP = restriction fragment length polymorphism. (A) A QTL likelihood interval is delineated using techniques described above. (B) Identification of recombinants in the QTL likelihood interval, and QTL fine mapping. (C) Chromosome walking. (D) Candidate gene isolation, and mutant complementation. See text for additional details of each step.

suggested, that QTLs represent clusters of genes which cumulatively (rather than individually) cause an observed phenotype (see Reference 32).

Several approaches have been described which can be employed to make the search for recombinants more efficient. PCR-based detection of markers (see Reference 33) in conjunction with microscale DNA extraction techniques (see References 34 and 35), can enable one worker to quickly and efficiently assay large populations for prospective recombinants. Further, PCR assays for several different loci might be multiplexed, and applied to pools of individuals[36] from different populations that carry different recombinant chromosome segments.

Ideally, recombinants will be identified from crosses between genetic stocks which are near isogenic for a small chromosome segment including a target QTL, thereby reducing the contribution of extraneous genetic variation to the error term for testing significance of the QTL. This is not a necessity but is a prudent precaution, especially for annual plants in which near-isogenic stocks can be developed rapidly using DNA marker-assisted selection. If one seeks to target multiple QTL regions, then near-isogenic stocks carrying different chromosome segments might be crossed to each other and recombinants simultaneously identified for two different chromosomal regions.

To characterize overlaps among recombinants in the QTL likelihood interval, DNA markers can often be drawn directly from a preexisting high-density map. In the absense of an adequate number of preexisting markers, the target region might be enriched for DNA markers by a number of techniques (see Chapter 2, this volume). One may be willing to accept a relatively low resolution of discrimination between recombinant stocks initially and obtain additional discrimination using subclones from megabase DNA elements during the course of a chromosome walk.

Phenotypic evaluation of numerous progeny from each recombinant is necessary to accurately determine the QTL genotype of each recombinant. The exact nature of progeny testing can be varied to accommodate the breeding system of the crop, and/or the gene action of the target QTL — in principle, progeny could be backcross/testcross, selfs, or even half-sib families derived from each individual recombinant.

A prudent approach to determine the minimum number of progeny which should be evaluated would be to use statistical power functions[37] including the estimated allele effect at the QTL (from the prior QTL likelihood interval mapping), together with an estimate of the magnitude of extraneous variation based on studying the recurrent parent in a similar test environment.

In principle, one could even genotype individual progeny to confirm co-segregation of marker and phenotype — however, if progeny testing is based on the near-isogenic line structure suggested above, this is probably unnecessary. If the target phenotype is expressed early in plant development, it may be reasonable to confirm the marker-phenotype association in key families that appear to contain recombination events near the target gene. In many cases, the phenotype will involve a reproductive organ directly, and be assayed early in plant development (in some cases, it cannot even be assayed in real-time). A compromise solution might be to retain small samples of tissue from individual plants in each family, and genotype selected key families after preliminary analysis of the phenotypes has been completed.

#### 11.2.3.1 Examples of New Information Gained by Substitution-Mapping of QTLs

Substitution mapping has been applied in several instances to shed new light on important questions. The technique was described in 1990 using, as an example, several introgressed chromosome segments of tomato, which conferred both desirable attributes and undesirable effects. In at least one instance, it was clear from substitution mapping that a reduction of fruit yield was caused by a gene independent from the nearby desirable QTL which increased soluble solids concentration of the tomato fruit.[5]

In a second case, a single small region of one chromosome has been transferred to maize from its wild relative teosinte, conferring a mutation which envelops the maize kernel with an indurate (hardened) glume.[38] This proved that a nearby candidate gene, "tunicate" (*Tu-1*) was *not* the gene which conditioned this phenotype. The new gene was designated *Tga-1* for "teosinte glume architecture." Ongoing experiments involve further dissection of this genomic region, and four additional genomic regions which control a suite of key differences between maize and teosinte, to resolve whether the manifold effects of these genomic regions are due to individual major genes with pleiotropic effects, or linked groups of genes with independent effects.[39]

Substitution mapping has been applied to several maize chromosome segments which are associated with heterotic increases in grain yield, and in at least one case heterosis has been deemed a result of dominant alleles at two different closely linked loci which were in repulsion phase in the homozygous parental stocks (see Chapter 14, this volume) — providing a concrete example of the classical proposal that heterosis might often be a result of multiple genetic loci, rather than a single locus at which a true heterozygote advantage was conferred.

Finally, molecular dissection of a region of rat chromosome 10 thought to carry a major hypertension gene, has revealed a complex of at least two genes. The use of random marker genetic screening methods initially showed that a 35 cM region of chromosome 10 of the Heidelberg strains of the stroke-prone hypertensive rat ($SHRSP_{HD}$) contained a major quantitative trait locus for blood pressure.[40,41] Subsequent, more detailed analysis of recombinant stocks in this chromosomal region demonstrated the presence of two QTLs, one associated with differences in basal blood pressure, and a second with blood pressure levels after exposure to excess dietary NaCl.[42]

## 11.3 SUMMARY

The issue of precision in genetic mapping is of growing importance, as technological advances now permit quantitative geneticists to ask questions about individual genetic loci affecting complex traits. Techniques now exist to "bridge" the gap in resolution between genetic mapping, and physical analysis of megabase DNA clones (see *Epilogue*). The increasing density of genetic maps, both directly and through comparative alignment of the chromosomes of different taxa, together with continuing improvements in megabase DNA cloning technology, suggest that high-precision genetic analysis of complex traits will become ever more routine. High-precision genetic mapping is likely to contribute substantially to basic objectives such as positional cloning of important genes and evaluating gene organization in divergent taxa, as well as to applied objectives such as DNA marker-assisted improvement of plants and animals.

## REFERENCES

1. Kowalski, S. D., Lan, T.-H., Feldmann, K. A., and Paterson, A. H., QTLs affecting flowering time in *Arabidopsis thaliana, Mol. Gen. Genet.*, 245, 548, 1994.
2. Meyerowitz, E. M., Structure and organization of the *Arabidopsis* nuclear genome, in *Arabidopsis,* Meyerowitz, E. and Somerville, C., Eds., Cold Spring Harbor Laboratory Press, Cold Spring Harbor, NY, 1994.
3. Paterson, A. H., Lin, Y. R., Li, Z., Schertz, K. F., Doebley, J. F., Pinson, S. R. M., Liu, S. C., Stansel, J. W., and Irvine, J. E., Convergent domestication of cereal crops by independent mutations at corresponding genetic loci, *Science,* 269, 1714, 1995.
4. Paterson, A. H., Lan, T.-H., Reischmann, K. P., Chang, C., Lin, Y.-R., Liu, S.-C., Burow, M. D., Kowalski, S. P., Katsar, C. S., DelMonte, T. A., Feldmann, K. A., Schertz, K. F., and Wendel, J. F., Toward a unified map of higher plant chromosomes, transcending the monocot-dicot divergence, *Nat. Genet.,* in press.
5. Paterson, A. H., Deverna, J. W., Lanini, B., and Tanksley, S. D., Fine mapping of quantitative trait loci using selected overlapping recombinant chromosomes in an interspecies cross of tomato, *Genetics,* 124, 735, 1990.
6. Lander, E. S. and Botstein, D., Mapping Mendelian factors underlying quantitative traits using RFLP linkage maps, *Genetics*, 121, 185, 1989; and Corrigendum, *Genetics,* 136, 705, 1994.
7. Paterson, A. H., Lander, E. S., Hewitt, J. D., Peterson, S., Lincoln, S. E., and Tanksley, S. D., Resolution of quantitative traits into Mendelian factors by using a complete map of restriction fragment length polymorphisms, *Nature*, 335, 721, 1988.
8. Knapp, S. J., Using molecular markers to map multiple quantitative trait loci: models for backcross, recombinant inbred, and doubled haploid progeny, *Theor. Appl. Genet.,* 81, 333, 1991.

9. Haley, C.S. and Knott, S. A., A simple method for mapping quantitative trait loci in line crosses using flanking markers, *Heredity,* 69, 315, 1992.
10. Jansen, R., Interval mapping of multiple quantitative trait loci, *Genetics,* 135, 205, 1993.
11. Zeng, Z.-B., Theoretical basis for separation of multiple linked gene effects in mapping quantitative trait loci, *Proc. Natl. Acad. Sci. U.S.A.,* 90, 10,972, 1993.
12. Lin, Y. R., Schertz, K. F., and Paterson, A. H., Comparative mapping of QTLs affecting plant height and flowering time in the Poaceae, in reference to an interspecific *Sorghum* population, *Genetics,* 141, 391, 1995.
13. Wright, S., *Genetics and the Evolution of Populations,* Chicago University Press, Chicago, 1968.
14. Hanson, W. D., Theoretical distribution of the initial linkage block lengths intact in the gametes of a population intermated for n generations, *Genetics,* 44, 839, 1959.
15. Hanson, W. D., The breakup of initial linkage blocks under selected mating systems, *Genetics,* 44, 857, 1959.
16. Miller, P. A. and Rawlings, J. O., Breakup of initial linkage blocks through intermating in a cotton breeding population, *Crop Sci.,* 7, 199, 1967.
17. Fredericksen, L. J. and Kronstad, W. E., A comparison of intermating and selfing following selection for heading date in two diverse winter wheat crosses, *Crop Sci.,* 25, 555, 1985.
18. Kwolek, T. F., Atkins, R. E., and Smith, O. S., Comparisons of agronomic characteristics in C0 and C4 of IAP3BR(M) random-mating grain sorghum population, *Crop Sci.,* 26, 1127, 1986.
19. Wells, W. C. and Kofoid, K. D., Selection indices to improve an intermating population of spring wheat, *Crop Sci.,* 26, 1104, 1986.
20. Tyagi, A. P., Correlation studies on yield and fiber traits in upland cotton (*Gossypium hirsutum* L.), *Theor. Appl. Genet.,* 74, 280, 1987.
21. Fatmi, A., Wagner, D. B., and Pfeiffer, T. W., Intermating schemes used to synthesize a population are equal in genetic consequences, *Crop Sci.,* 32, 89, 1992.
22. Hanson, W. D., Early generation analysis of lengths of heterozygous chromosome segments around a locus held heterozygous with backcrossing or selfing, *Genetics,* 44, 833, 1959.
23. Haldane, J. B. S. and Waddington, C. H., Inbreeding and linkage, *Genetics,* 16, 357, 1931.
24. Taylor, B., Recombinant inbred strains: use in gene mapping, in *Origins of Inbred Mice,* Morse, H., Ed., Academic Press, New York, 1978, 423–438.
25. Brim, C. A., A modified pedigree method of selection in soybeans, *Crop Sci.,* 6, 220, 1966.
26. Liu, S., Kowalski, S. P., Lan, T., Feldmann, K. A., and Paterson, A. H., Genome-wide high resolution mapping by recurrent intermating using *Arabidopsis thaliana* as a model, *Genetics,* 142, 247, 1996.
27. Burr, B., Burr, F. A., Thompson, K. H., Albertson, M. C., and Stuber, C. W., Gene mapping with recombinant inbreds in maize, *Genetics,* 118, 519, 1988.
28. Burr, B. and Burr, F. A., Recombinant inbreds for molecular mapping in maize: theoretical and practical considerations, *Trends Genet.,* 7, 55, 1991.
29. Burr, B., Burr, F. A., and Matz, E. C., Mapping genes with recombinant inbreds, in *The Maize Handbook,* Freeling, M. and Walbot, V., Eds., Springer-Verlag, New York, 1993, 249–254.
30. Beavis, W. D., Lee, M., Hallauer, A. R., Owens, T., Katt, M., and Blair, D., The influence of random mating on recombination among RFLP loci, *Maize Genet. Coop. Newsl.,* 66, 52–53, 1992, (nonrefereed newsletter).
31. Darvasi, A. and Soller, M., Advanced intercross lines, an experimental population for fine genetic mapping, *Genetics,* 141, 1199, 1995.
32. Michelmore, R. W. and Shaw, D., Character dissection, *Nature,* 335, 698, 1988.
33. Konieczny, A. and Ausubel, F. M., A procedure for mapping *Arabidopsis* mutations using co-dominant ecotype-specific PCR-based markers, *Plant J.,* 4, 403, 1993.
34. Wang, G., Wing, R., and Paterson, A. H., PCR amplification of DNA extracted from single seeds, facilitating DNA-marker assisted selection, *Nucl. Acids Res.,* 21, 2527, 1993.
35. Klimyuk, V., Carroll, B. J., Thomas, C. M., and Jones, J. D. G., Alkali treatment for rapid preparation of plant material for reliable PCR analysis, *Plant J.,* 3, 493, 1993.
36. Churchill, G. A., Giovannoni, J. J., and Tanksley, S. D., Pooled-sampling makes high-resolution mapping practical with DNA markers, *Proc. Natl. Acad. Sci. U.S.A.,* 90, 16, 1993.
37. Snedecor, G. W. and Cochran, W. G., Statistical Methods, 7th ed., Iowa State University Press, Ames, IA, 1980.

38. Dorweiler, J., Stec, A., Kermicle, J., and Doebley, J., Teosinte glume architecture. 1. A genetic locus controlling a key step in maize evolution, *Science,* 262, 233, 1993.
39. Doebley, J., Mapping the genes that made maize, *Trends Genet.*, 8, 302, 1992.
40. Hilbert, P., Lindpaintner, K., Beckmann, J. S., Serikawa, T., Soubrier, F., Dubay, C., Cartwright, P., De Gouyon, B., Julier, C., Takahasi, S., et al., Chromosomal mapping of two genetic loci associated with blood-pressure regulation in hereditary hypertensive rats, *Nature*, 353, 521, 1991.
41. Jacob, H. J., Lindpainter, K., Lincoln, S. E., Kusumi, K., Bunker, R. K., Mao, Y.-P., Ganten, D., Dzau, V. J., and Lander, E. S., Genetic mapping of a gene causing hypertension in the stroke-prone spontaneously hypertensive rat, *Cell,* 67, 213, 1991.
42. Kreutz, R., Hubner, N., James, M. R., Bihoreau, M., Gaugueir, D., Lathrop, G. M., Ganten, D., and Lindpainter, K., Dissection of a quantitative trait locus for genetic hypertension on rat chromosome 10, *Proc. Natl. Acad. Sci. U.S.A.,* 92, 8778, 1995.

# 12 Compilation and Distribution of Data on Complex Traits

*Douglas W. Bigwood*

## CONTENTS



## 12.1 INTRODUCTION

QTL data is among the most complex data existing in genetics. Complete reporting requires raw data, summary statistics, graphical representations, and a detailed explanation of experimental design and analysis. Data complexity, in itself, is not necessarily problematic. However, when combined with the fact that there is a lack of a consistency in quantitative trait loci (QTL) data reporting and terminology, information concerning complex traits is often difficult to utilize. One trend, however, is inescapable: data distribution will, in all likelihood, be predominately via the World Wide Web. The most important step anyone interested in QTL data can take is to get connected to the Internet and become familiar with the Web.

This chapter will begin with sections related to the reporting and formatting of QTL information, then present a survey of currently available information, identify means of finding new QTL information on the World Wide Web, and end with a discussion of future software development which will enhance the utility of QTL data.

## 12.2 REPORTING AND FORMATTING QTL DATA

In order for QTL data to be useful, it is important that careful attention is paid as to how the data is recorded, formatted, and ultimately presented to a user. Perhaps the most definitive article on the subject is reporting and accessing QTL information in USDA's Maize Genome Database by Byrne et al.[1] In the article, the authors present three key questions that should be answerable by querying a well-designed database. These are:

0-8493-7686-6/98/$0.00+$.50

1. Do QTL identified for a given trait in one population or environment correspond to those detected in other populations or environments?
2. For a specified chromosome segment in which a QTL was detected, what other traits have been associated with the same region, either through QTL studies, classical mapping of mutant phenotypes, or restricted fragment length polymorphism (RFLP) mapping of cDNAs? Responses to this type of query may offer hints of allelism, pleiotropic effects, or closely linked traits which might affect a marker-assisted selection strategy.
3. Given the increasing evidence of synteny and colinearity among grass family genomes, do QTL locations identified in one species correspond to QTL or other types of loci detected in corresponding regions of other species?

Currently, the paucity of QTL data makes answering the first question difficult, if not impossible, in many cases. It will remain difficult unless researchers report data in a consistent manner or the data is heavily curated. If this is done, then devising a scheme to automatically identify such relationships becomes feasible. As reported in the article, it is not necessary to force conformation to some mandatory schema, but to be aware of key information when reporting or reviewing QTL data.

Answering the second question requires an investigator to sift through a large amount of data to pull out significant relations unless a database is well designed and has an advanced query interface. Unfortunately, this is difficult, but this paper presents an excellent, detailed treatment of the subject which is too extensive to present here. One likely solution will be the development of complex displays which integrate many types of data and present potential relationships graphically. The displays will need to be flexible enough to allow the user to filter data in many different ways.

The last question is a particularly hard one to address because of the inconsistencies in data reporting and terminology. The terminology issue is lessened somewhat within a taxonomic group such as grasses, but remains problematic. Often, extensive human interpretation is necessary to identify, for example, which phenotypic descriptions identify things that are the same. Resolution of semantic issues will be necessary in order to automate comparative analysis among species. One potential solution is presented later in the section on future developments.

### 12.2.1 *Journal of Quantitative Trait Loci*

Possibly the best single source of new QTL information is the *Journal of Quantitative Trait Loci* (JQTL) which began publication in 1995. JQTL is sponsored by the Crop Science Society of America and is only available through the World Wide Web at the AGIS server (http://probe.nalusda.gov:8000/otherdocs/jqtl/index.html). Table 12.1 provides a representative list of some recent papers published in JQTL.

There are several advantages of electronic publication vis-a-vis printed publication. Many of these are exploited to their fullest. First, cost is significantly reduced due to the elimination of printing and distribution. This results in a reduced need to shorten total pages and allows the inclusion of tables, figures, and supporting data that might otherwise be eliminated. Several JQTL papers even include raw data. Second, when publishing via the World Wide Web, hypertext links greatly increase the facility with which a reader can retrieve and view related information. JQTL papers contain hypertext links to the Agricultural Genome Information System (AGIS) database objects (e.g., germplasm and locus information), AGRICOLA bibliographic records (including abstracts when present), and other papers when they exist online. Third, text can be separated from figures and tables, but these are instantly accessible with a single mouse click. They can even be brought up in a separate window and made available without incessant page flipping. Fourth, electronic text is easily indexed. JQTL articles are searchable by keyword. Figure 12.1 shows the anatomy of a JQTL document[2] and gives an example of the features mentioned above (at least as much as possible on a printed page).

**TABLE 12.1**
***Journal of Quantitative Trait Loci* — Table of Contents**

1. PLABQTL: A Program for Composite Interval Mapping of QTL
   H.F. Utz and A.E. Melchinger
2. Multiple Disease Resistance Loci and Their Relationship to Agronomic and Quality Loci in a Spring Barley Population
   Patrick Hayes, Doris Prehn, Hugo Vivar, Tom Blake, Andre Comeau, Isabelle Henry, Mareike Johnston, Berne Jones, Brian Steffenson, and C.A. St. Pierre
3. Chromosomal Regions Associated with Quantitative Traits in Oat
   Wilawan Siripoonwiwat, Louise S. O'Donoughue, Darrell Wesenberg, David L. Hoffman, Jos F. Barbosa-Neto, and Mark E. Sorrells
4. Evaluating Gene Effects of a Major Barley Seed Dormancy QTL in Reciprocal Backcross Populations
   Steve Larson, Glenn Bryan, William Dyer, and Tom Blake
5. Association of a Seed Weight Factor with the Phaseolin Seed Storage Protein Locus Across Genotypes, Environments, and Genomes in *Phaseolus-Vigna* spp.: Sax (1923) revisited
   William C. Johnson, Cristina Menéndez, Rubens Nodari, Epimaki M.K. Koinange, Steve Magnusson, Shree P. Singh, and Paul Gepts
6. Constructing Genetic Maps by Rapid Chain Delineation
   R.W. Doerge
7. Analysis of QTL Workshop I Granddaughter Design Data Using Least-Squares, Residual Maximum Likelihood and Bayesian Methods
   Pekka Uimari, Qin Zhang, Fernando Grignola, Ina Hoeschele, and Georg Thaller

## 12.3 SURVEY OF CURRENTLY AVAILABLE DATA AND/OR DATABASES

The AGIS contains the largest collection of genome databases. Many of these contain QTL data of varying amounts and detail. Searching can be accomplished in a number of ways (Figure 12.2) via the World Wide Web including simple or Boolean keyword searches using either WAIS or agrep, a search tool which allows fuzzy matches. Figure 12.3 shows a portion of the result of a WAIS search for QTL on the Soybase database. Each of the objects is retrievable by a mouse click on the object name. Query Builder and Query by Example provide interfaces for constructing more complex queries. Table-maker allows the user to retrieve data in tabular form using a simple forms-based interface. Figure 12.4 shows a table of loci with their map positions for all QTL studies in Soybase containing the word height. As in the WAIS example, any object can be retrieved with a single mouse click. A full-featured query language interface is also available. Some of the databases have added QTL interval data such that it can be displayed on a genetic map. Figure 12.5, a genetic map of chromosome 7 from RiceGenes, shows a QTL for blast resistance (qBlast-7-1) along with a linked locus (RG528 — an RFLP probe) which is highlighted. In addition to querying, the databases can be browsed on a class-by-class basis. The AGIS gopher server allows only WAIS searching.

A brief survey of the AGIS databases is presented below followed by a survey of databases available elsewhere which contain QTL data. Also, a list of miscellaneous resources related to QTL information is presented. The surveys include a list of traits covered (where applicable), the types of information provided, the best method(s) for searching for additional QTL data, and additional URLs where applicable. The URLs for AGIS are: http://probe.nalusda.gov and gopher://probe.nalusda.gov. AGIS contains contact information for each of the databases. In addition, for ACeDB-formatted databases, the data and database software can be downloaded via anonymous ftp from probe.nalusda.gov. Following the surveys is a brief guide to finding additional QTL information on the Internet.

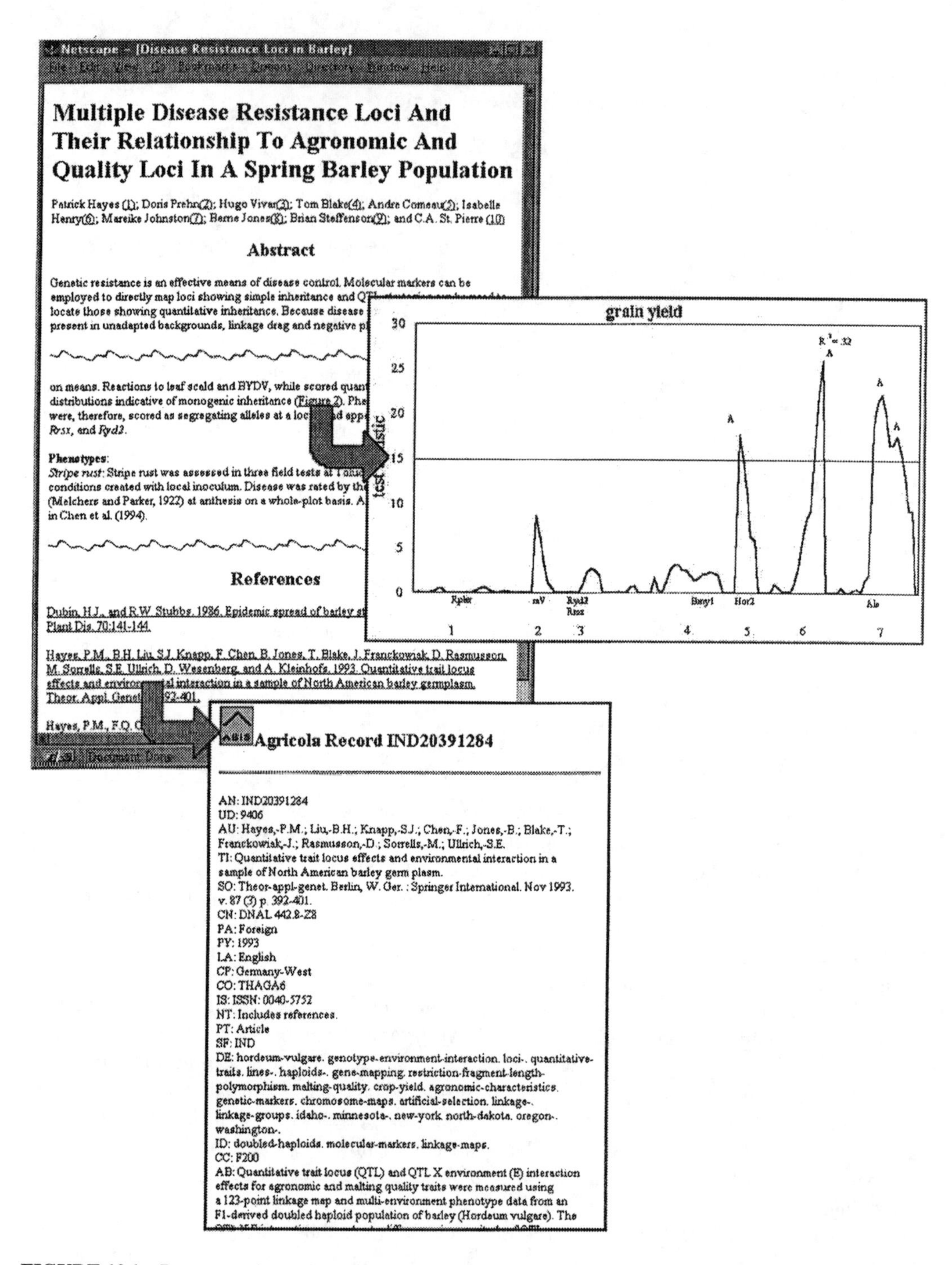


**FIGURE 12.1** Representative article from the *Journal of Quantitative Trait Loci* showing hypertext links to a figure and a reference. (See Reference 2.)

### 12.3.1 AGIS Databases

**MAIZE DB**

| | |
|---|---|
| Traits: | Starch content, protein content, and plants per embryonic embryo |
| Information: | Descriptions of traits, map locations, alleles, and statistical data |

AGIS **AGIS Search Form**

[ Start a new search ]

**Database group:** plant ( or reselect )

**Select a query method:**

ACEDB Query Language [ About ]
compose complex queries, requires knowledge of ACEDB query language syntax, searches a single database

ACEDB Query Builder [ About ]
build complex queries by choosing item names and operators from a list, searches a single database

ACEDB Query by Example [ About ]
fill in form for simple pattern matching queries, searches a single database

Fuzzy search (agrep) [ About ]
inexact and keyword matching using boolean expressions, searches multiple databases or files

WAIS search [ About ]
keyword matching using boolean expressions, searches multiple databases or files

ACEDB Table-maker [ About ]
form interface for generating tables from a single database

**Genome Informatics Group / 14 July 1995**

**FIGURE 12.2** Agricultural Genome Information System database search options.

Searching: Keyword search on QTL
Additional URL: http://teosinte.agron.missouri.edu

**RICEGENES**
Traits: Blast resistance
Information: Description of trait, map locations, detailed description of study, statistical data, graphical images of symptoms and maps
Searching: Browse QTL class (WWW) or keyword search on QTL (gopher)

**SOYBASE**
Traits: Canopy height, date of first flower, hard seededness, iron efficiency, leaf area, width and length, linoleate, linolenate, lodging, oil content, oleate content, palmitate content, plant height, protein content, beginning of seed development, seed filling period, seed pod maturity date, seed yield, cyst nematode resistance, stearate content, stem diameter and length
Information: Description of trait, statistical data
Searching: Browse QTL_Study class (WWW) or keyword search on QTL_study (gopher)

**GRAINGENES** (Wheat, Barley, Oats, and Other Small Grains)
Traits: Preharvest sprouting
Information: Description of trait, detailed description of study, statistical data, graphical images of autorads and maps
Searching: Browse QTL class (WWW), keyword search on QTL (WWW and gopher)

AGIS **WAIS Search Results**

---

**Searchstring = qtl**

**Databases searched = soybase**

---

**soybase:**

```
Score   Class: Object
-----   -------------
• 707     QTL_Study:  ?QTL_Study
• 455     Locus:  I
• 447     Locus:  BCI
• 447     Locus:  Sac007
• 437     Locus:  Pb
• 428     Locus:  G173
• 422     Locus:  G017.3-1
• 422     Locus:  K474-1
• 422     Locus:  K474-2
• 408     Locus:  A060-1
• 408     Locus:  K018
• 396     Locus:  A060-2
• 311     Trait:  Soybean cyst nematode resistance
• 309     Trait:  Seed filling period
• 287     Author:  Mansur, L.M.
• 286     Locus:  A245-1
• 282     Locus:  A343-1
• 282     Locus:  A385
• 275     QTL_Study:  Iron Efficiency_2
• 271     QTL_Study:  Lodging_2
• 269     Locus:  A109-1
• 268     Author:  Oliveira, A.
• 266     Trait:  Beginning Seed
• 262     QTL_Study:  Plant height_2
• 259     QTL_Study:  Seed yield_2
• 254     Author:  Orf, J.
• 250     QTL_Study:  Seed pod maturity_3
• 246     Author:  Cianzio, S.R.
• 239     QTL_Study:  Lodging_1
• 239     QTL_Study:  Plant height_1
• 237     Trait:  Leaf area
• 233     QTL_Study:  Canopy height_2
• 231     QTL_Study:  Leaf area_1
```

**FIGURE 12.3** Result of a WAIS keyword search for QTL in the Soybase database.

## 12.3.2 Other Genome and Genetic Databases

**GDB** (Human Genome Database)

| | |
|---|---|
| Information: | Citations referring to QTLs, Medline IDs where applicable |
| Searching: | Keyword search on *quantitative trait* in Abstract field of Citation table |
| Access: | http://gdbwww.gdb.org |

**MGD**

| | |
|---|---|
| Traits: | Dietary obesity, high affinity choline uptake, hypothermia due to alcohol sensitivity, morphine preference, skin tumor susceptibility, and tolerance to alcohol |
| Information: | Brief description of experiment, graphical representation of map location |
| Searching: | Select Type QTL from menu on Genetic Markers and Mouse Locus Catalog |
| Access: | http://www.informatics.jax.org/mgd.html |

AGIS **SoyBase**

**Table-maker : output**

[ help ]

| QTL_Study | Locus | Map | Position |
|---|---|---|---|
| Canopy height_1 | K390 | F | 4.7 |
| Canopy height_1 | R013-2 | D | 47.6 |
| Canopy height_2 | | | |
| Plant height_1 | | | |
| Plant height_2 | A060-1 | | |
| Plant height_2 | A060-2 | | |
| Plant height_2 | A397 | C | 305.6 |
| Plant height_2 | A397 | L1 | 77.4 |
| Plant height_2 | R079 | L14 | 42.1 |
| Plant height_2 | R079 | M | 19 |

**Query:** ACEDB Query Language | ACEDB Query Builder | ACEDB Query by Example | Fuzzy search (agrep) | WAIS search | ACEDB Table-maker

**Jump to:** All classes | All models

**Genome Informatics Group / 11 October 1995**

**FIGURE 12.4** Result of a Table-maker search for QTLs containing the word height.

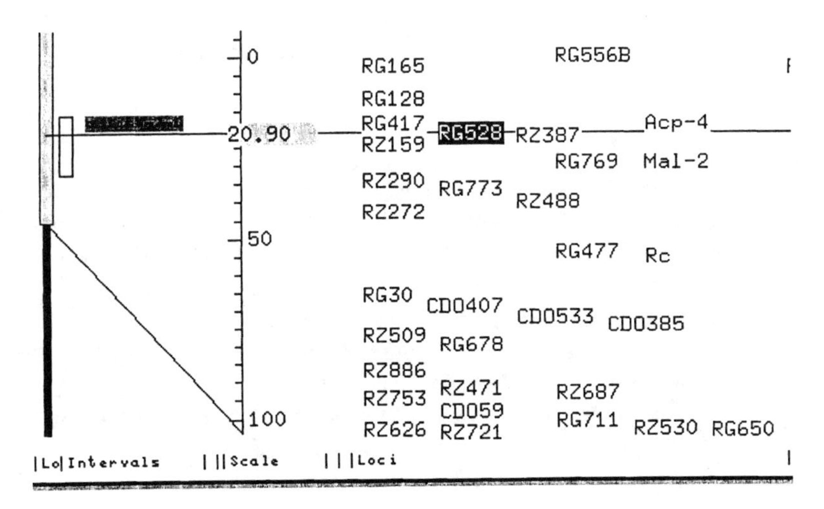


**FIGURE 12.5** A genetic map of chromosome 7 from RiceGenes showing a QTL for blast resistance (the open rectangle at left). A related locus (RG528) is highlighted.

### 12.3.3 Reference Databases

**MEDLINE** (Molecular Biology Subset)
Information: Citations and abstracts
Searching: Keyword search for QTL or quantitative trait locus/loci or quantitative traits
Access: http://www3.ncbi.nlm.nih.gov/Entrez; CD-ROM available from SilverPlatter Information, Inc.,100 River Ridge Drive, Norwood, MA 02062-5026, U.S.

**AGRICOLA (January 1989-Present)**
Information: Citations and abstracts
Searching: Keyword search for QTL or quantitative traits (ISIS) or quantitative trait locus/loci (plant genome subset)
Access: ISIS — (also includes the National Agricultural Library s Online Catalog) telnet://opac.nal.usda.gov
plant genome subset through 1993 — gopher://probe.nalusda.gov:7020/77/agricola.agidx; CD-ROM available from SilverPlatter Information, Inc.,100 River Ridge Drive, Norwood, MA 02062-5026, U.S.

**PLANT GENOME CONFERENCE ABSTRACTS**
Searching: Keyword search for QTL or quantitative trait locus/loci
Access: http://probe.nalusda.gov:8000/otherdocs/pg/index.html

### 12.3.4 Miscellaneous Resources on the Internet

**QTL MAPPING PAGE**
Description: Links to various QTL mapping documents on the World Wide Web maintained by Brad Sherman of the USDA Dendrome project
Access: http://s27w007.pswfs.gov/qtl/

**QUANTITATIVE GENETICS RESOURCES PAGE**
Description: An electronic supplement to the textbook *Fundamentals of Quantitative Genetics* by Mike Lynch and Bruce Walsh (page maintainer)
Access: http://nitro.biosci.arizona.edu/zbook/book.html

**QTL CARTOGRAPHER TUTORIAL**
Description: A tutorial for this QTL mapping software written by Christopher J. Basten, Bruce S. Weir, and Zhao-Bang Zeng
Access: http://www2.ncsu.edu/ncsu/CIL/stat_genetics/qtlcart/qtltutor.html

**MAPMAKER3 SOFTWARE DISTRIBUTION SITE**
Description: Distribution site for this mapping software which includes MAPMAKER/QTL. Produced by the MIT Whitehead Institute.
Access: http://www-genome.wi.mit.edu/ftp/distribution/software/mapmaker3/

**MSIM AND MQTL PAGES**
Description: Documents decribing these two software packages written by Nick Tinker which are for automated simulation of genetic markers and QTL and simplified composite QTL interval mapping
Access: http://gnome.agrenv.mcgill.ca/tinker/msim.htm
http://gnome.agrenv.mcgill.ca/tinker/mqtl.htm

**TABLE 12.2**
**Internet Search Services**

| Service | Access | Comments |
|---|---|---|
| Alta Vista | http://www.altavista.digital.com | One of the fastest and most comprehensive, allows searching of newsgroups |
| Yahoo! | http://www.yahoo.com | Groups sites by subject matter |
| Infoseek Guide | http://www.infoseek.com | Groups sites by subject matter |
| Lycos | http://www.lycos.com | Groups sites by subject matter |
| Excite | http://www.excite.com | Also searchable by concept, allows searching of newsgroups |
| Magellan | http://www.mckinley.com | Groups sites by subject matter |

**USDA, COOPERATIVE STATE RESEARCH, EDUCATION AND EXTENSION SERVICE HOME PAGE**

Description: Information about funding opportunities, programs, and grant awards at the USDA

Access: http://www.reeusda.gov/

#### 12.3.4.1 Finding Additional Information on the Internet

Table 12.2 lists several services which index information found by infobots on the Internet. All of these provide the ability to do keyword searching on a vast number of documents, most of which reside on the World Wide Web. Periodically searching one or more of these services will yield new QTL (and, of course, other types of) information soon after it becomes available. In addition, it is often worthwhile checking the resources listed above on a regular basis for new information or data retrieval methods. The discipline of bioinformatics is changing rapidly and new developments appear on almost a daily basis. Usenet newsgroups are also an important source of information concerning new developments. The newsgroups in the bionet hierarchy contain many postings related to biological information. Of particular interest is bionet.announce where most database and service providers announce new developments. Finally, some World Wide Web sites keep comprehensive lists of (and links to) resources for the molecular biologist. Some of the more extensive sites are Harvard Biological Laboratories (http://gogli.harvard.edu), Pedro's Biomolecular Research Tools (http://www.public.iastate.edu/~pedro/research_tools.html), and EBI's BioCatalog of molecular biology/genetics software (http://www.ebi.ac.uk/biocat/biocat.html).

## 12.4 FUTURE DEVELOPMENTS

The future of QTL data retrieval and assimilation will likely depend upon the development of user interfaces which can automatically draw together and integrate information from diverse sources. Unfortunately, like urban sprawl, the building of information resources is largely done without regard to neighbor resources, which makes this integration difficult. However, the effort required to build the software and to manually build semantic relationships among resources should provide a big payoff. One promising approach is the Biology Workbench which has been developed at the National Center for Supercomputing Applications.[3] The goal is to provide a user with a set of query mechanisms which can identify the set of suitable resources to search and combine the query results into a uniform report. These results can then be fed directly into the appropriate tool(s), which have been integrated into the workbench, for further analysis. Figure 12.6 shows the overall concept for the Biology Workbench which will ultimately include genome, metabolism, sequence, and structure information. Note that the user interface will be such that the details of the inner workings will be completely hidden. Similar work is underway elsewhere.

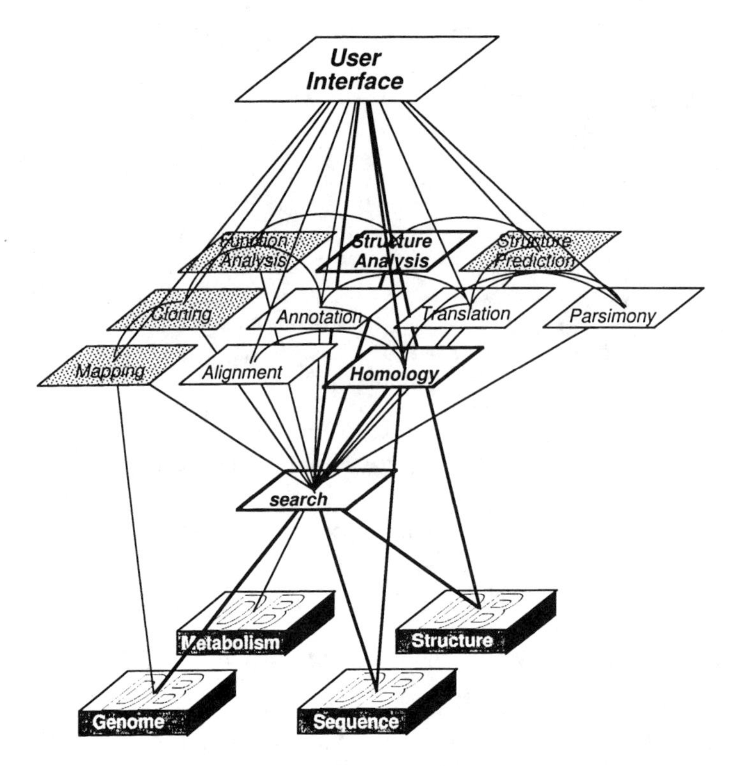


**FIGURE 12.6** Conceptual overview of NCSA's Biology Workbench. (Data from Reference 3 with permission.)

## ACKNOWLEDGMENT

I would like to thank Michael Shives for his help in the preparation of this manuscript.

## REFERENCES

1. Byrne, P. F., Berlyn, M. B., Coe, E. H., Davis, G. L., Polacco, M. L., Hancock, D. C., and Letovsky, S. I., Reporting and accessing QTL information in USDA's Maize Genome Database, *J. Quantitative Trait Loci*, 1, 3, 1995. (http://probe.nalusda.gov:8000/otherdocs/jqtl/jqtl1995-03/text11r.html)
2. Hayes, P., Prehn, D., Vivar, H., Blake, T., Comeau, A., Henry, I., Johnston, M., Jones, B., Steffenson, B., and St. Pierre, C. A., Multiple disease resistance loci and their relationship to agronomic and quality loci in a spring barley population, *J. Quantitative Trait Loci*, 2, 2, 1996. (http://probe.nalusda.gov:8000/otherdocs/jqtl/jqtl1996-02/jqtl22.html)
3. Jamison, C., Stupar, M., Fenton, J. M., Unwin, R., Jakobsson, E., and Subramaniam, S., The biology workbench — a WWW-based virtual computing environment for the macromolecular sequences and structures, unpublished manuscript.

# PART II

## CASE HISTORIES

# 13 Case History in Plant Domestication: Sorghum, An Example of Cereal Evolution

*Andrew H. Paterson, Keith F. Schertz, Yann-rong Lin, and Zhikang Li*

## CONTENTS



## 13.1 INDEPENDENT EVOLUTION OF MANY CEREAL CROPS PROVIDES A MODEL TO INVESTIGATE THE MOLECULAR BASIS OF DOMESTICATION

Most of the calories which feed humankind are derived from crops in the plant family Poaceae, the grasses. Diverse members of this large family have been independently selected for similar traits, by human civilizations in Africa, Asia, and the Americas, respectively. These independent

0-8493-7686-6/98/$0.00+$.50

episodes of selection resulted in the evolution of annual genotypes with large carbohydrate-rich grains that adhere to the plant from perennial ancestors which widely disperse their small seeds.

Detailed lists of traits which distinguish cultivated grain crops from their wild relatives or weedy intermediates have been compiled based on extensive study of morphology across many taxa.[1] Several common themes are apparent:

### 13.1.1 Reduction of Seed (Grain) Dispersal

Reduced seed dispersal is characteristic of virtually all cultivated grasses, and represents an obstacle to utilization of many potential new crops. However, the degree to which seed dispersal has been restricted is variable among different crops. For example, maize has evolved extraordinary restrictions on seed dispersal under the control of at least ten quantitative trait loci (QTL), not only reducing the tendency of the mature pistillate inflorescence to disarticulate or "shatter," but also tightly enveloping it in leaves. By contrast, Asian/African rice[2] has only an intermediate level of impedance to disarticulation, as a result of at least three QTLs.[2] The grains of Asian/African rice have usually been separated from the vegetative parts of the inflorescence by human hands, and the intermediate degree of shattering may reflect a preference of human populations for genotypes which provide a compromise between harvest efficiency and threshability.

### 13.1.2 Increased Seed Size and Reduced Seed Dormancy

In natural populations, production of large numbers of small seed with a high degree of dormancy confers "insurance" to a genotype, as both spatial and temporal distribution of progeny reduces the likelihood that a cataclysmic event will eliminate all from the gene pool.[3] By contrast, in annual crops, "fitness" of a genotype is determined by the number and vigor of seeds it contributes to the sole harvest. Large, vigorous seeds with no dormancy are likely to germinate more quickly than their neighbors, and compete successfully for growth-limiting resources such as light, moisture, and nutrients.

### 13.1.3 Synchronization of Seed/Grain Production

In natural populations, asynchrony of seed/grain maturity can be a selective advantage, reducing both the susceptibility of the genotype to climatic disasters, and the impetus for coevolution of pest populations with plant growth cycles.[3] In crops, typically only harvested once, breeders select for a single large burst of seed/grain production which matures before pest populations have reached damaging levels. In particular, reduction in the number of "tillers," axillary shoots or hypocotyl-derived buds which lag behind the primary inflorescence in their development, is a common feature of most domesticates.

An important aspect of synchronization is the allocation of photosynthate to seeds rather than to perennation organs such as rhizomes. Many wild or weedy grasses overwinter and spread by underground stems, or "rhizomes." Under adverse conditions in which a wild grass must make a "choice" between allocation of photosynthate to reproduction (seed) vs. persistence (rhizomes), persistence tends to be favored.[4] Elimination of rhizomes has not only redirected additional photosynthate to seeds, but also facilitated highly mechanized "row crop" production systems.

### 13.1.4 Reduction of Plant Stature (Height)

In natural populations, tall stature affords a competitive advantage for light, and increases the effectiveness of seed dispersal.[3] However, in agriculture, reduced height is necessary for machine harvest, and to avoid wind or other hazards. In most crops, series of height mutations, e.g., *Rht1*-*Rht10* in wheat[5] and *d1*-*d9* in maize,[6] have been preserved or induced and play a prominent role in breeding.

### 13.1.5 Coordination of Flowering with Photoperiod

In the semiarid tropics which represent the likely centers of origin for many important grain crops, short daylength serves as a cue by which plants coordinate seed development with the season of optimal rainfall.[3] However, in temperate latitudes, short-day flowering results in initiation of seed development dangerously late in the growing season, when solar radiation is declining and pest populations are high. Because many major crops derive from tropical ancestors, it has been necessary to select for photoperiod insensitive (day-neutral) mutations in order to adapt them to temperate agriculture. Some prominent examples of such mutations include *ma1* in sorghum,[7] and *se-1*, *se-2*, and *se-3* in rice.[8]

## 13.2 MAPPING DETERMINANTS OF SORGHUM DOMESTICATION

Over the past several years, we have focused considerable effort on investigating the inheritance of traits associated with the domestication of grain sorghum (*Sorghum bicolor* L.) from its wild relatives. Sorghum was a fortuitous choice for these studies, because of the availability of cross-compatible wild species which retain the morphology of non-grain producing grasses. The availability of detailed "comparative maps" showing the correspondence of sorghum chromosomes with those of many other Poaceae taxa, enabled us to evaluate the relationships between genomic locations of genes/QTLs in these different taxa.[2]

In choosing a population upon which to base our studies, we sought to cross an agronomically acceptable sorghum inbred,* with a relative that exemplified the morphological features common to wild grasses. The wild grass needed to be sexually compatible with sorghum, with the same chromosome number (2n = 20) and normal cytology. Further, there should have been little opportunity for gene flow (introgression) between the wild grass and sorghum. A search of the classical sorghum literature revealed that the logical candidate was *S. propinquum*, a strongly-rhizomatous perennial indigenous to the Pacific rim (particularly Indonesia and the Philippines). The allopatric geographical distributions of *S. propinquum* and *S. bicolor* (indigenous to Africa) indicated that the likelihood of recent gene flow was minimal. The possibility of gene flow cannot be absolutely ruled out due to overlap in the geographic distributions of both *S. bicolor* and *S. propinquum* with those of their probable interspecific hybrid, the polyploid *S. halepense* (johnson grass). Further, there is also the possibility of association between *S. propinquum* and the kaoliang sorghums of China.[9] Classical literature suggested that crosses between *S. bicolor* and *S. propinquum* were fertile, and exhibited normal cytology.[10]

A single plant of the cytosterile line Atx623 was pollinated by hand from a single plant of a *S. propinquum* accession obtained from ICRISAT, and the resulting F1 was selfed to produce a large population of F2 seeds. Seed dormancy was evident both in *S. propinquum* and its F2 progeny, but imbibition with micromolar concentrations of gibberrellin A3 stimulated ca. 70% of F1 and F2 seed to germinate. Most of these survived to maturity. No clear albinos or other gross aberrations were noted, however, *S. propinquum* and a subset of its hybrid progeny exhibited varying degrees of chlorosis and tip burn during early seedling development in the greenhouse. This disappeared quickly if seedlings were transplanted to the field, and gradually even in the greenhouse as plants grew older, and was assumed to represent a nutritional defect.

Despite the high fertility and fecundity of their interspecific hybrids, *S. bicolor* and *S. propinquum* showed an extraordinary degree of molecular divergence. Length polymorphisms were evident for about 70% of genomic restriction fragments or arbitrarily primed polymerase chain reaction (PCR) amplification products. A primary RFLP map comprised of 10 linkage groups putatively corresponding to the 10 sorghum chromosomes was assembled (see Chapter 1,

---

* Most commercial sorghum production in the U.S. is based upon hybrids between two inbred lines using cytoplasmic male sterility systems similar to those of maize.

Figure 1.1).[11] Because the sorghum chromosomes are small and indistinct and the sorghum plant is not tolerant of aneuploidy, it has not been possible to ascertain the relationship between linkage groups and chromosomes as done in other taxa. However, more than 1000 additional DNA markers have been added to the sorghum map[11a] and virtually all show linkage to the established map, indicating that the map covers all regions of the genome.

A subset of markers from the primary sorghum map, well-spaced at an average of 14-cM intervals across the chromosomes (see Figure 1.2 of Chapter 1), was applied to 370 interspecific F2 progeny grown in the field near College Station, TX. Growth and development of these plants was documented by measuring more than 30 phenotypes. Subjective phenotypes, relying on the perception of an investigator rather than an objective measurement, were evaluated independently by two or more investigators and average scores were used for data analysis. Putatively discrete phenotypes were accepted as 'genetic markers' if their addition to the genetic map did not significantly expand the length of the interval between the nearest flanking restriction fragment length polymorphism (RFLP) markers.* Three phenotypes, (non)shattering, brown vs. white testa, and purple hypocotyl, could be mapped as genetic markers. Other phenotypes were associated with particular chromosomal locations by chi-squared contingency tests (if measured on discrete scales) or interval mapping[12] if measured on continuous scales. Appropriate mathematical transformations of raw data were used as needed to normalize distributions of residual terms in the genetic model.

For most quantitative traits, the 370 F2 individuals were sufficient to resolve QTLs explaining about 4% or more of phenotypic variance in this population (however, see Chapter 10 for a detailed discussion of the possibility of false negative results in such a context). Genetic control of selected traits was as follows.

### 13.2.1 Plant Height

The average height of the main culm, tallest, and shortest flowering tillers for *S. bicolor* cv. "BTx623" was 109 (±6) cm, for *S. propinquum* was 396 (±40) cm, and for the F2 population was 290 (±99) cm. The phenotypic distribution of F2 progeny was bimodal, however a wide range of phenotypes within the two classes suggested the influence of multiple genes,[13] consistent with classical literature.[14] A total of six QTLs collectively accounted for 71.0% of phenotypic variation in plant height, and were distributed across five chromosomes (linkage groups; A, C, D, G [2], J). A single QTL on LG (linkage group) D explained 54.8% of phenotypic variation, with additive effect of 87.9 cm and dominance deviation of 63.9 cm. For five of the six (83%) height QTLs, the *S. propinquum* alleles exerted a positive additive effect (i.e., increased height). Among these five, four showed dominance or overdominance for increased height, and one was additive. The final QTL was "overdominant," with the heterozygote taller than either parent.

### 13.2.2 Flowering

The average time from planting to flowering of the main culm and (up to) the first five tillers was 115.5 (±7.8) d for the *S. bicolor* parent, 189 (±1.9) d for *S. propinquum,* and 149.7 (±37.7) d for the F2 population. Because 12 F2 progeny that had not yet flowered by frost (28 November 1992, 233 d) were excluded from the analysis, the data were slightly conservative (biased by reduced variation in flowering time). As was true of height, the phenotypic distribution of flowering dates for F2 progeny was bimodal, however a wide range of phenotypes within the two classes suggested the influence of multiple genes,[13] consistent with classical literature.[14] Three QTLs collectively accounted for 86.7% of the phenotypic variation in average days to flowering. For all three flowering QTLs, the *S. propinquum* alleles conferred late flowering. The *S. propinquum* alleles of *FlrAvgD1* and *FlrFstG1* were dominant, and of *FlrAvgB1* was recessive.

---

* Map expansion, due to incongruity between a phenotype and the flanking DNA markers, would suggest that additional genes were associated with the phenotype or that imperfect penetrance/expressivity were manifested.

*FlrAvgD1*, which alone could explain most of the phenotypic variance in flowering time, is of special interest. Previously, Quinby and Karper[7] suggested that the short-day vs. day-neutral dichotomy in crosses between temperate and tropical sorghums could be accounted for by a single genetic locus, which they named *maturity-1* (abbreviated *ma-1*). Since *FlrAvgD1* is the only one among these QTL which could account for the dichotomy, and since it exhibits the further property discovered by Quinby and Karper that it is closely linked to a locus with a major effect on the height of the sorghum plant,[14] we have accepted *ma-1* as the proper name for *FlrAvgD1*. Moreover, we have shown that the *ma-1* locus appears instrumental in regulation of flowering across virtually all *S. bicolor* races,[13] and probably in many other grass taxa.[2]

### 13.2.3 Seed Size

Although late flowering precluded production of mature seed by *S. propinquum* in the field, greenhouse-grown seed of *S. propinquum* are typically about 10% of the mass of *S. bicolor* seed. A total of nine QTLs, located on eight linkage groups (A, B [2], C, D, E, F, I, J) collectively accounted for 51.7% of phenotypic variation in seed size, with individual QTLs explaining 5.3 to 11.9% of variation. In all cases, the *S. propinquum* allele conferred the reduced seed size. The mode of gene action of *S. propinquum* alleles ranged widely, from largely dominant to largely recessive.[11a]

### 13.2.4 Seed Number

A total of four QTLs, located on different linkage groups (A, B, C, H) collectively account for 19.1% of phenotypic variation in seed number with individual QTLs ranging from 4.2 to 6.8%. In three cases, dominant *S. propinquum* alleles increased seed number and in the remaining case a largely recessive *S. propinquum* allele reduced seed number.[11a]

### 13.2.5 Tiller number

*S. propinquum* is abundantly tillering with a single crown often producing 100 or more tillers in the first growing season. By contrast, cultivated *S. bicolor* genotypes rarely produce more than 2 to 3 tillers even when grown at a very low density (plants in our study were 1 m apart). A total of four QTLs, located on LGs C, D, H, and J, accounted for 23.7% of phenotypic variation in the number of tillers at 8 weeks after seeding (prior to flowering).[15] The *S. propinquum* allele at each of these four loci was associated with increased tillering. Of the loci (LGs C, H), two showed largely dominant gene action, one (LG J) showed largely additive gene action, and one (LG D) showed largely recessive gene action. The LG C tillering QTL corresponded very closely to one of the QTLs affecting rhizomatousness, with largely overlapping 1-LOD likelihood intervals, and maximum-likelihood peaks ca. 7 cM apart. It was proposed[15] that a single gene at this locus may regulate the number of vegetative initials available to differentiate either into tillers or into rhizomes, consistent with developmental literature,[16] and that additional independent genes may be involved in determining the fate of each initial.

### 13.2.6 Rhizomes

*S. propinquum* is abundantly rhizomatous, and is the probable source of the rhizomatous trait of "Johnson Grass" (*S. halepense*).[15] By contrast, no *S. bicolor* genotype, either cultivated or wild, has been unequivocally demonstrated to produce rhizomes.

Three distinct regions of LG C accounted for 21.8% of phenotypic variance in the number of above-ground rhizome-derived shoots. In all cases, the *S. propinquum* alleles conferred enhanced rhizomatousness.[15] While no chromosomes other than LG C accounted for detectable variation in above-ground rhizome-derived shoots, the extent of subterranean rhizomes was influenced by additional QTLs on LGs B, D, F, G, H, and I.[14] These accounted for an additional 31% of variance in the extent of below-ground rhizomes, beyond the 14% accounted for by the two LG C QTLs.

The *S. propinquum* allele increased rhizomatousness in all cases except LG D, where the *S. propinquum* homozygote showed a marginally significant (LOD 3.09) reduction in rhizomatousness. Of the eight QTLs, four showed simple additive gene action, two showed largely dominant gene action, and one showed largely recessive gene action. Finally, one QTL (on LG B) was "overdominant" with the heterozygous genotypes showing greater rhizomatousness than either parental homozygote.

## 13.3 COMPARATIVE ANALYSIS OF DOMESTICATION

Comparative genetic maps, using common DNA probes to show the relative alignment of chromosomes in different taxa, enable one to make approximate comparisons of the locations of QTLs in species which cannot be crossed to one another. Such analyses are particularly interesting in the grasses, because many taxa have been independently domesticated for similar purposes — therefore, one can evaluate the possibility of a common genetic basis underlying domestication. Moreover, detailed comparative maps have been assembled for most of the leading cultivated grasses.

QTLs associated with domestication tend to fall at corresponding locations in different grass taxa, much more often than would be expected by chance.[2] One possible explanation of this finding is that some QTLs in different taxa may be the result of independent mutations in corresponding genes. To date, we have shown a high degree of correspondence among QTLs affecting flowering time, plant height, seed size, and seed dispersal ("shattering"), in the genomes of sorghum, maize, and rice. As comparative data becomes available for additional phenotypes, and additional taxa, it seems likely that many additional examples will be found.

Correspondence of QTLs provides additional support for the hypothesis that many complex phenotypes may have a relatively simple genetic basis. Classical quantitative genetic theory has suggested that complex traits such as seed size may be influenced by a virtually infinite number of genes, each with a very small effect — consistent with the gradualistic model of evolution which emerged in the mid-20th century. By contrast, over the past 20 years 'punctuational' models, invoking more rapid selection for fewer genes with larger effects, have gained support.

The nonrandom distributions of QTLs we have observed tend to support punctuational models for phenotypic evolution. If mutation in a virtually infinite number of genes could confer the phenotypes studied, the correspondence we observed would be unlikely to occur. However, it is important to acknowledge that the power of mapping experiments to detect QTLs is clearly finite, and many QTLs of small effect may escape detection (see Chapter 10, this volume). It remains for future investigators to evaluate whether the pattern of correspondence extends to these smaller QTLs, probably using modified experimental designs (see Chapter 15, this volume). In particular, an important area for further research is to evaluate levels of correspondence among QTLs segregating in elite gene pools of domesticates.

## 13.4 PATTERNS OF GENE ACTION IMPLICATE SELECTION FOR LOSS-OF-FUNCTION ALLELES AS AN IMPORTANT COMPONENT OF DOMESTICATION

Rapid genetic changes such as domestication are often thought to be associated with selection for loss-of-function mutations. From first principles, it is simpler to disrupt the function of a gene, by as little as a single base substitution, than to recruit genes to novel functions. Such loss-of-function mutations are indicated clearly by cases of "dominance" of one allele over another, and may even account for additivity of alleles.

QTL mapping data for most of the traits we have studied, supports the view that domestication has selected for new mutant alleles in many crop gene pools (Table 13.1). The wild (*Sorghum propinquum*) alleles were dominant for the discrete "shattering" locus, four (80%) QTLs affecting

**TABLE 13.1**
**Action of *Sorghum propinquum* Allele at QTLs for Traits Related to Domestication**

| | | Mode of gene action[a] | | | |
|---|---|---|---|---|---|
| Trait | No. of Genes/QTLs | Dom | Add | Rec | Overdom |
| Shattering | 1 | 1 | 0 | 0 | 0 |
| Height | 6 | 4 | 1 | 0 | 1 |
| Flowering | 3 | 2 | 0 | 1 | 0 |
| Seed size | 9 | 2 | 1 | 6 | 0 |
| Tiller number | 4 | 1 | 2 | 1 | 0 |
| Rhizomatousness | 8 | 2 | 4 | 1 | 1 |
| Overall | 31 | 12 | 8 | 9 | 2 |

[a] Dom = dominant; Add = additive; Rec = recessive; and Overdom = overdominant.

plant height, two (67%) QTLs affecting flowering, three (75%) QTLs affecting seed number, and two (50%) QTLs affecting tillering. Dominant *S. propinquum* alleles also outnumbered recessive alleles for rhizomatousness, however four loci showed additive gene action.

One trait, seed size, was a curious exception, with six (67%) of the nine *S. propinquum* alleles for reduced seed size being recessive to the corresponding *S. bicolor* alleles.

## 13.5 APPLICATIONS OF INFORMATION ABOUT PLANT DOMESTICATION

While studies of plant domestication contribute much to the basic understanding of evolution and development, it is less obvious how such studies can contribute to improved agricultural productivity. Arguments against the utility of such studies might emphasize the fact that crossing a weed with a crop will only show "what we have already gained," rather than offering opportunities to make further gains.

In our view, there are at least three important applications of genetic information from crop domestication studies, which supplement the valuable genetic/developmental information which they yield.

### 13.5.1 Improvement of Prospective New Crops

Modern agriculture is based almost entirely on less than 50 major crops. These few plants represent only a tiny fraction of the potential genetic diversity resulting from 200 million years of plant evolution. Many as-yet wild species have novel agronomic, nutritional, or biochemical attributes which offer opportunities to reduce dependence on synthetic products, diversify farm income, and provide sustainable and profitable alternatives to high-input agricultural systems.

The domestication process for new crops might be accelerated by using cloned genes associated with domestication in existing crops, to engineer the required mutants. Such an approach might be much faster and more economical than imposing new episodes of selection, to identify new mutations in many of the same genes already shown to account for key aspects of domestication.[2] Isolation of genes associated with key steps in crop domestication would offer the potential to quickly engineer such mutations into new gene pools using antisense mRNA technology[17] or a similar approach. For example, domestication of many new seed crops would be greatly facilitated by suppression of shattering, e.g., wild rice,[18,19] birdsfoot trefoil,[20a] castor[20,21a] oilseed spurge,[21] Vernonia,[22] and others.

### 13.5.2 New Sources of Variation for Improvement of Other Crops

Exceptions to the high level of correspondence among "domestication QTLs" may provide opportunities for crop improvement. For example, the order in which mutations happened to occur may influence the selective advantage afforded subsequent mutations.[2] The African domesticators of sorghum may simply have been fortunate to find a mutant in a critical step leading to grain abscission (*Sh1*) which "turned off" the pathway, accounting for ~100% of phenotypic variance in crosses between shattering and nonshattering types. By contrast, the American domesticators of maize may not have been so lucky, but still succeeded in reducing disarticulation by "pyramiding" mutations with smaller effects on several distinct steps. Moreover, independent and random occurrence of mutations in paralogous genes may have formed new alleles with very different phenotypic consequences. For example, putatively paralogous non-shattering mutations on maize chromosomes 3 and 8 explain grossly different portions of phenotypic variance, in the same population.[2]

Such incongruities among QTLs, contrasting with the overall picture of correspondence, may point to opportunities for genetic engineering of improved productivity.[2] For example, transformation of rice with a maize chromosome 2 allele which explains 23.6% of phenotype variance in seed mass might improve rice seed mass, since the corresponding chromosomal region of rice has not been associated with variation in seed mass.[2] Such use of genetic variation which transcends species boundaries may afford qualitative improvements in quantitative traits which are manipulated slowly by classical techniques and have been refractory to biotechnology.

### 13.5.3 Ongoing Interactions between Crops and Weeds

Some crops continue to interact with their wild ancestors, either through competition for growth-limiting resources such as light, water and nutrients or through genetic exchange. Because many dispersal mechanisms have been eliminated from crop gene pools, genetic analysis of crop × weed hybrids is a starting point for molecular cloning of genes associated with "weediness."[15]

## 13.6 SUMMARY

A cross between cultivated sorghum and its wild relative has permitted molecular dissection of many aspects of crop domestication and molecular mapping of a host of genes/QTLs playing important roles in growth and development of grasses. Populations derived from crosses between cultivated germplasm and wild relatives offer many exciting opportunities for botanical research with relevance to evolution and development as well as both classical and entrepreneurial approaches to crop improvement. Such populations, often made at the request of enthusiastic molecular biologists to expedite identification of DNA markers for genetic mapping, offer many exciting opportunities for botanical research which have not yet been adequately exploited.

## REFERENCES

1. Harlan, J. R., De Wet, J. M. J., and Price, E. G., Comparative evolution of cereals. *Evolution*, 27, 311, 1973.
2. Paterson, A. H., Lin, Y. R., Li, Z., Schertz, K. F., Doebley, J. F., Pinson, S. R. M., Liu, S. C., Stansel, J. W., and Irvine, J. E., Convergent domestication of cereal crops by independent mutations at corresponding genetic loci, *Science,* 269, 1714, 1995.
3. Harper, J. L., *Plant Population Biology,* Academic Press, London, 1977.
4. Oyer, E. B., Gries, G. A., and Rogers, B. J., The seasonal reproduction of Johnson Grass plants, *Weeds,* 7, 13, 1959.
5. Gale, M. D., Dwarfing genes in wheat, in *Progress in Plant Breeding,* Vol. 1, Russell, G., Ed., Butterworths, London, 1985, 1–35.

6. Coe, E. H. and Neuffer, M. G., Gene loci and linkage map of corn (maize) (*Zea mays* L.) (2N=20), in O'Brien, S. J., Ed., *Genetic Maps: Locus Maps of Complex Genomes,* 6th ed., Cold Spring Harbor Laboratory Press, Cold Spring Harbor, New York, 1993, 6.157–6.189.
7. Quinby, J. R. and Karper, R. E., The inheritance of three genes that influence time of floral initiation and maturity date in milo, *J. Am. Soc. Agron.*, 37, 916, 1945.
8. Kinoshita, T., and Takahashi, M., The one hundredth report of genetical studies on rice plant — Linkage studies and future prospects, *J. Fac. Agr. Hokkaido Univ.,* 65, 1, 1991.
9. Sauer, J., *Historical Geography of Crop Plants: A Select Roster,* CRC Press, Boca Raton, FL, 1993.
10. Doggett, H., Sorghum, in *Evolution of Crop Plants,* Simmonds, N. W., Ed., Longman, Essex, UK, 1976, 112–117.
11. Chittenden, L. M., Schertz, K. F., Lin, Y., Wing, R. A., and Paterson, A. H., RFLP mapping of a cross between Sorghum bicolor and S. propinquum, suitable for high-density mapping, suggests ancestral duplication of Sorghum chromosomes, *Theor. Appl. Genet.*, 87, 925, 1994.
11a. Paterson, A. H. et al., Unpublished results, 1997.
12. Lander, E. S. and Botstein, D., Mapping Mendelian factors underlying quantitative traits using RFLP linkage maps, *Genetics*, 121, 185, 1989; and Corrigendum, *Genetics,* 136, 705, 1994.
13. Lin, Y. R., Schertz, K. F., and Paterson, A. H, Comparative mapping of QTLs affecting plant height and flowering time in the Gramineae, in reference to an interspecific *Sorghum* population, *Genetics,* 141, 391, 1995.
14. Quinby, J. R. and Karper, R. E., Inheritance of height in sorghum, *Agron. J.,* 46, 211, 1954.
15. Paterson, A. H., Schertz, K. F., Lin, Y. R., Liu, S. C., and Chang, Y. L., The weediness of wild plants: molecular analysis of genes responsible for dispersal and persistence of johnsongrass (*Sorghum halepense* L. Pers.), *Proc. Natl. Acad. Sci. U.S.A.*, 92, 6127, 1995.
16. Gizmawy, I., Kigel, J., Koller, D., and Ofir, M., Initiation, orientation, and early development of primary rhizomes in *Sorghum halepense* (L.) Pers., *Ann. Bot.,* 55, 343, 1985.
17. Bourque, J. E., Antisense strategies for genetic manipulations in plants, *Plant Sci.,* 105, 125, 1995.
18. Hayes P. M., Stucker R. E., and Wandrey G. G., The domestication of American wild rice, Zizania-Palustris, *Econ. Bot.,* 43, 203, 1989.
19. Brungardt, S., Growing wild rice can be a shattering experience. Minnesota science — Agricultural Experiment Station, University of Minnesota, 43, 4, 1988.
20. Domingo, W. E. and Crooks, D. M., Investigations with the castor-bean plant. III. Fertilizers, clipping, method of planting, and time of harvest. *J. Am. Soc. Agron.*, 37, 910, 1945.
20a. Murphy, R. P., Personal communication, 1983.
21. Pascual, M. J. and Correal, E., Mutation studies of an oilseed spurge rich in vernolic acid, *Crop Sci.,* 32, 95, 1992.
21a. Auld, D., Personal communication, 1995.
22. Massey, J. H., Harvesting *Vernonia anthelminthica* (L.) WILLD to reduce seed shattering losses, *Agron. J.,* 63, 812, 1971.

# 14 Case History in Crop Improvement: Yield Heterosis in Maize

*Charles W. Stuber*

## CONTENTS



## 14.1 INTRODUCTION

When I began my career in the study of the inheritance of quantitative traits, there were two primary options for a researcher in this area to pursue: 1) develop new theory or enhance the theory already available, or 2) test the theory in appropriate empirical investigations. I attempted a few theoretical approaches,[1-3] but soon decided that quantitative theory was not my forte. Scientists such as C. C. Cockerham, R. E. Comstock, W. D. Hanson, and O. Kempthorne could develop the statistical approaches, while I would concentrate on empirical studies.

My earlier studies in maize focused on attempts to measure the relative effects of epistasis in comparison with additive and dominance effects, not only for predicting population improvement but also as a component in heterosis or hybrid vigor. However, the inability to control many of the vagaries of the environment led me to look for some type of trait that could be measured in the laboratory and that was correlated with the traits measured in the field. I still remember a discussion with Dr. George F. Sprague in the late 1960s in which I asked him whether I should consider some type of laboratory investigations that might help to better understand the inheritance of quantitative traits and might also assist in improvement of such traits. (Dr. Sprague was Investigations Leader for the USDA-ARS national corn and sorghum research programs and was my supervisor at that time.) His comment was "Charlie, I think it is a good idea, but I strongly encourage you to continue with your field research. Your publication record could suffer a severe drought period without an active field research program." It was good advice.

At that time, the laboratory traits that seemed most amenable for this purpose were isozymes, electrophoretic variants of enzymes. In our first attempt to relate isozymes to quantitative traits, such as grain yield in maize, changes of allelic frequencies at four isozyme loci were monitored over several cycles of recurrent selection for increased yield.[4] Although the evidence was not overwhelming, the few significant associations detected between isozyme marker loci and grain yield provided the encouragement for further research in this area. It should be noted that the acronym, QTL (for quantitative trait locus), which is in common usage today, was not coined until about 1975.

With the few encouraging results cited above, it appeared that the use of isozyme loci as markers for studying quantitative traits in maize might be a viable approach, however, in the early 1970s very few isozyme loci had been mapped in maize. Also, the electrophoretic technology was not developed for efficiently characterizing numerous isozyme loci on the large populations of plants required for quantitative inheritance studies. Dr. M. M. Goodman and I began a very fruitful collaboration which has resulted in the mapping of more than 40 isozyme loci using techniques whereby several enzyme systems can be characterized on a single starch gel.[5-7] Dr. Goodman's interests in the use of this technology focused largely on evolutionary studies in maize and the characterization of genetic relationships among racial collections of maize. My research has focused on the study of the genetic basis of phenomena such as heterosis and genotype by environment interaction, and on the use of marker technology for enhancing plant breeding efficiency.

## 14.2 EARLY MARKER INVESTIGATIONS IN MAIZE

During the 1970s and early 1980s, several pioneering studies were conducted in maize that focused on associating marker genotypes with quantitative trait performances.[8-10] In several of these earlier studies, changes of allelic frequencies at a large number of isozyme marker loci were monitored over successive cycles of long-term selection in several populations of maize.[11-14] Changes of allelic frequencies at numerous loci were shown to be highly correlated with changes in several morphological and reproductive traits in maize, including the selected trait, grain yield. The impetus for the more recent activity in the use of genetic markers (isozymes and DNA-based markers) for identifying and mapping QTLs was provided by these investigations.

In addition, our laboratory and several others conducted investigations in which marker [isozyme or restricted fragment length polymorphism (RFLP)] diversity of inbred lines was correlated with performance (usually grain yield) in single-cross hybrids.[9,10] A major objective of these studies was to evaluate the use of markers for prediction of hybrid performance from crosses among untested inbred lines. The number of markers used in these studies varied from fewer than 11 isozymes to 230 RFLPs. These investigations showed that genetic distances based on marker data agreed well with pedigree data for assigning lines to heterotic groups.[15-19] However, in those studies that included field evaluations, it was concluded that isozyme and RFLP genotypic data were of limited usefulness for predicting the heterotic performance between unrelated inbred maize lines.[9,10]

Several factors contributed to the limited predictive value of marker data. In those studies using only isozyme genotypic data, the small number of isozyme loci assayed had effectively marked only a small fraction of the genome. Thus, only a limited proportion of the QTLs contributing to the hybrid response would be sampled. Also, it cannot be assumed that allelic differences at marker loci equate to allelic differences at linked QTLs or vice versa. For a limited number of markers to be effective as predictors for hybrid performance, the effects (including types of gene action) of the linked QTL "alleles" must be ascertained. Even with the large number of RFLP markers (230) used in the study reported by Smith et al.,[20] many of the conditions outlined by Bernardo[21] for effective prediction of hybrid performance based on molecular marker heterozygosity undoubtedly were not met.

## 14.3 MAPPING QTLs CONTRIBUTING TO HETEROSIS IN MAIZE

Heterosis (or hybrid vigor) has been a major contributor to the success of the commercial maize industry and is often an important component of the breeding strategies of many crop and horticultural plants. The term 'heterosis' was coined and first proposed by G. H. Shull in 1914,[22] and normally is defined in terms of F1 superiority over some measure of the performance of one or both parents. Genetic explanations for this phenomenon include: (1) true overdominance (i.e., single loci for which two alleles have the property that the heterozygote is superior to either homozygote), (2) pseudo-overdominance as proposed by Crow[23] (i.e., closely linked loci at which alleles have dominant or partially dominant advantageous effects are in repulsion phase linkage), and (3) certain types of epistasis.[24]

Our marker-facilitated research program at Raleigh, North Carolina, has identified and mapped QTLs associated with hybrid performance in 15 F2 populations derived from seven elite inbred lines and five inbred lines with a partial exotic component (Latin American, expected to be 50%). Although the early studies used only isozymes as markers, some studies used both isozymes and RFLPs.[25-28] The primary focus has been on grain yield, however, measurements recorded on individual plants in the field evaluations included dimensions, weights, and counts of numerous vegetative and reproductive plant parts as well as silking and pollen shedding dates. In the studies of (CO159 × Tx303)F2 and (T232 × CM37)F2, nearly 1900 plants were genotyped and evaluated for more than 80 quantitative traits in each population.

Results from these F2 investigations showed that QTLs affecting grain yield, and most of the other quantitative traits, were generally distributed throughout the genome, however, some chromosomal regions tended to contribute greater effects than others to trait expression. For example, major factors associated with the expression of grain yield were detected in the vicinity of *Mdh4, Adh1,* and *Phi1* on chromosome 1L; *Dia1* on chromosome 2S; *Mdh3* and *Pgd2* on chromosome 3L; *Amp3, Mdh5,* and *Pgm2* on chromosome 5S; *Idh1* on chromosome 8L; and *Acp1* on chromosome 9S. Not all of the chromosome regions were well marked for those studies using only isozymes, and presumably major factors also may have been segregating in regions devoid of marker loci in these studies.

A re-evaluation of the (CO159 × Tx303)F2 population was conducted using both RFLPs and isozymes as genetic markers.[28] By increasing the number of markers from 17 to 114, more accurate localization of QTLs was possible. Marker loci associated with grain yield (and several other traits) generally corresponded well with the earlier results where comparisons were possible. However, a number of previously unmarked genomic regions were found to contain factors with large effects on certain traits. Some of the detected genetic factors affected several yield 'component' traits whereby they counter-balanced each other, thus producing no net effect on overall grain yield.[28]

The documented number of maize populations evaluated for grain yield QTLs probably exceeds 50 (more than 20 have been studied in the author's research program), and each population has shown a unique distribution of genetic factors significantly associated with the yield trait. Certain chromosomal regions (such as 1L, 5S, and 6L) have shown QTLs in a preponderance of the reported investigations. Other regions have shown significant associations with yield only occasionally. However, at least one grain yield QTL has been reported on each of the 20 chromosome arms of maize.

The magnitudes of effects associated with specific QTLs has varied greatly among documented investigations. In the study of the (CO159 × Tx303)F2 and (T232 × CM37)F2 populations, the number of plants measured in each population (1776 and 1930, respectively) was great enough to detect factors contributing as little as 0.2% ($p < 0.01$) of the phenotypic variation in several yield-related traits.[25,26] In one study, nearly 35% of the estimated genetic variation for grain yield was attributed to a major QTL in the vicinity of the isozyme marker *Amp3* on chromosome 5.[29]

## 14.4 THE B73 × MO17 HYBRID STORY

Although we had conducted QTL mapping, and even some QTL manipulating studies in a number of populations, many of the parental lines chosen for the studies did not represent the elite materials that were being used in commercial maize production. In our initial studies, parental lines were chosen because they differed for many morphological traits (we wanted to maximize phenotypic variability) and because they differed for alleles at the isozyme marker loci that we were able to assay (we wanted to maximize the number of polymorphic marker loci). Consequently, there were many skeptics who were not convinced that the marker-facilitated technology had utility in the 'real' breeding community. Therefore, we decided to focus on a hybrid, B73 × Mo17, which had been widely used for several years in commercial production, and even today, derivatives of the parental lines are still widely used. During the past 8 years, much of our research has been based on materials generated from B73 and Mo17 or developed in combination with these lines.

The lines B73 and Mo17 are not only widely used and highly productive, they also represent the two most widely used 'heterotic groups' in U.S. maize-breeding programs. (Crosses made between lines within a 'heterotic group' usually exhibit relatively less heterosis than crosses made between lines from different groups.) The two most widely used groups in the U.S. are usually referred to as 'Iowa Stiff Stalk Synthetic' (BSSS) and 'Lancaster' or 'Lancaster Sure Crop'. A large number of the commercially grown hybrids are produced from crossing two inbred lines, one derived from each of these groups. Examples of lines derived from BSSS are B73, B37, and A632; examples from the Lancaster group are Mo17, Oh43, and Va35.

The initial study based on the inbred lines B73 and Mo17 was designed to explore the genetic bases of two important phenomena, heterosis and genotype-by-environment interaction. With the use of 76 marker loci (9 isozyme and 67 RFLP, which were linked to about 95% of the genome), QTLs contributing to heterosis for grain yield were mapped on 9 of the 10 maize chromosomes using 264 backcross families.[29] For the mapping analyses, backcrosses to B73 were analyzed separately from the backcrosses to Mo17, and for those QTLs associated with grain yield, the heterozygotes showed a higher phenotypic value than the respective homozygotes (with only one exception). These results, therefore, suggested not only overdominant gene action (most likely pseudo-overdominance), but also that the detected QTLs contributed significantly to the expression of heterosis in the B73 × Mo17 hybrid. Although the field evaluations were made in six diverse environments (four in North Carolina, one in Iowa, and one in Illinois), there was little evidence for QTL-by-environment interaction.

Data from the above study have been re-analyzed recently using a modification of the North Carolina experimental design III.[30] The latter analysis agrees with the earlier report and strongly suggests the presence of multiple linked QTLs on most chromosomes that have significant effects on grain yield. However, the results differ from the earlier report in that the design III analysis (a very powerful tool for evaluating dominance variation) favors the hypothesis of dominance of favorable genes to explain the observed heterosis. However, overdominance cannot be ruled out.

### 14.4.1 Fine-Mapping

Reducing the size of the regions identified as containing major QTLs through 'fine-mapping' is an important requirement for marker-assisted selection (MAS) and has been envisioned as the initial step for identifying single genes that ultimately could be manipulated using transformation (recombinant DNA) technology.[9] For example, in the mapping study based on the cross of B73 × Mo17 discussed above, a region near the isozyme marker *Amp3* on chromosome 5 accounted for about 20% of the phenotypic variation for grain yield. This one large region was dissected into at least two smaller QTLs.[31] Effects at these two QTLs appear to act in a dominant manner, each showing significance in one testcross used in the study but not the other. Therefore, the results indicate that the genetic factors are in repulsion phase linkage, thus supporting the dominance theory of heterosis. One other segment in the targeted region on chromosome 5 showed a significant association with yield, but the effect was not consistently expressed and may be spurious. The largest of the three segments was mapped to a 27.5-cM interval near the marker *Amp3*.[31]

A region on chromosome 4 was also targeted for fine-mapping and a previously identified QTL in the vicinity of RFLP marker *UMC15* has been placed into a 3.1-cM segment.[32] This QTL appears to show additive gene action. In the fine-mapping study, another QTL on chromosome 4 was identified about 50 cM from the *UMC15* marker. This QTL, in the vicinity of *UMC66*, appears to show overdominant gene action. However, we suspect that there may be two loci linked in repulsion phase as was found on chromosome 5.

### 14.4.2 Mapping in Stress Environments

Although there is little evidence for QTL by environment interaction in most investigations of traits such as grain yield in maize, these studies usually have been conducted under near optimum growing conditions. Therefore, the breeder has little information for making decisions on the selection of appropriate environments for evaluating QTLs that might enhance breeding strategies under less than optimal growing conditions. We have conducted a major investigation designed to evaluate and compare several aspects of QTL mapping under both optimum and stress environments.

Experimental materials used for this study were derived from the set of F3 lines used in the heterosis study reported by Stuber et al.[29] From these F3 lines, 208 F6 recombinant inbred (RI) lines were developed by single seed descent. The lines were genotyped with about 110 RFLP and 8 isozyme markers and were backcrossed to the two parental lines, B73 and Mo17. The progenies from these backcrosses have been evaluated in four field environments — two years and two locations — each with a 2 × 2 × 2 factorial arrangement. The factors were low and high soil moisture (severe drought stress and no moisture stress), low and high planting density (14,500 plants/acre and 29,000 plants/acre) and low and high nitrogen (about 50 lb/acre and 200 lb/acre) as outlined below:

| | | |
|---|---|---|
| Low moisture | Low density | Low nitrogen |
| | | High nitrogen |
| | High density | Low nitrogen |
| | | High nitrogen |
| High moisture | Low density | Low nitrogen |
| | | High nitrogen |
| | High density | Low nitrogen |
| | | High nitrogen |

Results from the RI line backcrosses were compared with earlier results using F3 line backcrosses when both were evaluated under nonstress environmental conditions. The comparisons showed that QTL mapping did not differ in the two generations of lines generated from the B73 × Mo17 cross. When comparing results from the RI line backcrosses evaluated in individual high stress environments with those evaluated in nonstress environments, there were nearly tenfold differences in grain yield. However, the results showed that no new QTLs (affecting traits such as grain yield) were detected under such high stress conditions.[29a] This agrees with earlier results showing little QTL by environment interaction, and implies that breeders and geneticists can rely on mapping data from favorable environments for breeding for materials adapted to stress environments.

### 14.4.3 Enhancement of B73 and Mo17 Lines

Earlier mapping studies in our program suggested that two elite inbred lines, Tx303 and Oh43, contained genetic factors that might contribute to the heterotic response of the B73 × Mo17 single-cross hybrid. Therefore, when the initial B73 × Mo17 study was conducted,[29] a companion study was made to identify and locate the putative genetic factors in Tx303 and Oh43. Planned mean comparisons among backcrosses and testcrosses in the two studies were used to identify six chromosomal segments in Tx303 that (if transferred into B73) would be expected to enhance the B73 × Mo17 hybrid response for grain yield. Likewise, another six segments were identified in Oh43 for transfer into Mo17 that would also be expected to enhance the B73 × Mo17 hybrid response.[33]

Three backcross generations were used for transfer of the identified segments into the target lines, B73 and Mo17. BC2 families were analyzed for marker genotypes which were then used to select individuals for the third backcrosses. Individuals were selected if they had the desired marker genotype in the vicinity of the donor segment to be transferred and the recipient line's genotype in the remainder of the genome. At least one marker was assayed on each of the 20 chromosome arms, and usually two markers were genotyped in the vicinity of the donor segment being transferred. Two selfing generations followed the third backcross. At each backcrossing and selfing stage, marker genotyping of plants was conducted in the same manner as for the BC2 families. However, if a marker locus became fixed in a line, that marker was not evaluated in that line in succeeding generations and only segregating loci were analyzed. This reduced laboratory analyses considerably.

After the second selfing generation, 141 BC3S2 modified B73 lines were identified for testcrossing to the original Mo17. Likewise, 116 BC3S2 modified Mo17 lines were targeted for testcrossing to the original B73. These 257 testcross hybrids were evaluated in replicated field plots in three locations in North Carolina.[34] Of the modified B73 × original Mo17 testcrosses, 45 (32%) yielded more than the check hybrid (normal B73 × normal Mo17) by at least one standard deviation. Only 15 (11%) yielded less than the check. Evaluations of modified Mo17 × original B73 testcrosses showed that 51 (44%) yielded more grain than the normal check hybrid and only 10 (<9%) yielded less than the check hybrid. The highest yielding hybrids exceeded the check by 8 to 11% (9 to 12 bushels/acre).

Based on these initial evaluations, the better performing modified lines were selected for intercrossing and were designated as "enhanced" lines. Of the "enhanced" B73 lines, 15 were chosen for crossing with 18 "enhanced" Mo17 lines producing 93 hybrids that were evaluated in replicated field trials in North Carolina in 1993.[33] At a planting density of about 29,000 plants/acre, only 15 of the test hybrids produced less grain than either of 2 check hybrids (normal B73 × normal Mo17 and a high yielding commercial hybrid, Pioneer 3165). However, six exceeded the checks by two standard deviations or more. The two highest yielding "enhanced" B73 × "enhanced" Mo17 hybrids exceeded the checks by more than 15% (22 to 24 bushels/acre).

These "enhanced" B73 × "enhanced" Mo17 hybrids were evaluated again in 1994, with some additional reciprocal crosses, which increased the number of test hybrids to 149.[33] Of these hybrids, 20 exceeded the check hybrids (same as in 1993) by two standard deviations or more, with the 3 best exceeding the checks by 12 to 15% (18 to 24 bushels/acre). Although the rankings changed slightly over the 2 years of testing, the average yields over both years for the best yielding hybrids exceeded the check hybrids by 8 to 10% (10 to 16 bushels/acre). More importantly, the parental lines that showed superior general combining ability in 1993 also showed similar performance in 1994. Evaluations of the highest yielding "enhanced" B73 × "enhanced" Mo17 hybrids were again conducted in 1995 in North Carolina and results corroborated those of the previous 2 years. In addition, nine of these hybrids were evaluated at four locations in central Iowa by Pioneer Hi-Bred International. Of the nine, three exceeded the grain yields of the best Pioneer commercial check hybrids by 6 to 10 bushels/acre. These results were surprising because all of the selections were based on evaluations in North Carolina.

Results from the introgression of the targeted segments from Tx303 into B73 and from Oh43 into Mo17 have demonstrated that marker-facilitated backcrossing can be successfully employed to manipulate and improve complex traits such as grain yield in maize. Not all of the six targeted segments have been successfully transferred into a single modified B73 or modified Mo17 line. There appears to be some indication that there may be no advantage in transferring more than two to four segments. In fact, there is some indication that there could be a disadvantage. Increasing the number of transferred segments may be replacing the recipient genome with an excessive amount of linked donor chromosomal segments that could cause a deleterious effect. Also, epistatic interactions between a larger number of introgressed segments may result in a negative effect (see Chapter 15, this volume). In addition, favorable epistatic complexes in coupling phase (e.g., between recurrent parent alleles) could be disrupted. Further evaluations are necessary to determine the effects of larger numbers of transferred segments.

### 14.4.4 Breeding Scheme Using Near-Isogenic Lines (NILs)

Although the results discussed above showed that the enhancement of lines B73 and Mo17 was successful, the procedure for development of the "enhanced" lines (NILs) was very inefficient and would not be recommended for a practical breeding program. That procedure required that the targeted segments (containing the putative QTLs) be identified in the donor lines prior to transfer to the recipient lines. In our maize program at Raleigh, we have outlined, and have tested, a marker-based breeding scheme for systematically generating superior lines without any prior identification of QTLs in the donor source(s). The identification and mapping of QTLs in the donor is a bonus derived from the evaluation of the NILs generated. Choice of the donor usually will be based on prior knowledge of its potential for providing favorable genetic factors, and, in maize, may involve appropriate heterotic relationships.

The procedure involves the generating of a series of NILs by sequentially replacing segments of an elite line (the recipient genome) with corresponding segments from the donor genome. The objective is for the group of NILs generated to contain, collectively, the complete genome of the donor source, with each NIL containing a different segment from the donor. Marker-facilitated backcrossing, followed by marker-facilitated selfing to fix the introgressed segments, is used to monitor the targeted segments from the donor and to recover the recipient genotype in the remainder of the genome. The number of backcrosses required will depend on the number of evaluations that can be made in the marker laboratory.

In maize, the NILs are then crossed to an appropriate tester(s) to create hybrid testcross progeny that are evaluated in replicated field trials (with appropriate checks) for the desired traits. The superior performing testcrosses will be presumed to have received donor segments that contain favorable QTLs. Therefore, QTLs are mapped by function, which should be an excellent criterion for QTL detection. The breeding scheme not only creates "enhanced" lines that are essentially identical to the original elite line, but it also provides for the identification and mapping of QTLs as a bonus with no additional cost. Obviously, the scheme is based on having a reasonably good marker map with alternate alleles in the donor and recipient lines.

This breeding strategy should be an excellent procedure for tapping into the potential of exotic germplasm. Furbeck[35] used this procedure to develop a small sample of NILs using the elite line, Mo17, as the recipient. An exotic population, derived from the accessions Cristal (MGIII) a Brazilian racial collection and Arizona (AYA41) a Peruvian collection, was used as the donor. Figure 14.1 shows some of the significant introgressed segments and traits involved. Using this procedure, positive segments from the exotic (such as the segment associated with *Phi1* on chromosome 1) are immediately available in a adapted background (Mo17) for further breeding use. Also, and equally important, segments with negative effects (such as those associated with *Dia1* and *Dia2*, both –9.5 bushels/acre effects) are eliminated. Moreover, because each segment was incorporated independently, the detection of positive effects is not biased by negative adjacent segments (e.g., *Phi1*, +10.5 bushels/acre, and *Dia2*, –9.5 bushels/acre, on chromosome 1). It should be noted that Eshed and Zamir[36] have effectively used this breeding procedure to extract favorable genetic factors from a wild species of tomato.

## 14.5 CONCLUSIONS

Molecular-marker technology has been demonstrated to be effective for identifying and mapping QTLs in maize, as well as in many other plant species, and for studying phenomena such as heterosis and genotype by environment interaction. The positive results from marker-facilitated selection and introgression studies should encourage the use of this technology by commercial breeders for transferring desired genes between breeding lines. Appropriate use of markers should increase the precision and efficiency of plant breeding, as well as expedite the acquisition of favorable genes from exotic populations or from wild species.

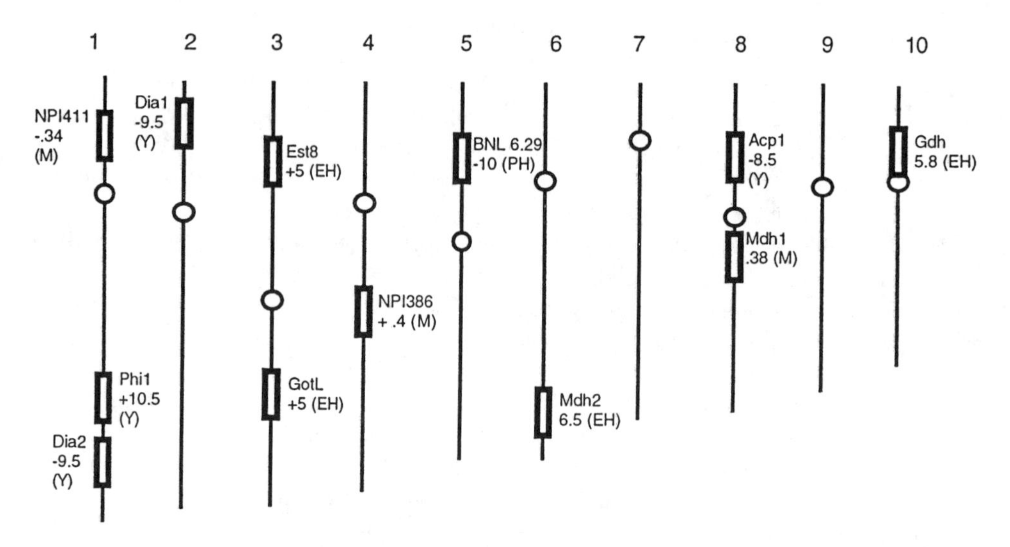


**FIGURE 14.1** Maize chromosome map showing locations of positive and negative effects on several traits derived from segments transferred from an exotic population (Cristal × Arizona) into the elite inbred line Mo17. Y = yield (bushels/acre); M = moisture (%); EH = ear height (cm); PH = plant height (cm).

Recent studies showing the high degree of homology and synteny between sorghum and maize genomes[37] should greatly enhance the efficiency of mapping quantitative traits in both species. Also, comparative mapping with other monocots has demonstrated many examples of conserved gene order and functions,[38-40] which should prove very beneficial in the identification and mapping of useful genes in maize.

## REFERENCES

1. Stuber, C. W. and Cockerham, C. C., Gene effects and variances in hybrid populations, *Genetics,* 54, 1279, 1966.
2. Stuber, C. W., Estimation of genetic variances using inbred relatives, *Crop Sci.*, 10, 129, 1970.
3. Brim, C. A. and Stuber, C. W., Application of genetic male sterility to recurrent selection schemes in soybeans, *Crop Sci.,* 13, 528, 1973.
4. Stuber, C. W. and Moll, R. H., Frequency changes of isozyme alleles in a selection experiment for grain yield in maize (*Zea mays* L.), *Crop Sci.*, 12, 337, 1972.
5. Goodman, M. M., Stuber, C. W., Newton, K., and Weissinger, H. H., Linkage relationships of 19 enzyme loci in maize, *Genetics,* 96, 697, 1980.
6. Goodman, M. M. and Stuber, C. W., Maize, in *Isozymes in Plant Genetics and Breeding, Part B,* Tanksley, S. D. and Orton, T. J., Eds., Elsevier Scientific, Amsterdam, 1983, 1.
7. Stuber, C. W., Wendel, J. F., Goodman, M. M., and Smith, J. S. C., Techniques and scoring procedures for starch gel electrophoresis of enzymes from maize (*Zea mays* L.), N.C. Agric. Res. Serv., N. C. State Univ., *Tech. Bull,* No. 286, 1988, 87 pp.
8. Stuber, C. W., Biochemical and molecular markers in plant breeding, in *Plant Breed. Rev.*, 9, 37, 1992.
9. Stuber, C. W., Breeding multigenic traits, in *DNA-Based Markers in Plants,* Phillips, R. and Vasil, I., Eds., Kluwer Academic Publishers, 1994, 97.
10. Stuber, C. W., Heterosis in plant breeding, in *Plant Breed. Rev.*, 12, 227, 1994.
11. Guse, R. A., Coors, J. G., Drolsom, P. N., and Tracy, W. F., Isozyme marker loci associated with cold tolerance and maturity in maize, *Theor. Appl. Genet.,*76, 398, 1988.
12. Kahler, A. L., Associations between enzyme marker loci and agronomic traits in maize, *Proc. 40th Annu. Corn and Sorghum Industry Research Conf.,* American Seed Trade Assoc., 40, 66, 1985.
13. Pollak, L. M., Gardner, C. O., and Parkhurst, A. M., Relationships between enzyme marker loci and morphological traits in two mass selected maize populations, *Crop Sci.,* 24, 1174, 1984.

14. Stuber, C. W., Moll, R. H., Goodman, M. M., Schaffer, H. E., and Weir, B. S., Allozyme frequency changes associated with selection for increased grain yield in maize (*Zea mays* L.), *Genetics,* 95, 225, 1980.
15. Melchinger, A. E., Lee, M., Lamkey, K. R., Hallauer, A. R., and Woodman, W. L., Genetic diversity for restriction fragment length polymorphisms and heterosis for two diallel sets of maize inbreds, *Theor. Appl. Genet.,* 80, 488, 1990.
16. Melchinger, A. E., Lee, M., Lamkey, K. R., and Woodman, W. L., Genetic diversity for restriction fragment length polymorphisms: relation to estimated genetic effects in maize inbreds, *Crop Sci.,* 30, 1033, 1990.
17. Godshalk, E. B., Lee, M., and Lamkey, K. R., Relationship of restriction fragment length polymorphisms to single-cross hybrid performance of maize, *Theor. Appl. Genet.,* 80, 273, 1990.
18. Livini, C., Ajmone-Marsan, P., Melchinger, A. E., Messmer, M. M., and Motto, M., Genetic diversity of maize inbred lines within and among heterotic groups revealed by RFLPs, *Theor. Appl. Genet.,* 84, 17, 1992.
19. Zehr, B. E., Dudley, J. W., and Chojecki, J., Some practical considerations for using RFLP markers to aid in selection during inbreeding of maize, *Theor. Appl. Genet.,* 84, 704, 1992.
20. Smith, O. S., Smith, J. S. C., Bowen, S. L., Tenborg, R. A., and Wall, S. J., Similarities among a group of elite maize inbreds as measured by pedigree, $F_1$ grain yield, grain yield heterosis, and RFLPs, *Theor. Appl. Genet.,* 80, 833, 1990.
21. Bernardo, R., Relationship between single-cross performance and molecular marker heterozygosity, *Theor. Appl. Genet.,* 83, 628, 1992.
22. Hayes, H. K., Development of the heterosis concept, in *Heterosis,* Gowen, J. W., Ed., Iowa State College Press, Ames, IA, 1952, 49.
23. Crow, J. F., Dominance and overdominance, in *Heterosis*, Gowen, J. W., Ed., Iowa State College Press, Ames, IA, 1952, 282.
24. Schnell, F. W. and Cockerham, C. C., Multiplicative vs. arbitrary gene action in heterosis, *Genetics,* 131, 461, 1992.
25. Edwards, M. D., Stuber, C. W., and Wendel, J. F., Molecular marker— facilitated investigations of quantitative trait loci in maize. I. Number, distribution, and types of gene action, *Genetics,* 116, 113, 1987.
26. Stuber, C. W., Edwards, M. D., and Wendel, J. F., Molecular marker facilitated investigations of quantitative trait loci in maize. II. Factors influencing yield and its component traits, *Crop Sci.,* 27, 639, 1987.
27. Abler, B. S. B., Edwards, M. D., and Stuber, C. W., Isoenzymatic identification of quantitative trait loci in crosses of elite maize inbreds, *Crop Sci.,* 31, 267, 1991.
28. Edwards, M. D., Helentjaris, T., Wright, S., and Stuber, C. W., Molecular-marker-facilitated investigations of quantitative trait loci in maize. IV. Analysis based on genome saturation with isozyme and restriction fragment length polymorphism markers, *Theor. Appl. Genet.,* 83, 765, 1992.
29. Stuber, C. W., Lincoln, S. E., Wolff, D. W., Helentjaris, T., and Lander, E. S., Identification of genetic factors contributing to heterosis in a hybrid from two elite maize inbred lines using molecular markers, *Genetics,* 132, 823, 1992.
29a. Stuber, C. W., LeDeaux, J. R., and Graham, G. I., Unpublished data, 1997.
30. Cocherham, C. C. and Zeng, Z-B., Design III with marker loci, *Genetics,* 143, 1437, 1996.
31. Graham, G. I., Wolff, D. W., and Stuber, C. W., Characterization of a yield quantitative trait locus (QTL) in maize by fine mapping, *Crop Sci.,* 37, to be published.
32. Graham, G. I., Mapping quantitative trait loci using specialized populations, Ph.D. thesis, North Carolina State University, Raleigh, (Diss. Abstr.), 1996.
33. Stuber, C. W., Success in the use of molecular markers for yield enhancement in corn, *Proc. 49th Annu. Corn and Sorghum Industry Research Conf.,* American Seed Trade Assoc., 49, 232, 1994.
34. Stuber, C. W. and Sisco, P. H., Marker-facilitated transfer of QTL alleles between elite inbred lines and responses in hybrids, *Proc. 46th Annu. Corn and Sorghum Industry Research Conf.,* American Seed Trade Assoc., 46, 104, 1991.
35. Furbeck, S. M., The development and evaluation of molecular-marker derived near isogenic lines to study quantitative traits in maize, Ph.D. thesis, North Carolina State University, Raleigh, (Diss. Abstr. DA9330287), 1993.
36. Eshed, Y. and Zamir, D., An introgression line population of *Lycopersicon pennellii* in the cultivated tomato enables the identification and fine-mapping of yield-associated QTL, *Genetics,* 141, 1147, 1995.

37. Whitkus, R., Doebley, J., and Lee, M., Comparative genome mapping of sorghum and maize, *Genetics*, 132, 119, 1992.
38. Ahn, S. and Tanksley, S. D., Comparative linkage maps of rice and maize genomes, *Proc. Natl. Acad. Sci. U.S.A.*, 90, 7980, 1993.
39. Ahn, S., Anderson, J. A., Sorrells, M. E., and Tanksley, S. D., Homoeologous relationships of rice, wheat, and maize chromosomes, *Mol. Gen. Genet.*, 241, 483, 1993.
40. Lin, Y.-R., Schertz, K. F., and Paterson, A. H., Comparative analysis of QTLs affecting plant height and maturity across the Poaceae, in reference to an interspecific sorghum population, *Genetics*, 141, 391, 1995.

# 15 Case History in Germplasm Introgression: Tomato Genetics and Breeding Using Nearly Isogenic Introgression Lines Derived from Wild Species

*Dani Zamir and Yuval Eshed*

## CONTENTS



## 15.1 INTRODUCTION

Tomato (*Lycopersicon esculentum,* 2n = 24) is a self-pollinated vegetable crop which is widely cultivated. Tomato's well-endowed genetic resources include 877 monogenic mutations and more than 1000 accessions representing eight wild species.[1] More than 1000 RFLP markers, spanning 1200 cM, have been placed on the tomato linkage map, providing the basis for resolving quantitative traits into discrete Mendelian factors.[2,3] The self-pollinated nature of the cultivated tomato enables the construction of populations that segregate for two alleles only at each locus, thereby simplifying analyses of the associations between markers and quantitative traits. Most quantitative trait loci (QTL)-mapping studies in tomato have been conducted on progenies of interspecific crosses because within *L. esculentum* there is very low DNA-marker variation.[4] Another reason for attempts to map QTL originating from exotic germplasm is the realization that the gene pool of the cultivated tomato, as is the case for many other crop plants, contains only a small fraction of the variation available within the genus.[5]

0-8493-7686-6/98/$0.00+$.50

In this case history, we review some of the major questions concerning genetic and breeding aspects of tomato. We demonstrate that nearly isogenic lines (NILs) representing the entire genome provide a more comprehensive view of the inheritance of traits showing continuous variation than conventional segregating populations. NILs for QTL (QTL-NILs) constructed in elite genetic backgrounds enhance the rate of variety development through the union of QTL identification and plant breeding.

## 15.2 NILS FOR THE ANALYSIS OF POLYGENIC TRAITS

### 15.2.1 HOW MANY GENES AFFECT TRAITS SHOWING CONTINUOUS VARIATION?

Estimates of the number of loci underlying quantitative variation vary greatly depending on the experimental approach used to answer this basic question. For example, using biometric techniques in a 76-generation divergent-selection experiment in maize, Dudley[6] estimated the number of effective factors controlling percent kernel protein to be 122. A much lower estimate of the number of QTL affecting this trait was obtained through marker analysis of the same population, which revealed 22 significant marker loci, some of them tightly linked.[7] The differences between these estimates reflect the polygenic and oligogenic models for the nature of quantitative variation.[8] These models are evaluated using the tomato system.

Fruit mass (FM) and fruit total soluble solids, measured in degrees Brix (B), have been analyzed in many interspecific tomato crosses. These traits show striking differences between the parental species: FM of the cultivated varieties ranged from 50 to 70 g with values of 4-5°B as compared to the wild species which were characterized by small fruits (1 to 5 g and 10 to 15°B). The minimum number of loci affecting FM in an interspecific tomato cross was estimated, using Wright's method, at between 10 and 11;[9,10] recently, Zeng et al.[11] suggested that this estimator is on the order of one third of the actual number of genes.

Several population structures and interspecific crosses were used to examine associations between markers covering the entire genome and the traits FM and Brix (Table 15.1). The simplest populations, F2 or BC, consist of plants which segregate on average for a large portion of the wild-species genome, and both marker and phenotypic data were collected on the same individuals. The estimated number of marker-identified QTL affecting FM in three such populations was between 6 and 7 (Table 15.1). A different population structure used to estimate numbers of QTL is recombinant inbreds (RIs) descended from the F2 of *L. esculentum* with *L. cheesmanii*. Plants of the RIs also contained a large portion of the wild-species genome but provided the advantage of being a permanent mapping resource which could be phenotyped in different years. Using this population, the number of FM QTL was estimated at 12; this higher estimate could have arisen from the overall homozygous nature of the lines which enabled a better detection of the additive components of the variation, and/or from the more accurate phenotyping of the individuals based on replicated evaluations. In the above populations, the mapping resolution for detecting a QTL is approximately 20 cM[16,17] and therefore, using a simplistic model, the maximum number of detectable QTL in a genome of 1200 cM is approximately 60.

The *L. pennellii* introgression line (IL) population was designed to generate QTL-NILs. It consists of 50 lines, each containing a single homozygous RFLP-defined wild-species chromosome segment. Together the lines provide complete coverage of the genome and a set of NILs to their recurrent parent — the processing tomato cultivar M82 (Figure 15.1).[18] The genetic assumption underlying the identification of QTL using the NILs was that any phenotypic difference between an IL and its nearly isogenic control is due to a QTL that resides on the chromosome segment introgressed from *L. pennellii*.[15] The minimum number of $p < 0.05$-significant QTL affecting a trait in the ILs was calculated on the basis of the following assumptions: (1) each IL affecting a quantitative trait carries only a single QTL; (2) two overlapping introgressions with a significant

**TABLE 15.1**
**The Number of Significant Effects ($p < 0.05$) of Wild Species QTL on FM and B**

| Wild species | Population structure | Population size | No. of FM QTL | No. of B QTL | Ref. |
|---|---|---|---|---|---|
| *Lycopersicon chmielewskii* | BC1 | 237 | 6 | 4 | 3 |
| *L. cheesmanii* | F2 | 350 | 7 | 4 | 12 |
| *L. pimpinellifolium* | BC1 | 257 | 7 | 3 | 13 |
| *L. cheesmanii* | RI | 97(6 reps) | 12 | 14 | 14 |
| *L. pennellii* | IL | 50(6 reps) | 18 | 23 | 15 |

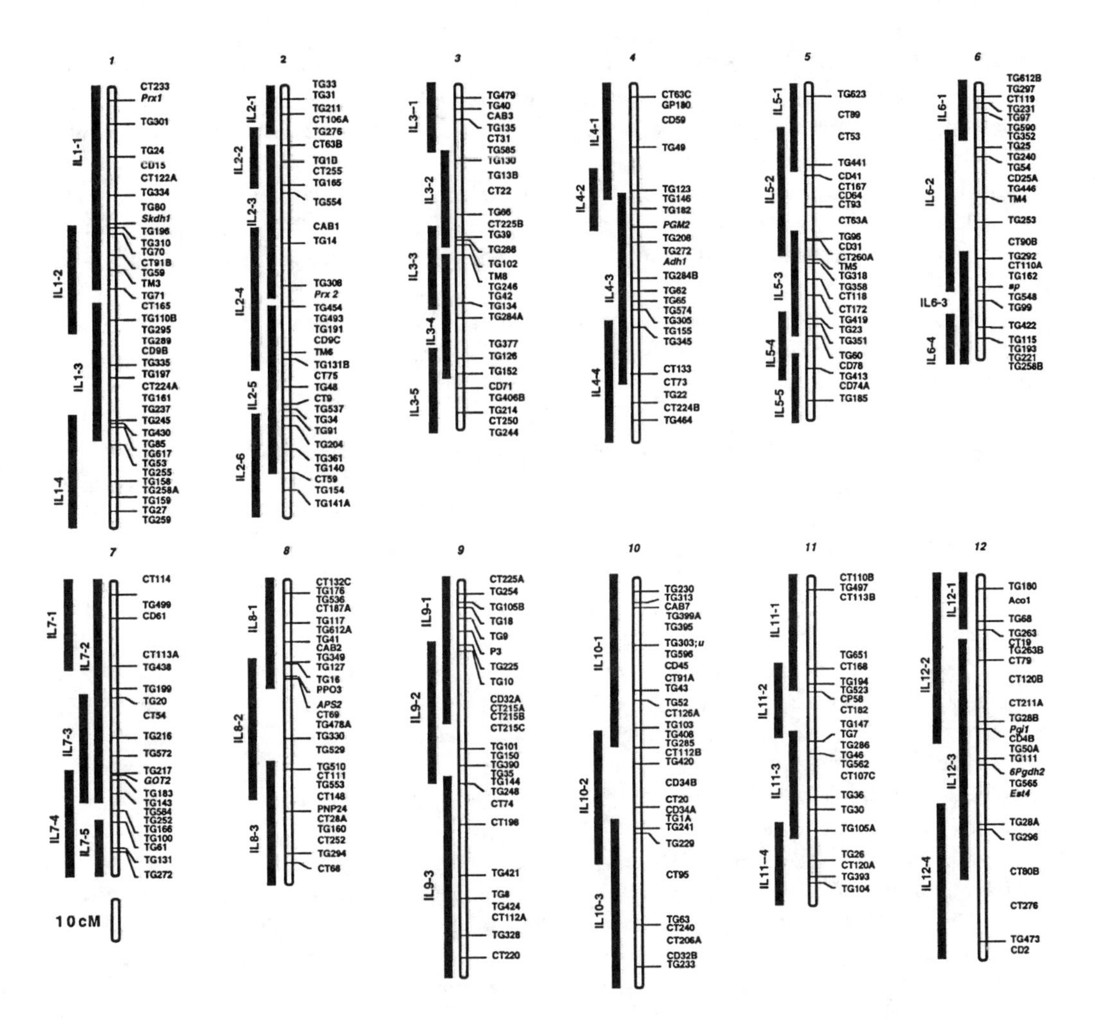


**FIGURE 15.1** Chromosomal locations, sizes and identities of the 50 *Lycopersicon pennellii* ILs. The genetic map was constructed on the basis of 119 BC1 plants as described by Eshed et al.[19] Mapped markers are connected to the chromosome with a line and markers not assayed on the BC1 map are placed according to their approximate positions based on Tanksley et al.[2] Each line was probed with all markers, and the ones showing wild-species alleles are marked with bars to the left of the chromosome.

effect on a trait (in the same direction relative to the control) carry the same QTL. Therefore, in the ILs the maximum number of detectable QTL is approximately 30. Despite this limitation, twice as many FM QTL were identified as compared to the previous populations (Table 15.1). The sensitivity of the ILs for identifying QTL was even more pronounced for B, where two to six times as many QTL were identified as compared to the other populations.

Using the *L. pennellii* ILs, QTL were mapped to various chromosome segments originating from the wild species. However, the effects associated with an introgressed segment could be due to the existence of one or more loci. A 60-cM segment on the long arm of chromosome 2 was responsible in the homozygous condition for a 60% reduction in FM relative to the control, M82. This chromosomal region apparently harbors FM QTL which are common to a number of wild tomato species.[20] Fine-mapping analysis of recombinant lines for that region identified three linked loci with a similar effect on FM; two of them were placed on an interval of 3 cM. Finer mapping may reveal additional FM QTL in these regions.[21]

Quantitative effects which appear to be associated with a single locus were inferred from cases of rare transgressive segregation. Using the ILs, 18 QTL for FM were identified but in only two cases (IL7-5 and IL12-1-1 with introgression sizes of 15 and 4 cM, respectively) alleles of the small-fruited wild species were associated with larger fruits; these effects were consistent in trials conducted in different years and genetic backgrounds.[21] Because transgression for FM in the entire 1200-cM genome was limited to these regions, it is unlikely that more than a single factor resides on the two aforementioned introgressions. A similar rationale can be developed for IL2-1 (introgression length of 16 cM) which carries the only QTL, out of 23 identified, responsible for reduced B relative to M82. This same line is also unique in its effects on plant weight and yield, suggesting the existence of a single pleiotropic factor.

Several features of the IL population contributed to its efficiency in detecting QTL, even when only a few replicates of each genotype were evaluated: (1) The lines contain single RFLP-defined introgressions, some of which produce effects of relatively large magnitude in which all the quantitative genetic variation between the NILs is associated with the introgressed segment;[22] (2) the permanent nature of the lines enabled testing of the introgression effects in different years. The results showed high reproducibility of the effects of the QTL mapped to the different introgressed chromosome segments;[21,22] (3) elimination of the "overshadowing effect" of major QTL on the ability to detect minor QTL (a major QTL contributes to large phenotypic variation, thereby masking the effects of other QTL segregating in the same population); (4) elimination of epistatic interactions between unlinked QTL; and finally, (5) the simple statistical procedure that relies on comparison to a common control and is therefore less affected by the need for experimentwise error. The resolving power of the nearly isogenic approach for detecting QTL is supported by studies which could not identify specific QTL in a whole-genome segregating population, but uncovered them after elimination of most of the genetic variation which was not associated with the studied loci. For example, Paterson et al.[23] detected five QTL for B in lines carrying a small portion of the *L. chmielewskii* genome whereas none of them had a significant effect in a BC1 analysis of the same cross.[3]

Overall, these results indicate that conventional segregating populations underestimate the effective number of QTL as compared to QTL-NILs. In view of the additional evidence revealed by the analysis of QTL epistasis (to be discussed later), we conclude that numerous loci with effects of different magnitudes underlie quantitative variation. Our estimate of the number of QTL affecting FM is consistent with the biometrical estimates as interpreted by Zeng et al.[11]

### 15.2.2 What Types of Gene Actions Are Revealed by QTL Studies?

Gene action of the QTL detected using the IL population was determined by comparing the homozygous ILs to their hybrids with the recurrent parent. FM and B QTL were intermediate between additivity and dominance;[15] this mode of inheritance is in agreement with results obtained by analysis of an F2 generation.[12] In contrast, fruit yield (Y) was strongly associated with overdominance whereby some of the heterozygous ILs had higher values relative to their corresponding homozygous parents. Detailed mapping analysis of a chromosome 1 introgression which showed overdominance for Y suggested the existence of two loci in *cis* with opposite effects. This result was therefore consistent with the pseudo-overdominance model for heterosis.[24] However, for the other heterotic introgressions, including *dw-1*,[25] the issue of the mode of gene action for heterosis is still unresolved. It is interesting to note that the wild species used for the tomato mapping studies were highly inferior to the cultivated variety with respect to Y, yet chromosome segments from these species contribute to the increased Y of commercially grown varieties.[26] This transgressive segregation is frequent for Y and for seedling morphological traits, whereas for FM and B, transgression was rare.[12,15,25]

### 15.2.3 How Reproducible Is the Effect of an Identified QTL?

Mendelian factors underlying quantitative traits in an interspecific tomato cross were compared in F2 and F3 generations of the same population.[12] Of 11 FM QTL identified in both generations in a trial conducted in California, six were significant both in F2 and F3. Of the five B QTL, two were significant in both generations. Differences between generations can result from interactions with the environment and/or may indicate that the resolution power of such populations is limited to QTL with large effects. In contrast, of 33 yield-associated QTL identified in a 2-year trial of selected ILs, 28 were significant in both experiments.[21]

### 15.2.4 Is the Difference between QTL-NILs Associated with the Introgressed Segment?

The use of the *L. pennellii* ILs to identify QTL is based on RFLP results which indicated that each line contains a single wild-species introgression. However, some of the lines may include small unidentified introgressions, and these segments may be responsible for the observed phenotypic effects. To test whether the difference between the IL and its nearly isogenic control lies solely in the introgressed segment, a simple experiment was performed using eight selected ILs.[22] An F2 resulting from a cross between each IL and M82 was subjected to RFLP analysis, and plants homozygous for the cultivated-tomato chromosome segment were compared quantitatively to M82. In no case were any differences detected, indicating that the observed phenotypic differences (which were verified using the plants carrying the *L. pennellii* introgressions) are due to the mapped chromosome segment.

### 15.2.5 Are QTL Conserved between Different Species?

To answer this question we compared the mapping results of the experiments presented in Table 15.1 for B and FM. Because different RFLP markers were used in the different studies, we aligned the QTL positions on the basis of the saturated tomato linkage map[2] (Figure 15.2). An additional problem in comparing these populations is the different criteria used to identify the QTL: interval mapping[3,12] vs. ANOVA.[13] We positioned the QTL in a region where the LOD score exceeded two, or $p < 0.01$. The three studies identified almost the same number of QTL for FM (six to seven) and B (three to four; Table 15.1). Moreover, two FM QTL on chromosomes 1 and 11 were identified in the three populations and two QTL, on chromosomes 2 and 6, in two populations. For B, QTL on chromosomes 3 and 6 were identified by the three studies. Overall, half of the QTL identified were common in the three studies involving three different wild species, carrying both red and green fruits.

The small number of QTL identified and their conservation across species reflects what we call major QTL. The identified QTL on chromosome 6 (Figure 15.2) represents the self-pruning (*sp*) mutation which controls plant habit in tomato; processing-tomato varieties are determinate

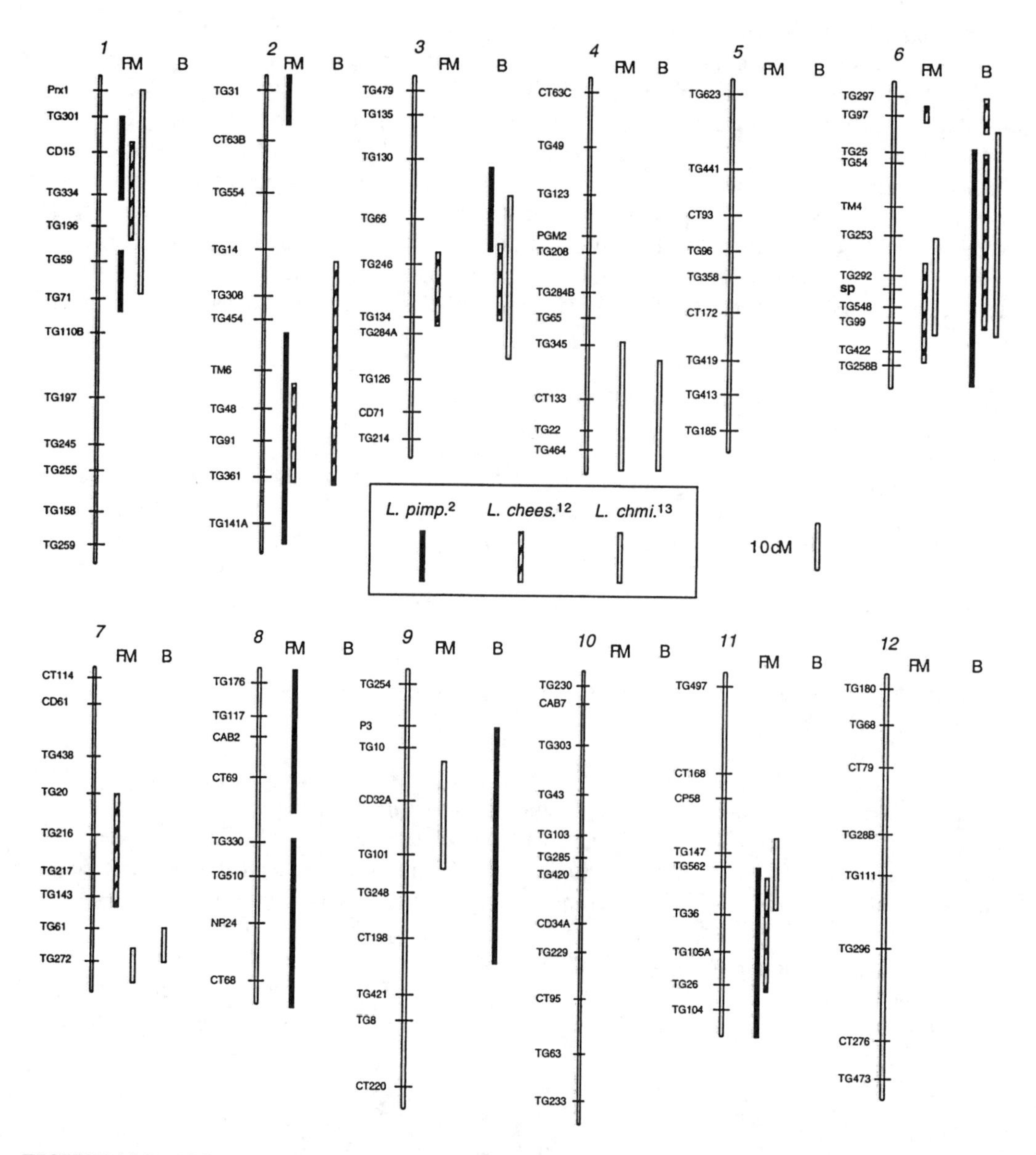


**FIGURE 15.2** Map positions of FM and B QTL identified in whole genome conventional segregating generations involving three different wild tomato species.[2,12,13]

due to homozygosity for the recessive allele and as a result have one or two leaves between inflorescences; this is in contrast to three leaves in the indeterminate wild species. The effect of this mutation on B was demonstrated by Emery and Munger[27] using NILs, and is associated with different sink-source relationships between the genotypes.

Similar map positions of major QTL among different crosses, species and even taxa have led to the postulation that many QTL are conserved.[28,29] A possible interpretation of these results is that major QTL, which in some cases can be considered qualitative Mendelian genes, are the ones efficiently identified using conventional segregating populations. The use of different experimental designs, such as ILs, should enable the detection of minor QTL where less conservation is expected.

### 15.2.6 How Prevalent Is QTL Epistasis?

An organism's phenotype results from the combined action of numerous Mendelian genes affecting qualitative and quantitative traits. Epistatic interactions of major genes have often been inferred from modified segregation ratios. For genes affecting quantitative traits, epistasis has been defined as the deviation from the sum of the independent effects of the individual genes.[30] Epistasis between QTL assayed in populations segregating for an entire genome has been found at a frequency close to that expected by chance alone.[21] In contrast, recent studies focusing on NILs, in which most genetic variation is associated with the studied QTL, have detected epistasis more frequently. Interaction between two QTL for number of spikelets in a maize-teosinte cross was not significant in the F2 generation but was highly significant when a combination of NILs for these QTL were evaluated.[31] The power of the nearly isogenic approach for detecting QTL epistasis was also demonstrated for genes affecting bristle number in *Drosophila*.[32,33] In F2 generations, some of the genotypic classes are represented at a lower frequency than others, and therefore a very large population would have to be grown in order to detect QTL interactions.[16] In NIL experiments, the frequency of two-locus genotypes can be balanced, and replicated measurements of identical genotypes can be assayed.

Unlike QTL-mapping studies in conventional segregating populations which have generally uncovered little evidence for epistasis, we showed, using the tomato NILs, that QTL epistasis is a significant component in determining phenotypic values.[21] We analyzed interactions in a half-diallel crossing scheme generated among 10 selected homozygous ILs. The 45 derived double heterozygotes were evaluated in the field for four yield-associated traits, along with the 10 single heterozygotes and M82. Of 180 (45 × 4) tested interactions, 28% were epistatic ($p < 0.05$) on both linear and geometric scales. Although the frequency of epistasis was high, additivity was the major component in the interaction of pairs of QTL. Nevertheless, the detected epistasis showed a consistent trend which was less than additive, i.e., the effect of the double heterozygotes was of a smaller magnitude than the sum of the effects of the corresponding single heterozygotes. The resolving power of the tomato NILs for a study of continuous traits was seen previously, when more QTL affecting FM and B were identified as compared to those found using other conventional segregating interspecific populations (Table 15.1). Moreover, we detected a higher-than-expected frequency of epistatic combinations between introgressions which, independently, had no significant effect on the measured trait. This observation indicates that the number of QTL is even higher than previously estimated.

The diminished additivity of QTL is consistent with our initial unexplained observation, revealed by comparing the QTL identified using the entire IL population with the parental species phenotypes (Figure 15.3). M82 had a value of 4.3°B, whereas the F1 interspecific hybrid with *L. pennellii* had a value of 10.2°B. This difference (5.9°B) is attributed to genes originating from the wild species. Summation of all the independent effects of the QTL (using the minimal estimate defined previously) in a heterozygous condition (IL × M82) yielded a value of 12.5, which is more than twice as high as the difference between the F1 interspecific hybrid and M82. A more pronounced effect of the less-than-additive trend was revealed when the effects of all the heterozygous ILs on B were added together, regardless of their significance. The cumulative effect on B, after correcting for the overlap in genome representation by the ILs (480 cM out of 1200 cM; Figure 15.1) was 16.3, almost three times higher than expected based on the M82 and F1 values. A similar less-than-additive trend was detected for FM. This calculation suggests that diminished additivity is intensified when a higher number of segregating QTL are involved.

The significant less-than-additive epistasis detected through the di-introgression study and the suggestion that epistatic effects are enhanced when more QTL are involved has an important bearing on QTL-mapping studies. In conventional segregating populations, where each individual carries a number of QTL, less-than-additive epistasis would reduce the effect of each individual QTL,

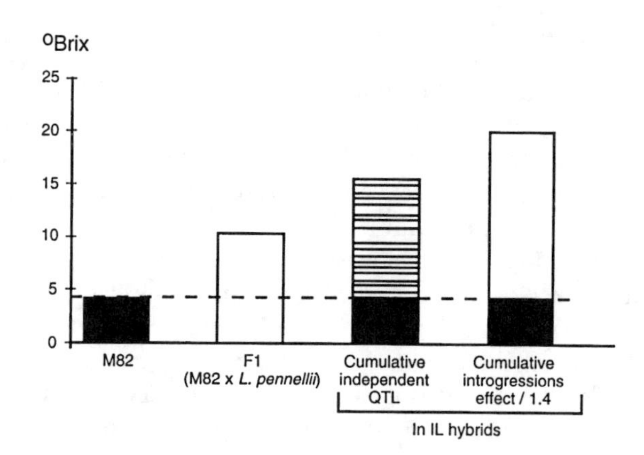


**FIGURE 15.3** Comparison of B values of M82, its F1 hybrid with *Lycopersicon pennellii*, the cumulative B expected based on additivity of the significant QTL identified independently using the *L. pennellii* ILs in a heterozygous state, and the cumulative B value obtained for all heterozygous ILs and corrected for overrepresentation of the genome by the overlapping regions.

rendering the identification of loci with minor effects difficult. When viewed from an evolutionary perspective, less-than-additive epistasis would contribute to developmentally buffered organisms through canalization of the phenotype. In this scenario, "loss" of an allele affecting a fitness trait would have a minimal effect on the phenotype.

## 15.3 NILS FOR BREEDING FOR COMPLEX TRAITS

### 15.3.1 How to Identify QTL for Horticultural Yield

Modern crop cultivars carry only a small fraction of the variation available in land races and wild species. The narrow genetic basis of many cultivated crops results from their pattern of domestication, which was often initiated from a small number of individuals, and from the standardization and uniformity requirements imposed by intensive agriculture.[5] Exotic germplasm is therefore considered a major reservoir of genetic variation for the improvement of crop plants, a realization that has led to the establishment of numerous collections which are maintained and distributed through gene banks.[34]

Wild germplasm has been used extensively as a source of agriculturally important monogenic traits, such as disease resistance; however its exploitation for more complex polygenic traits, such as yield, has been limited. Phenotypic distribution of yield shows continuous variation, and is affected by a number of contributing genes and the environment. In early traditional segregating generations (F2/F3, BC) resulting from crosses of wild-species and cultivated varieties, the plants contain a large portion of the wild-species genome; this often leads to fertility problems and interferes with evaluation of yield potential. An additional difficulty in estimating horticultural yield consists of the need for multiple individuals in a tested plot, the practical design in commercial fields. Since the ILs harbor only a small portion of the *L. pennellii* genome (an average of 2.75%), their phenotype generally resembles that of the cultivated variety and yield can be estimated in field trials in plots and compared to the nearly isogenic processing variety M82. The effect of the introgression was also assayed in a heterozygous condition, by comparing the hybrids of M82 with the ILs to their parents.

The processing-tomato industry requires varieties that are adapted to a single machine harvest and produce high fruit yield per unit area combined with high levels of total soluble solids (mainly sugars and organic acids) in the fruit. The use of such varieties renders the manufacturing of concentrates more cost effective via a reduction in expenses associated with transportation and water evaporation at the processing plant. In tomato, as in many other crops, yield and quality parameters are inversely related.[35] Therefore, we calculated horticultural yield in our studies as the product of B and Y (BY). This parameter provides an estimate of the weight of soluble solids produced per unit area and is directly related to the amount of processed product (e.g., tomato paste) that can be produced by each genotype.

To identify QTL associated with horticultural yield (BY), replicated plot trials of the tested genotypes need to be conducted. The first step in the utilization of the *L. pennellii* ILs for mapping such QTL was to screen the entire population in a nonreplicated plot trial. The results of this survey were compared to those obtained in a replicated trial of the same genotypes grown as single plants at wide spacing.[22] Fruit characteristics (FM and B) were similar between the two stands, but Y was generally different. The eight lines that outperformed the control in the plot survey were subjected to detailed analysis the following year. The effects of these introgressions, measured on single plants, were reproducible relative to the previous year's results. In a replicated plot trial of these ILs and their hybrids involving two genetic backgrounds, BY in seven of the eight hybrids was 7 to 13% higher than that of their nearly isogenic controls. Combining the two introgressions with the largest contribution to horticultural yield in plots resulted in a 20% increase relative to the control in the third year.

Utilization of wild-tomato germplasm for breaking the negative relationship between Y and B was demonstrated in a pioneering study by Rick.[36] Through repeated backcrossing of a *L. chmielewskii* accession to *L. esculentum* and selection for large fruits with high B values, novel germplasm was developed which was later subjected to RFLP analysis with the aim of identifying the introgressed wild-species chromosome segments.[37] Two independent introgressions on chromosome 7 were identified which were associated with increased B values.[38] Physiological studies aimed at unraveling the physiological mechanisms associated with the introgressions suggested that the effects are dependent on the translocation of photosynthates and water during fruit ripening.

The general insensitivity of the identified QTL to genetic background, as demonstrated in several studies,[15,22,26] most probably results from the large genetic distance between the parental species, and the very limited use which has been made by plant breeders of wild germplasm for the general improvement of the cultivated crop. Except for Y and B, the main focus of the IL studies, we identified wild-species alleles that contributed to firmer fruit, better internal color, a higher level of carotenoids and vitamins, earlier setting and less stem retention. In all these cases, the identified *L. pennellii* alleles acted in a direction opposite to that expected based on performance of the parental lines.

### 15.3.2 How Can We Expedite the Rate of Variety Development?

The main disadvantage of IL populations is the length of time and the large amount of work required for their development. We estimate that after complete genotyping of a BC1 population and selection of the appropriate lines providing coverage of the genome with minimal overlaps, two additional cycles of backcrosses accompanied by genotypic and phenotypic selection are required. In a selfed BC3 population, the founders of an IL population can be identified.

Advanced backcross (AB) QTL analysis has recently been suggested as a method for the discovery and transfer of agronomically valuable QTL from wild germplasm into elite breeding lines.[39] The strategy is to conduct the QTL and marker analyses in BC2 or BC3 generations which have previously been subjected to selection against deleterious genes originating from the wild source. The rationale behind the use of these generations is to lower the proportion of the donor-parent genome in each of the lines and to avoid the masking effect of deleterious wild recessive alleles. Using such an approach, even early generation interspecific populations can be assayed for yield in plot trials. Based on simulation studies it was suggested that AB breeding can be effective for the identification of additive, partially dominant, dominant and overdominant QTL. This approach was used to identify QTL and transfer alleles originating from *L. pimpinellifolium* capable of enhancing a wide range of traits important for processing tomatoes.[40] Using marker data, it was possible to select QTL-NILs in a single generation after the identification of the putative QTL. The short time between the mapping experiment and production of an elite line isogenic for the gene of interest should encourage wider utilization of exotic germplasm through AB-QTL breeding.

## REFERENCES


1. Rick, C. M. and Chetelat, R. T., TGRC stock lists, *Rep. Tom. Genet. Coop.*, 45, 53, 1995.
2. Tanksley, S. D., Ganal, M. W., Prince, J. C., de Vicente, M. C., Bonierbale, M. W., Broun, P., Fulton, T. M., Giovanonni, J. J., Grandillo, S., Martin, G. B., Messeguer, R., Miller, J. C., Miller, L., Paterson, A. H., Pineda, O., Roder, M. S., Wing, R. A., Wu, W., and Young, N. D., High density molecular linkage maps of the tomato and potato genomes: biological inferences and practical applications, *Genetics*, 232, 1141, 1992.
3. Paterson, A. H., Lander, E. S., Hewitt, J. D., Peterson, S., Lincoln, S. E., and Tanksley, S. D., Resolution of quantitative traits into Mendelian factors, using a complete linkage map of restriction fragment length polymorphisms, *Nature*, 335, 721, 1988.
4. Miller, J. C. and Tanksley, S. D., RFLP analysis of phylogenetic relationship and genetic variation in the genus *Lycopersicon*, *Theor. Appl. Genet.*, 80, 437, 1990.
5. Ladizinsky, G., Founder effect in crop-plant evolution, *Econ. Bot.*, 39, 191, 1985.
6. Dudley, J. W., Seventy-six generations of selection for oil and protein precentage in maize, in *Proc. International Conference on Quantitative Genetics*, Pollak, E., Kempthorne, O. and Bailey, T. B., Jr., Eds., Iowa State Univ. Press, Ames, IA, 1977, 459.
7. Goldman, I. L., Rocheford, T. R., and Dudley, J. W., Quantitative trait loci influencing protein and starch concentration in the Illinois Long Term Selection maize strains, *Theor. Appl. Genet.*, 87, 217, 1993.
8. Barton, N. H. and Turelli, M., Evolutionary quantitative genetics: how little do we know?, *Annu. Rev. Genet.*, 23, 337, 1989.
9. Lande, R., The minimum number of genes contributing to quantitative variation between and within populations, *Genetics*, 99, 541, 1981.
10. Powers, L., The nature of the series of environmental variances and the estimation of the genetic variance and geometric means in crosses involving species of *Lycopersicon*, *Genetics*, 27, 561, 1942.
11. Zeng, Z. B., Houle, D., and Cockerham, C. C., How informative is Wright's estimator of the number of genes affecting a quantitative character?, *Genetics*, 126, 235, 1990.
12. Paterson, A. H., Damon, S., Hewitt, J. D., Zamir, D., Rabinowitch, H. D., Lincoln, S. E., Lander, E. S., and Tanksley, S. D., Mendelian factors underlying quantitative traits in tomato: comparison across species, generations and environments, *Genetics*, 127, 181, 1991.
13. Grandillo, S. and Tanksley, S. D., QTL analysis of horticultural traits differentiating the cultivated tomato from the closley related species *Lycopersicon pimpinellifolium*, *Theor. Appl. Genet.*, in press.
14. Goldman, I. L., Paran, I., and Zamir, D., Quantitative trait locus analysis of a recombinant inbred line population derived from a *Lycopersicon esculentum* × *Lycopersicon cheesmanii* cross, *Theor. Appl. Genet.*, 90, 925, 1995.
15. Eshed, Y. and Zamir, D., Introgression line population of *Lycopersicon pennellii* in the cultivated tomato enables the identification and fine mapping of yield associated QTL, *Genetics*, 141, 1147, 1995.
16. Tanksley, S. D., Mapping polygenes, *Annu. Rev. Genet.*, 27, 205, 1993.
17. Darvasi, A., Weinreb, A., Minke, V., Weller, J. I., and Soller, M., Detecting marker-QTL linkage and estimating QTL gene effect and map location using a saturated genetic map, *Genetics*, 134, 943, 1993.
18. Eshed, Y. and Zamir, D., A genomic library of *Lycopersicon pennellii* in *L. esculentum*: a tool for fine mapping of genes, *Euphytica*, 79, 175, 1994.
19. Eshed, Y., Abu-Abied, M., Saranga, Y., and Zamir, D., *Lycopersicon esculentum* lines containing small overlapping introgressions from *L. pennellii*, *Theor. Appl. Genet.*, 83, 1027, 1992.
20. Alpert, K., Grandillo, S., and Tanksley, S. D., *fw*2.2: a major QTL controlling fruit weight is common to both red and green-fruited tomato species, *Theor. Appl. Genet.*, 91, 994, 1995.
21. Eshed, Y. and Zamir, D., Less-than-additive epistatic interactions of quantitative trait loci in tomato, *Genetics*, 143, 1807, 1996.
22. Eshed, Y., Gera, G., and Zamir, D., A genome-wide search for wild-species alleles that increase horticultural yield of processing tomatoes, *Theor. Appl. Genet.*, 93, 877, 1996.

23. Paterson, A. H., DeVerna, J. W., Lanini, B., and Tanksley, S. D., Fine mapping of quantitative trait loci using selected overlapping recombinant chromosomes, in an interspecific cross of tomato, *Genetics,* 124, 735, 1990.
24. Crow, J. F., Dominance and overdominance, in *Heterosis*, Gowen, J. W., Ed., Iowa State College Press, Ames, IA, 1952.
25. De-Vicente, M. C. and Tanksley, S. D., QTL analysis of transgressive segregation in an interspecific tomato cross, *Genetics,* 134, 585, 1993.
26. Eshed, Y. and Zamir, D., Introgressions from *Lycopersicon pennellii* can improve the soluble solids yield of tomato hybrids, *Theor. Appl. Genet.*, 88, 891, 1994.
27. Emery, G. C. and Munger, M. H., Effect of inherited differences in growth habit on fruit size and soluble solids in tomato, *J. Am. Soc. Hort. Sci.*, 95, 410, 1970.
28. Fatokun, C. A., Menancio-Hautea, D. I., Danesh, D., and Young, N. D., Evidence for orthologous seed weight genes in cowpea and mung bean based on RFLP mapping, *Genetics,* 132, 841, 1992.
29. Lin, Y., Schertz, K. F., and Paterson, A. H., Comparative analysis of QTLs affecting plant height and maturity across Poacceae, in reference to an interspecific sorghum population, *Genetics,* 141, 391, 1995.
30. Falconer, D. S., *Introduction to Quantitative Genetics*, John Wiley & Sons, New York, 1989.
31. Doebley, J., Stec, A., and Gustus, C., *Teosinte branched1* and the origin of maize: evidence for epistasis and the evolution of dominance, *Genetics,* 141, 333, 1995.
32. Spickett, S. G. and Thoday, J. M., Regular responses to selection. 3. Interactions between located polygenes, *Genet. Res.*, 7, 96, 1966.
33. Long, A. D., Mullaney, S. L., Reid, L. A., Fry, J. D., Langley, C. H., and Mackay, T. F. C., High resolution mapping of genetic factors affecting abdominal bristle number in *Drosophila melanogaster, Genetics,* 139, 1273, 1995.
34. Esquinace-Alcazar, J. T., Plant genetic resources, in *Plant Breeding, Principles and Prospects,* Hayward, M. D., Bosemark, N. O., and Ramagosa, I., Eds., Chapman and Hall, 1993, 33.
35. Stevens, M. A. and Rick, C. M., Genetics and breeding, in *The Tomato Crop, a Scientific Basis for Improvement,* Atherton, J. G. and Rudich, J., Eds., Chapman and Hall, New York, 1987, 35.
36. Rick, C. M., High soluble-solids content in large-fruited tomato lines derived from a wild green-fruited species, *Hilgardia,* 42, 493, 1974.
37. Tanksley, S. D. and Hewitt, J., Use of molecular markers in breeding for soluble solids content in tomato — a re-examination, *Theor. Appl. Genet.,* 75, 811, 1988.
38. Azanza, F., Kim, D., Tanksley, S. D., and Juvik, J. A., Genes from *Lycopersicon chmielewskii* affecting tomato quality during fruit ripening, *Theor. Appl. Genet.*, 91, 495, 1995.
39. Tanksley, S. D. and Nelson, J. C., Advanced backcross QTL analysis: a method for simultaneous discovery and transfer of valuble QTLs from unadapted germplasm into elite breeding lines, *Theor. Appl. Genet.,* 92, 191, 1996.
40. Tanksley, S. D., Grandillo, S., Fulton, T. M., Zamir, D., Eshed, Y., Petiard, V., Lopez, J., and Beck-Bunn, T., Advanced backcross QTL analysis in a cross between an elite processing line of tomato and its wild relative *L. pimpinellifolium, Theor. Appl. Genet.,* 92, 213, 1996.

# 16 Case History in Genetics of Long-Lived Plants: Molecular Approaches to Domestication of a Fast-Growing Forest Tree: *Populus*

*H.D. Bradshaw, Jr.*

## CONTENTS



## 16.1 INTRODUCTION

Forest trees are fascinating organisms of considerable significance ecologically, aesthetically, and economically. Most of the earth's terrestrial biomass is found in forest trees. Tree-dominated ecosytems shelter our richest stores of plant and animal diversity. Native forests have for centuries provided us with lumber, fiber, fuel, and food. However, many natural forests have been eliminated permanently by land clearing for agriculture or grazing, or compromised by harvesting timber at unsustainable levels. Indirect effects of human activities, such as fire suppression, have also changed the character of previously "wild" forests. It is apparent that the cutting of managed forests must be reduced to sustainable rates and that the few remaining vestiges of virgin forest should be preserved intact. How will these conservation goals be met in the face of increasing worldwide demand for forest products? One part of the solution is to increase wood production on a relatively small, intensively managed land base. Intensive management includes the establishment of breeding programs designed to domesticate forest trees for maximum economic yield in a nearly agricultural setting. Although maize and wheat breeders have a 5,000 to 10,000 year head start on domestication, we forest geneticists hope that our efforts at capturing the genetic diversity still abundant in most

0-8493-7686-6/98/$0.00+$.50

forest trees will be accelerated by recent advances in molecular genetics (e.g., genome and QTL mapping) and by an interdisciplinary approach to achieving a basic understanding of tree growth and development.

## 16.2 WHY PEAS AND NOT TREES?

Gregor Mendel brought three critical concepts to his study of trait transmission. First, he designed his experiment to answer a specific question about the inheritance of discrete characters. Second, he analyzed his data quantitatively instead of by simple description. Third, and most important, he chose an experimental system (garden peas, *Pisum sativum*) with self-compatible true-breeding (i.e., homozygous) lines differing in discrete phenotypic characters, a short generation interval, and a physical size well suited to planting in an environmentally homogeneous monastery garden. Many other agricultural crops and model organisms (e.g., maize, *Drosophila*, or *Arabidopsis*) might have served Mendel nearly as well. By Mendel's criteria, forest trees would have been a poor choice indeed. Most forest tree species are dedicated outcrossers with high levels of individual heterozygosity. Inbreeding depression can be severe. There are only a handful of discrete, monogenic, visible traits known in the best-studied trees, and none in most species. Mendel was a patient man, but would he have been willing to wait 2 years between pollination and seed germination, as is necessary in pines? Would he have been able to sustain his interest in genetics for the additional 4 to 20 years required to produce the F2 generation? Where would he have found the space to plant thousands of trees in a uniform "garden"? I suspect that Mendel had no difficulty in deciding between peas and trees for his famous experiments. The wonder is that anyone has worked on forest genetics since!

The role of genetics is seldom apparent when observing natural populations of forest trees. Trees growing in nature are found in heterogeneous environments, often in uneven-aged stands, and in admixture with other species. For traits of practical interest, such as stem volume growth, there is no obvious way to judge whether an exceptionally large tree became large by virtue of a superior genotype, or by fortuitously occupying a favorable environment. In other words, the heritability of many commercially important traits is low when estimated in typical extensive forestry sites replete with stumps, boulders, competing vegetation, and uncontrolled herbivory. The perennial growth habit of forest trees adds further complications. The environment is variable across time as well as space, leading to significant and unpredictable genotype × environment (G × E) interactions. For example, a particular tree genotype may grow well in wet years but be a poor competitor in dry years, while its neighbor tree might have the opposite response. Since growth is a cumulative phenomenon, the peculiarities of the weather in the year of seedling establishment could have effects on growth that persist for years and poorly reflect the phenotype of the same tree had it been planted in an "average" year. The long lifespan of trees gives pests and pathogens (with their short life cycles) many opportunities to overcome the tree's genetic defenses. Consider also that most properties of wood, which is the principal commodity produced by forest trees, are very different in juvenile and mature trees. The juvenile-mature correlation is similarly low for other traits, such as growth, making selection of desirable genotypes for breeding and propagation a slow process.

## 16.3 COPING WITH TREES

How do breeders and geneticists cope with these recalcitrant trees? The most successful approach has been to take advantage of the major virtue of forest trees, i.e., their high levels of genetic diversity, and to then apply the most powerful genetic analysis methods available to understand and exploit this variation. For example, tree breeders were the first to establish common garden tests of geographically diverse populations, a methodology later adopted by ecological geneticists, such as Turesson and Clausen, working on herbaceous plants. In forestry, these common garden trials are called "provenance tests" and are designed to identify particularly desirable seed sources. According to Wright[1] the oldest well documented provenance test was planted with Scots pine

(*Pinus sylvestris*) seedlings in 1821 by Philippe de Vilmorin on his estate in France. The first test included 30 seed sources from Russia, Latvia, Germany, Scotland, and France, and showed that height, diameter, bole form, needle length, and bud color were heritable. Since that time, provenance trials have been carried out in many tree species, giving insight into the evolution of adaptive variation in nature and its practical consequences for forestry. Forest geneticists were quick to develop or adopt other new techniques in genetics, including statistical genetics, population genetics with allozyme and DNA markers, regeneration of transgenic plants, and molecular biology.

With the advent of general methods for mapping quantitative trait loci (QTLs),[2,3] forest geneticists and tree breeders have embraced this new technology as a basic research tool[4-7] while retaining a healthy skepticism of its practical benefits.[8] For several reasons, the promise of QTL mapping has a powerful allure for those of us working with forest trees. A deep understanding of quantitative trait variation would permit a rational domestication program to be implemented for some tree species. From an applied tree breeding perspective, the biggest intermediate-term reward from QTL mapping would be the development of marker-assisted selection (MAS) for superior genotypes, indirectly selecting for the desired phenotype using information from genetic markers flanking QTLs. Since MAS can be performed at the seedling stage, years or decades before the selected trait might be manifested in the adult phenotype, the savings in time, space, and testing effort are obvious. Because of the long juvenile period and generation interval of most forest trees, it is arguable that MAS has more to offer in forestry than in agriculture.[9] For example, if MAS proves useful not just in identifying superior clones for propagation in the current generation, but also in choosing superior parents with complementary multilocus genotypes for the next generation, 10 years or more might be saved in each breeding cycle. In addition to MAS, there are significant short-term and long-term benefits to tree breeding expected from QTL mapping. Markers developed for genome mapping have already proven useful for assessment of clonal identity[10] and for the study of natural populations upon which we depend for germplasm.[11] In the long run we expect that the genes underlying QTLs will be cloned and manipulated directly in transgenic trees, circumventing many of the breeding steps now necessary to produce trees with specific QTL allele configurations. As a basic research tool, QTL mapping in trees has been and will continue to be productive in two ways. First, QTL mapping can be viewed as a natural extension of the statistical framework which has characterized much of forest genetics and all of tree breeding for the past 50 years. Whereas phenotypic variation in trees previously had been partitioned into components of populations-within-species, families-within-populations, and clones-within-families, an additional level of resolution is added by explaining variation among clones in terms of segregating QTL alleles. At the scale of the QTL, we are much closer than ever before to biological reality, i.e., the genes responsible for observed variation. The discovery of QTLs with large effects on tree phenotypes has also stimulated a re-examination of assumptions often made in traditional statistical analyses, including the notion that continuous phenotypic distributions are most often the result of polygenic inheritance of the "infinitesimal" type. The second favorable outcome of QTL mapping in forest trees has been the eagerness with which scientists in many disciplines have welcomed the opportunity to work with genetically well-characterized pedigrees. Pathologists, physiologists, molecular biologists, and ecologists immediately grasped the added explanatory power which could be brought to their experiments within mapped families. As a broad, detailed knowledge of these intensively studied families develops, we can expect new insights into fundamental processes of tree growth and development and how these processes might be manipulated during domestication to enhance the quantity and quality of forest products.

## 16.4 *POPULUS* AS A MODEL FOREST TREE

The term "forest tree" may be operationally useful, but, botanically speaking, it is an unnatural amalgamation of angiosperms and gymnosperms, deciduous and evergreen, hermaphroditic, monoecious, and dioecious, temperate and tropical. Given the range of organisms lumped together as "forest trees", perhaps it is overly optimistic (or presumptuous) to designate any one model system capable of serving as a generic "forest tree". With that caveat in mind, however, the angiosperm genus *Populus* satisfies many of the requirements of a model system, at least from the molecular geneticist's point of view.

While many aspects of biology are common to all plants, and hence may be studied as well, and more easily, in *Arabidopsis*, some anatomical features, physiological processes, and developmental pathways must be studied in a woody plant. Therefore, the most important attribute of *Populus* (poplars, cottonwoods, and aspens) as a model system is that all *Populus* are trees. They can form a dominant overstory, reaching heights of more than 60 m. Though individual stems are not particularly long lived by tree standards (perhaps 200 to 300 years), vegetative clones of quaking aspen (*P. tremuloides*) may be as old as 10,000 years and cover an area of 80 ha, making them among the largest living things on earth. Like all trees, and unlike herbaceous plants, *Populus* undergo perennial, cumulative, woody growth from apical, lateral, and cambial meristems. They become dormant in autumn, cold-hardy in winter, and renew growth in the spring. If we wish to study the genetic control of wood formation or vegetative dormancy, it seems logical to begin work in a plant which has these features.

Among tree genera, *Populus* has some compelling advantages as a model system.

- Most species are fast-growing, with annual height increments of 3 to 5 m in operational plantations. This rapid growth has encouraged intensive culture of *Populus* in the temperate zone for at least the past 400 years. In the Pacific Northwest of North America, above-ground dry biomass yields of 20 to 30 Mg/ha/yr (5 to 10 times the yield of native conifer forests) have attracted industrial interest in *Populus* for energy, pulp, engineered wood products (e.g., oriented strandboard), veneer, and solid wood. Fast growth means that rotations (i.e., the interval between planting and harvest) are short, generally 4 to 18 years. With short rotations, meaningful testing for commercially relevant phenotypes such as growth and wood quality can be conducted within 2 to 4 years of breeding. Willows (*Salix*) and *Eucalyptus* are two other angiosperm genera known for rapid growth, but typical rotation lengths for conifers range from 12 to 60 years.
- The genus *Populus* is genetically diverse, with about 30 species worldwide. Among species, there is considerable variation in adaptive traits such as crown architecture, water use efficiency, and disease resistance. There may also be substantial variation within species such as quaking aspen (*P. tremuloides*), which is the most widely distributed tree in North America. Among competing model systems, willows and *Eucalyptus* have about 10 times as many species as *Populus*.
- The genetic diversity among species may be captured by interspecific hybridization. Hybrids frequently show heterosis for growth and other traits, and are planted commercially on a large scale. Because all members of the genus are diploid (2n = 38), the hybrids are usually fertile and F2 progenies segregating for a wide variety of traits can be produced. Interspecific hybridization is seldom used in operational forest tree breeding except in *Populus* and *Eucalyptus*, although heterosis for growth has been observed in numerous other taxa.[12]
- Sexual propagation of *Populus* is usually straightforward. Most species are strictly dioecious, and all are normally wind pollinated. Pollen can be stored for long periods of time. Controlled crosses may be performed on detached female branches in the greenhouse, with each pollination yielding hundreds of seeds within 6 to 8 weeks. Seedlings grow to a height of 2 to 3 m in their first year. Most other trees are more difficult to breed than *Populus*, either because the pollen doesn't store well (often the case with insect-pollinated species), the female tree must be climbed repeatedly in order to perform the pollinations, each pollination yields only one or a few seeds, seed development is slow, or some combination of these factors.
- Vegetative propagation is easy for most species and hybrids. Most industrial plantations are established from 30-cm unrooted cuttings (i.e., "sticks"), so clonal genetic experiments

reflect the performance expected from commercial planting stock. Both mature and juvenile *Populus* trees are readily cloned from dormant cuttings, making it possible to design genetic tests with parental, F1, and F2 trees growing side-by-side. Replicated clonal trials give accurate estimates of the phenotype and any environmental variance; hence, broad-sense heritabilities are calculated to set an upper bound on the proportion of phenotypic variance which can be explained by QTLs. Genetic tests may be replicated in time and space, or destructively sampled (such as for wood quality traits) without losing the genotype. Most importantly, cloned multigeneration pedigrees serve as a "portable" genetic resource for scientists in many disciplines. Cuttings from a family whose molecular marker genotypes are known at hundreds of loci may be shipped to investigators all over the world, leading to the accumulation of QTL patterns for many traits. Clonal propagules in *Populus* are the functional equivalent of recombinant inbred lines in annual plants. While *Populus*, *Salix*, and *Eucalyptus* plantations generally are clonal, almost no conifers are propagated clonally for commercial planting on a large scale, but that will probably change as technologies (rooted cuttings, micropropagation *in vitro*, somatic embryogenesis) are implemented to capture the nonadditive fraction of genetic variance.

- The generation interval in *Populus* is fairly short (for a tree), with seed-to-seed times of 4 to 5 years under favorable conditions.
- The physical size of the *Populus* genome is rather small. A diploid *Populus* nucleus contains about 1.1 pg of DNA,[13] equivalent to a haploid genome complexity of 500 Mb (similar to that of tomato and rice, about 5 times larger than *Arabidopsis,* and 6 times smaller than maize). These 500 Mb are distributed over 19 chromosomes with a combined genetic length estimated to be 2400 to 2800 cM,[14] giving an average correspondence between the physical and genetic lengths of approximately 200 kb/cM (about the same value as in *Arabidopsis*). By way of contrast, the physical size of the genome of the best-studied conifer, loblolly pine (*Pinus taeda*), is 40 times as great as that of *Populus*, and the average relationship between physical length and genetic distance is about 8000 kb/cM. The favorable ratio between genetic and physical length in *Populus* makes it a rational choice for map-based cloning of genes of special importance to forest trees.
- Transgenic systems are well-developed in *Populus*. *Populus* was the first forest tree to be genetically engineered,[15] and continues to be the most widely used woody plant for transgenic work. This is particularly relevant if map-based cloning is to become the method of choice for gene isolation, since cloned candidate genes from the mapped region can be tested directly for function (or complementation) in transgenic *Populus*. Optimism about the future of transgenic approaches to the genetic improvement of *Populus* has led to the formation of the Tree Genetic Engineering Research Cooperative at Oregon State University. None of the willow, *Eucalyptus*, or conifer systems have as facile a transformation/regeneration system, although there has been steady improvement in this area.
- Perhaps the greatest strength of *Populus* as a model system is the breadth of existing knowledge from many scientific disciplines. For many of the reasons listed above, *Populus* has become a focal point for tree physiologists, anatomists, biochemists, molecular biologists, pathologists, and ecologists. In each of these scientific fields it has been recognized that *Populus* offers an unprecedented (in forest trees) control of genetic stocks by using clonal propagules, so the *Populus* research community has given a warm reception to the pedigrees, genetic markers, and linkage maps developed for QTL identification.

## 16.5 MAJOR QUESTIONS IN *POPULUS* GENETICS

*Populus*, along with other forest trees, is in the earliest stages of domestication. The choice of breeding and selection schemes presently made will have long-term impacts on the success of domestication and the conservation of wild germplasm. In order to implement a rational plan for domestication, several questions about the genetics of hybrid *Populus* must be addressed.

- What is the genetic basis of heterosis for stem volume growth observed in many interspecific hybrids and which breeding strategy may best exploit this?
- Are morphological and physiological traits under oligogenic or polygenic control and how may these traits be manipulated to improve productivity?
- What is the genetic basis of adaptation to climate and photoperiod and what is the genetic mechanism responsible for genotype × environment interaction?
- What is the genetic basis of disease resistance, both of the qualitative/vertical and quantitative/horizontal types?

To answer these questions, QTL mapping has been used as part of a collaboration among tree breeders, growers, physiologists, and pathologists.

## 16.6 MAPPING PEDIGREES IN HYBRID *POPULUS*

While a linkage map has been made for an intraspecific cross in *Populus tremuloides* (quaking aspen),[16] most linkage mapping has been done in interspecific hybrid pedigrees.[14,17-19] There are two reasons for this. First, interspecific hybrids are the typical commercial planting stock in many parts of the world, so mapping results derived from F1 families can be applied directly to operational breeding. Segregation within F1 families depends on heterozygosity in the parental trees, but since average heterozygosity values in *Populus*, as in most trees, are 20% or higher, usually there is ample variation with which to work. Thus far, only qualitative loci for disease resistance have been mapped in *Populus* F1 hybrids,[18,19] but QTL mapping work in interspecific F1 hybrid families of *Eucalyptus*[6] demonstrates the feasibility of mapping true quantitative trait loci in the F1 generation. Second, because there is substantial genetic differentiation among species within *Populus*, and this variation can be forced to segregate widely in the F2 generation produced from mating interspecific F1 hybrids, QTLs affecting adaptive and commercial traits such as heterotic stem volume growth, stem form, crown geometry, leaf phenology, and disease resistance can be mapped.[5]

Most of the QTL mapping effort in *Populus* has been focused on a single three-generation pedigree. The parental generation consists of a female *P. trichocarpa* (black cottonwood clone 93-968) from a maritime climate in western Washington state and a male *P. deltoides* (eastern cottonwood clone ILL-129) from a continental climate in Illinois. These two trees were mated in 1981 to produce the F1 family 53, and in 1988 two of the F1 hybrids (53-246 and 53-242) were sibmated to generate the F2 family 331. This mating was repeated in 1990 to increase the number of F2 offspring to 379. All three generations of the mapping pedigree are maintained in stoolbeds, where the new growth is cut back to the ground each winter to encourage the resprouting of multiple stems suitable for making hardwood cuttings for establishing experimental plantings. The parental and F1 trees are also kept in a clone arboretum where they are allowed to grow large enough to flower, so that additional crosses may be done when necessary.

The map used for QTL mapping in F2 family 331 consists of about 250 restriction fragment length polymorphism (RFLP) markers and 100 random amplification of polymorphic DNA (RAPD markers), and covers roughly two-thirds of the ~2600 cM length of the *Populus* genome.[14]

## 16.7 QTL MAPPING IN INTERSPECIFIC *POPULUS* HYBRIDS

Amid considerable skepticism that QTLs with measurable phenotypic effects exist in forest trees, but secure in the knowledge that such QTLs did exist in other interspecific plant hybrids, a pilot-scale clonal trial consisting of the two parental genotypes, two F1 genotypes, and 54 F2 genotypes was planted in 1990. Each clone was represented by six ramets distributed in two-tree plots across three blocks. Obviously, with so few F2 offspring only QTLs with large magnitudes of effect can be detected, and the estimates of QTL phenotypic effect are unlikely to be very accurate.[20] Another limitation of this initial study was that the QTL analysis was performed as though the F2 was the result of a founding cross between inbred lines; in this case, two species of *Populus*. That is, F2 marker data were encoded as simply homozygous for the *P. trichocarpa* allele (TT), homozygous for the *P. deltoides* allele (DD), or heterozygous (TD) in spite of the fact that the parental trees are themselves highly heterozygous and often the RFLP data revealed three or four marker alleles segregating in the F2. We felt that the strong genetic differentiation between *P. trichocarpa* and *P. deltoides*, which are in separate sections within the genus and are adapted to almost nonoverlapping environments, made it plausible that the major QTLs involved in heterosis are essentially fixed for different alleles in each species. A further complication of the genetic analysis is inbreeding depression, since the F2 is the product of a sibmating.[21] Fortunately, the tremendous level of segregating variation in the F2 of an interspecific hybrid makes QTL mapping possible even with the small progeny size and overly simplified genetic model employed.

Growth, stem form, leaf, and branch traits were measured after 1 and 2 years in the field. By the end of the second growing season the taller trees were over 5 m in height. Broad-sense heritabilities for growth, form, and phenology traits were high (0.62 to 0.98), reflecting the uniformity of the agricultural environment typical of hybrid poplar culture and the very large genetic variance produced by interspecific hybridization. For nearly every trait measured at least onc QTL with a significant LOD score (≥2.9) was found, and each of these QTLs accounted for at least 30% of the phenotypic variance.[5] For the highest heritability trait, date of bud burst in the spring (important for tolerance to late spring frosts), a five-QTL model could account for 85% of the total genetic variance in the trait. By any reasonable standard, such QTLs of large effect can be considered "major" genes, and call into question the assumption that "complex" traits such as growth and adaptation must be the result of many genes of small and equal effect. Genome-wide marker heterozygosity was uncorrelated with growth in the F2, further supporting the notion that neither heterosis nor inbreeding depression are determined by the cumulative effect of many loci distributed throughout the *Populus* genome. Heterosis for stem volume in the F1 can be explained by the complementary dominance of a few favorable QTL alleles for increased stem height from the *P. trichocarpa* parent and for increased stem diameter from the *P. deltoides* parent. Perhaps the most interesting finding to emerge from this pilot study is that QTLs controlling stem diameter (basal area) growth are found clustered with QTLs affecting leaf traits on sylleptic branches.[5] Sylleptic branches arise from axillary buds that have not undergone a winter dormancy and are much more abundant on *P. trichocarpa* than on *P. deltoides*. Sylleptic branches proliferate in response to light, water, and nutrients, and allocate most of their fixed carbon to the growing stem. Sylleptics are ephemeral, self-pruning when they become shaded as the crown moves upward with each growing season. Tree physiologists have recognized the importance of increased sylleptic branch leaf area for the superior stem volume growth of hybrid poplar,[22-24] and QTL mapping has made the genetic connection between the traits clearer.

The pilot-scale trial with 54 F2 genotypes is, as of this writing, in its seventh year of growth. While we continue to use this trial for demonstration purposes and for some mapping of major QTLs for disease resistance,[25] most of the current QTL mapping effort is directed at two much larger plantations established in contrasting growing environments east and west of the Cascade Range in Oregon. In these QTL validation trials, there are 375 F2 genotypes from the F2 family 331, along with the parental and F1 clones, replicated as two-tree plots in three or four blocks at both locations. F2 populations of this size are adequate to search for QTLs of small effect and/or low heritability traits. West of the Cascades the climate is maritime, with cool moist winters and warm dry summers. The plantation is adjacent to large natural stands of *P. trichocarpa* along the lower Columbia River. East of the Cascades, the climate is continental with cold winters and hot summers, and a total annual precipitation of less than 30 cm. Here, the QTL mapping plantation is adjacent to natural vegetation such as sagebrush, and to drip-irrigated stands of hybrid poplar destined for fiber production. Preliminary results of QTL mapping in these new plantations suggest that some QTLs will be important in both environments, but that others will contribute to genotype × environment interactions as seen in other plants.[26]

Collaborative efforts among physiologists, pathologists, and tree breeders have expanded with the installation of these large replicated trials. In addition to the "traditional" growth and morphological characters measured in the pilot-scale plantation, new traits are being mapped in F2 family 331, such as dormancy and cold hardiness, water-use efficiency, osmotic adjustment, stem hydraulic conductance, plant growth regulator concentration, leaf nutrient levels, wood quality, and disease and insect resistance.

## 16.8 WHAT HAVE WE LEARNED FROM QTL MAPPING IN *POPULUS*?

One important outcome of QTL mapping in *Populus* is that for the first time there is objective evidence that "major" genes may be responsible for some of the variation observed in quantitative traits, such as stem volume growth, in a forest tree. In retrospect, this outcome should perhaps have been expected[8] in light of similar experiments carried out in interspecific hybrids of herbaceous plants.[2,27] "Major" QTLs for growth and wood quality have now been found in *Eucalyptus* hybrids[6] and intraspecific crosses of loblolly[4] and maritime[7] pine, and it seems probable that many more forest-tree QTLs will be identified when suitable experiments are performed.

QTL mapping in *Populus* has also demonstrated the value of close cooperation among scientists in different disciplines. The pedigrees constructed for QTL mapping have become "standards" for physiology and pathology research. The sharing of common genetic stocks and protocols has promoted the formation of a network of information about all aspects of hybrid poplar biology (http://poplar1.cfr.washington.edu). As knowledge about F2 family 331 continues to accumulate, we anticipate that QTL "coincidence mapping" will lead to a better fundamental understanding of tree growth and development. It is likely that an effort to sequence and map all or most of the expressed genes in *Populus* will be undertaken in the next few years, perhaps culminating in the placement of candidate genes near QTLs of interest.

## 16.9 HOW WILL QTL MAPS CHANGE POPLAR BREEDING?

The discovery of "major" QTLs, often with nonadditive effects, might seem to present a challenge to the orthodox "infinitesimal" model of quantitative trait variation assumed by most quantitative genetics theory and by many tree breeders. It would be a mistake to utter this challenge too loudly at the present time, since most of the QTL mapping performed in plants so far has been done with a "stacked deck", that is, in pedigrees known to be segregating widely for the traits to be studied. I suspect that when the population-level complexities of allele number, frequency, magnitude of effect, and interaction with other loci and the environment are taken into account, the "oligogenic" (within-family) and "polygenic" (across the whole breeding population) models of quantitative trait inheritance will look very similar. The most obvious extension of QTL mapping would be to examine patterns of QTL "activity" in an entire breeding population, or a mating design drawn from it. Such an approach would give a much more complete picture of the distribution of genetic variation at the population level where most quantitative genetics work and all of tree breeding are done.

QTL maps could play an important role in choosing among long-term breeding strategies in *Populus*. Currently, only F1 hybrids are in widespread commercial production, and there has been little interest in advanced-generation (F2, backcross) breeding. F1 breeding is limited by the slow pace of reciprocal recurrent selection schemes when applied to trees, and by the fact that only two species at a time can be involved in the production of a hybrid clone. It is possible to treat the entire *Populus* genus as one large gene pool, and QTL mapping could be used to identify the best alleles for each trait from each species (recognizing that within-species variation exists for many of these traits, as well). Using MAS, a composite genotype of the best QTL alleles could be synthesized to test the merits of an advanced-generation breeding strategy. This composite genotype represents a "molecular ideotype" analogous to the morphological ideotypes developed for crop plants to maximize yield in monoculture.

The ultimate extension of QTL mapping to breeding may come when the genetic and physical maps of the *Populus* genome can be reconciled, and QTLs can be cloned and manipulated directly. With its small genome and efficient transgenic system, *Populus* is well positioned to take a lead role in the positional cloning of genes of special interest in forest trees.

## ACKNOWLEDGMENTS

All the *Populus* QTL mapping experiments were done in collaboration with Reini Stettler, with whom I have had many productive discussions regarding this chapter. Molecular marker data and phenotypic measurements were collected by colleagues too numerous to mention (you know who you are). This work was supported by the USDA National Research Initiative Competitive Grants Program, the U.S. Department of Energy Biofuels Feedstock Development Program, the Washington Technology Centers, the Consortium for Plant Biotechnology Research, Boise Cascade Corporation, James River Corporation, and the Poplar Molecular Genetics Cooperative.

## REFERENCES

1. Wright, J. W., *Introduction to Forest Genetics*, Academic Press, New York, 1976.
2. Paterson, A. H., Lander, E. S., Hewitt, J. D., Peterson, S., Lincoln, S. E., and Tanksley, S. D., Resolution of quantitative traits into Mendelian factors by using a complete linkage map of restriction fragment length polymorphisms, *Nature*, 335, 721, 1988.
3. Lander, E. S. and Botstein, D., Mapping Mendelian factors underlying quantitative traits using RFLP linkage maps, *Genetics*, 121, 185, 1989.
4. Groover, A., Devey, M., Fiddler, T., Lee, J., Megraw, R., Mitchell-Olds, T., Sherman, B., Vujcic, S., Williams, C., and Neale, D., Identification of quantitative trait loci influencing wood specific gravity in an outbred pedigree of loblolly pine, *Genetics*, 138, 1293, 1994.
5. Bradshaw, H. D., Jr. and Stettler, R. F., Molecular genetics of growth and development in *Populus*. IV. Mapping QTLs with large effects on growth, form, and phenology in a forest tree, *Genetics*, 139, 963, 1995.
6. Grattapaglia, D., Bertolucci, F. L., and Sederoff, R. R., Genetic mapping of QTL controlling vegetative propagation in *Eucalyptus grandis* and *E. urophylla* using a pseudotestcross strategy and RAPD markers, *Theor. Appl. Genet.*, 90, 933, 1995.
7. Plomion, C., Durel, C. E., and O'Malley, D. M., Genetic dissection of height in maritime pine seedlings raised under accelerated growth conditions, *Theor. Appl. Genet.*, 93, 849, 1996.
8. Strauss, S. H., Lande, R., and Namkoong, G., Limitations of molecular-marker-aided selection in forest tree breeding, *Can. J. For. Res.*, 22, 1050, 1992.
9. Williams, C. G. and Neale, D. B., Conifer wood quality and marker-aided selection: a case study, *Can. J. For. Res.*, 22, 1009, 1992.
10. Castiglione, S., Wang, G., Damiani, G., Bandi, C., Bisoffi, S., and Sala, F., RAPD fingerprints for identification and for taxonomic studies of elite poplar (*Populus* spp.) clones, *Theor. Appl. Genet.*, 87, 54, 1993.
11. Keim, P., Paige, K. N., Whitham, T. G., and Lark, K. G., Genetic analysis of an interspecific hybrid swarm of *Populus*: occurrence of unidirectional introgression, *Genetics*, 123, 557, 1989.
12. Zobel, B. and Talbert, J., *Applied Forest Tree Improvement*, John Wiley & Sons, New York, 1984.
13. Bradshaw, H. D., Jr. and Stettler, R. F., Molecular genetics of growth and development in *Populus*. I. Triploidy in hybrid poplars, *Theor. Appl. Genet.*, 86, 301, 1993.
14. Bradshaw, H. D., Jr., Villar, M., Watson, B. D., Otto, K. G., Stewart, S., and Stettler, R. F., Molecular genetics of growth and development in *Populus*. III. A genetic linkage map of a hybrid poplar composed of RFLP, STS, and RAPD markers, *Theor. Appl. Genet.*, 89, 167, 1994.

15. Fillatti, J., Sellmer, J., McCown, B., Haissig, B., and Comai, L., *Agrobacterium*-mediated transformation and regeneration of poplar, *Mol. Gen. Genet.*, 206, 192, 1987.
16. Liu, Z. and Furnier, G. R., Inheritance and linkage of allozymes and RFLPs in trembling aspen, *J. Heredity*, 84, 419, 1993.
17. Newcombe, G., Bradshaw, H. D., Jr., Chastagner, G. A., and Stettler, R. F., A major gene for resistance to *Melampsora medusae* f.sp. *deltoidae* in a hybrid poplar pedigree, *Phytopath.*, 86, 1996.
18. Villar, M., Lefevre, F., Bradshaw, H. D., Jr., and Teissier du Cros, E., Molecular genetics of rust resistance in poplars (*Melampsora larici-populina* Kleb./*Populus* sp.) by bulked segregant analysis in a 2 × 2 factorial mating design, *Genetics*, 143, 531, 1996.
19. Cervera, M. T., Gusmao, J., Steenackers, M., Peleman, J., Storme, V., Vanden Broeck, A., Van Montagu, M., and Boerjan, W., Identification of AFLP molecular markers for resistance against *Melampsora larici-populina* in *Populus*, *Theor. Appl. Genet.*, 93, 733, 1996.
20. Beavis, W. D., The power and deceit of QTL experiments: lessons from comparative QTL studies, in *Proc. of the Corn and Sorghum Industry Res. Conf.*, American Seed Trade Association, Washington, D.C., 1994, 250.
21. Bradshaw, H. D., Jr. and Stettler, R. F., Molecular genetics of growth and development in *Populus*. II. Segregation distortion due to genetic load, *Theor. Appl. Genet.*, 89, 551, 1994.
22. Hinckley, T. M., Ceulemans, R., Dunlap, J. M., Figliola, A., Heilman, P. E., Isebrands, J. G., Scarascia-Mugnozza, G., Schulte, P. J., Smit, B., Stettler, R. F., van Volkenburgh, E., and Wiard, B. M., Physiological, morphological, and anatomical components of hybrid vigor in *Populus*, in *Structural and Functional Responses to Environmental Stresses*, Kreeb, K. H., Richter, H., and Hinckley, T. M., Eds., SPB Academic,The Hague, The Netherlands, 1989, 199.
23. Ceulemans, R., Scarascia-Mugnozza, G., Wiard, B. M., Braatne, J. H., Hinckley, T. M., Stettler, R. F., Isebrands, J. G., and Heilman, P. E., Production physiology and morphology of *Populus* species and their hybrids grown under short rotation. 1. Clonal comparisons of 4-year growth and phenology, *Can. J For. Res.*, 22, 1937, 1992.
24. Hinckley, T. M., Braatne, J., Ceulemans, R., Clum, P., Dunlap, J., Newman, D., Smit, B., Scarascia-Mugnozza, G., and van Volkenburgh, E., Growth dynamics and canopy structure, in *Ecophysiology of Short Rotation Forest Crop*, Mitchell, C. P., Ford-Robertson, J. B., Hinckley, T., and Sennerby-Forsse, L., Eds., Elsevier Applied Science, New York, 1992, 1.
25. Newcombe, G. and Bradshaw, H. D., Jr., Quantitative trait loci conferring resistance in hybrid poplar to leaf spot caused by *Septoria populicola*, *Can. J. For. Res.*, in press.
26. Paterson, A. H., Damon, S., Hewitt, J. D., Zamir, D., Rabinowitch, H. D., Lincoln, S. E., Lander, E. S., and Tanksley, S. D., Mendelian factors underlying quantitative traits in tomato: comparison across species, generations, and environments, *Genetics*, 127, 181, 1991.
27. Keim, P., Diers, B. W., and Shoemaker, R. C., Genetic analysis of soybean hard seededness with molecular markers, *Theor. Appl. Genet.*, 79, 465, 1990.

# 17 Case History in Animal Improvement: Mapping Complex Traits in Ruminants

*Michel Georges*

## CONTENTS



0-8493-7686-6/98/$0.00+$.50

## 17.1 INTRODUCTION

With the advent of recombinant DNA methods and the emergence of *genomics* as a new, very powerful discipline for genome analysis, it is now conceivable to identify the genes upon which artificial selection acts. Besides the fundamental interest in understanding the molecular biology underlying production traits such as muscular development, milk yield and composition, disease resistance, etc., it is primarily the prospect of more efficient marker assisted selection (MAS) that has spurred the present enthusiasm for this area of research. The most popular approach towards the identification of production genes fits into the general "positional (candidate) cloning" strategy; i.e., clone the genes based on their map position after their initial localization by linkage analysis.[1]

The majority of economically relevant production traits in livestock, however, are complex in nature. As pointed out by Lander and Schork,[2] trait "complexities" arise whenever the simple Mendelian correspondence between phenotype and genotype breaks down. This can be due to a variety of factors including environmental and developmental noise, locus heterogeneity, oligo- or polygenic inheritance, and even nonconventional transmission mechanisms. While mapping single-gene traits has become routine with the available tools, the real challenge awaiting geneticists is the dissection of complex traits into their Mendelian components. In this chapter we report our experience with three different production traits in ruminants, all characterized by varying degrees of complexity.

## 17.2 THE *MH* GENE ON BOVINE CHROMOSOME 2 CAUSES DOUBLE-MUSCLING IN SEVERAL CATTLE BREEDS

### 17.2.1 Double-Muscling in Cattle: A Complex Trait

In cattle as in most other mammals, muscularity is rightfully considered a continuously distributed, quantitative phenotype. Interestingly, however, cattle breeders have traditionally recognized a distinct group of animals which they refer to as "double-muscled" as a result of their unique generalized muscular hypertrophy. This trait has been valued very differently depending on the economic context. In dairy cattle, where double-muscling occasionally appears in offspring of perfectly normal looking animals, it is considered a genetic defect. While double-muscling is actively selected in some beef breeds, it is avoided in others because of the associated management difficulties such as a high rate of dystocia (i.e., calving problems). It should be noted that in beef breeds in particular, the distinction between double-muscled and conventional animals becomes ambiguous, double-muscled individuals appearing as the extremes of a continuously distributed phenotype rather than a distinct entity.

While the hereditary nature of double-muscling was suggested earlier, the precise mode of transmission has remained controversial.[3] Depending on the study and population in which the study was performed, double-muscling has been reported to follow monogenic (recessive or dominant), oligogenic or even polygenic inheritance. Incomplete penetrance and variable expressivity were usually quoted in these studies. It appeared therefore that double-muscling as traditionally defined by breeders might in fact encompass a mixture of causally heterogeneous entities.

### 17.2.2 Experimental Crosses Involving Belgian Blue Double-Muscled Individuals Reveal a Locus with Major Effect on Muscular Development

One of the most thorough series of studies on double-muscling is that of Hanset and colleagues[4-7] in the Belgian Blue Cattle breed. Objective criteria of muscular development, such as dressing-out percentage, lean and fat percentage, plasma and red-cell creatine and creatinine concentrations, measured on nearly 150 randomly selected animals reared in standardized conditions, clearly revealed abnormal, bimodal distributions of these phenotypes, therefore objectively confirming the visual classification in double-muscled and conventional types traditionally performed by breeders. Resolving the phenotypic distribution into two component normal populations with common variance using a maximum likelihood procedure, revealed mean difference of three to four standard deviations depending on the trait, pointing towards the segregation of an allele with major effect on muscular development with a population frequency close to 50%.[4] The most convincing evidence in favor of such a major gene, however, came from experimental crosses involving double-muscled Belgian Blue sires and Holstein Friesian dairy cows, i.e., animals with very poor muscular development. While F1 offspring showed a phenotypic distribution very similar to their Holstein Friesian dams, backcrossing these F1's to double-muscled sires produced a bimodal backcross (BC) generation, clearly pointing towards the Mendelian segregation of a recessive *mh* (muscular hypertrophy) allele.[5]

Histological examination of double-muscled tissues in this breed actually revealed that what was commonly referred to as muscular hypertrophy is actually a muscular hyperplasia, i.e., an increase in the number of muscle fibers rather than an increase in their individual diameter.[6] Moreover, it was shown that while double-muscling — as its name implies — is traditionally considered a muscle-specific phenotype, other tissues are equally if not more affected, usually by a decrease in their relative mass.[7]

### 17.2.3 Microsatellite Mapping of the *mh* Locus to Bovine Chromosome 2

The same kind of experimental crosses were subsequently used to perform a whole genome scan using a microsatellite-based marker map. To perform the linkage analysis, animals were classified as double-muscled or conventional. Very significant LOD scores were obtained on chromosome 2 (>17), and multipoint linkage analysis positioned the *mh* locus at the centromeric end of this chromosome, at 2 cM from the nearest microsatellite marker: TGLA44. The corresponding chromosomal region accounted for all the variance of the trait assumed to be fully penetrant in this experiment.[8]

To verify the role of this locus in the distinction between double-muscled and conventional animals within the Belgian Blue Cattle breed, pedigrees were gathered in the general population, composed of conventional sires and their double-muscled offspring. Only individuals with unambiguous phenotypes were included in the analysis. Three such pedigrees, representing 61 offspring jointly, were genotyped for the chromosome 2 markers yielding a LOD score >12 without obligated recombinant for the closest TGLA44 marker. These results indicated that the same locus underlying the segregation of double-muscling in the expermental crosses also accounts for the segregation of that trait in the general Belgian Blue population.[8]

Comparing marker allele frequencies between *mh* and wild-type chromosomes sampled in the Belgian Blue population using the maximum likelihood procedure developed by Terwilliger et al.,[9] revealed a clear association between *mh* and TGLA44 alleles. As no such evidence was found for most distant chromosome 2 markers, population stratification could be excluded as the cause for this association, pointing towards genuine linkage disequilibrium. As linkage disequilibrium is expected for closely linked markers only, these results confirmed the proximity between the TGLA44 and *mh* loci.[10]

### 17.2.4 Towards Interbreed Identical-by-Descent (IBD) Fine-Mapping of the *mh* Locus

Having identified the locus responsible for the double-muscling phenotype in Belgian Blue, will allow us to more efficiently address the issue of genetic homo- or heterogeneity of this entity among cattle breeds. One of the other breeds in which double-muscling has been abundantly described is the Spanish Asturiana breed. Early segregation analysis performed in this breed suggested a dominant rather than recessive mode of inheritance, apparently pointing towards a distinct genetic determinism when compared to Belgian Blue. Nevertheless, samples were gathered in this population from animals that showed a clearly conventional phenotype and that had sired double-muscled offspring. A total of 28 such offspring were collected from 7 sires. The ensuing pedigree material was genotyped with chromosome 2 microsatellites and subjected to linkage analysis. LOD scores >5 were obtained, and multipoint linkage analysis pointed towards a virtually identical map position of what is more than likely the same *mh* locus. Linkage disequilibrium analysis performed as previously done in the Belgian Blue population revealed a similar association, although extending over a longer chromosome segment. This observation might reflect the more recent intense selection for the double-muscling phenotype in this population. It should be noted that different TGLA44 alleles were associated with *mh* in both populations.[10]

These data strongly suggest locus homogeneity of double-muscling in Belgian Blue and Asturiana. It is unclear at this point, however, whether allelic homogeneity also applies. Finding different TGLA44 alleles associated with *mh* in both populations does not preclude allelic homogeneity, as the identified linkage disequilibrium might have been established by selection or drift after segregation of the two chromosome lines leading to Belgian Blue and Asturiana, respectively. The development of more closely linked markers will be needed to resolve this issue.

It should be noted that allelic homogeneity of double-muscling in different breeds might provide an interesting alternative towards fine-mapping the *mh* locus prior to positional cloning. As pointed out by Boehnke, recombinants can quickly become the primary factor limiting the resolution of genetic fine-mapping.[11] More than 200 informative meioses might be needed to define a 1 cM segment containing the gene of interest. Such numbers may be very difficult if not impossible to obtain with species like cattle. An alternative therefore would be to identify an identical-by-descent (IBD) chromosome segment flanking the *mh* locus among breeds sharing an IBD causal mutation in that gene. Historical records suggest that the spread of the double-muscling phenotype among continental breeds might be linked to the migration of Shorthorn individuals in the early 19th century. It is not unlikely, therefore, that double-muscling in at least a subset of these breeds involves the same mutation coalescing 50 generations back. This mutation would then be flanked by an ancestral chromosome segment of 2/[50*b] cM, where b corresponds to the number of breeds sharing this mutation.[10] Careful selection of "recombinant" individuals in the respective breeds might help to reduce the size of the shared segment even more.

### 17.2.5 Evidence for Allelic Heterogeneity: Is the *mh* Locus a QTL for Meat Production?

In an experimental *Bos taurus* × *Bos indicus* cross designed to map quantitative trait locus (QTL) affecting a variety of phenotypes including growth and carcass characteristics, preliminary evidence was found for a QTL affecting the adult weight of the animal. Although the maximum likelihood position of this QTL did not coincide exactly with the *mh* location, the respective support intervals are clearly overlapping.[10a] It is intriguing therefore to speculate that the same gene might underlie both phenotypic effects. As both phenotypes are fairly distinct, however, this would likely imply allelic heterogeneity at the corresponding locus.

## 17.3 QTLs with Large Effects on Milk Yield and Composition Are Still Segregating in Highly Selected Dairy Cattle Populations

### 17.3.1 Milk Yield and Composition Are Archetypal Quantitative Traits

Breeding schemes in dairy cattle are based on selection indices combining breeding values for a number of traits.[12] The economically dominant ones, however, are milk yield and composition (particularly fat and protein yield and percentage) as these are directly related to the potential yields of butter and cheese.

Analysis of milk yield and composition in commercial dairy populations reveals continuous, quasinormal distributions. In the absence of any conclusive evidence for the segregation of major genotypes, milk production traits have been treated using biometrical methods assuming an underlying infinitesimal model.[12] The resemblance observed between relatives reflects narrow-sense heritabilities estimated around 0.25 for yield traits (milk, fat, and protein) and 0.50 for percentage traits (fat and protein). Genetic correlations were estimated around 0.8 among yield traits, 0.6 among percentage traits, –0.3 between milk yield and percentage traits, and between 0.1 and 0.2 between component yield and percentage.[13] Individual breeding values are commonly estimated using Best Linear Unbiased Predictors obtained with an individual animal, and superior animals used as progenitors in subsequent generations.[14] Over the last 20 years, annual milk production per cow has increased in the U.S. from 4500 to 6800 kg, primarily as a result of the genetic improvement of the herd.[13]

### 17.3.2 Dairy Breeding Schemes Offer a Unique Niche for MAS

The development of artificial insemination and semen cryopreservation has offered the possibility to apply extreme selection pressure on the sire side through the extensive use of elite bulls. The identification of genetically superior sires, however, required the implementation of large-scale progeny-testing schemes. From candidate sires preselected on the basis of the estimated breeding values of their elite parents, 50 to 200 daughters are being produced. The genetic merit of the candidates is then inferred with high accuracy from the performances of their milking daughters: a genetic evaluation procedure requiring a minimum of 5 years at an estimated cost of $35,000 per bull. Thousands of bulls are annually tested following this procedure resulting in the selection of the top 10% to be extensively used in the general population. Combined with extensive data-collection campaigns and sophisticated statistical methods to dissect phenotypes into genetic and nongenetic components, this scheme has allowed for spectacular genetic progress despite the associated increase in generation interval. Alternative schemes have been envisaged, in which sires are selected on the performances of their full-sisters (produced by multiple ovulation and embryo transfer) rather than daughters, but have not yet found widespread application.[15]

The need for the progeny-test primarily stems from the differentiation among full-sibs due to Mendelian sampling, i.e., the fact that full-sibs inherit a different sample of genes from their respective parents. This makes it difficult to predict the breeding value of an individual based on the estimated breeding values of its parents only. Identifying the loci contributing to Mendelian sampling, however, would allow a preselection of candidate bulls based on the QTL-alleles they have inherited from their parents. The prospect of marker assisted preselection of young dairy bulls prior to progeny-testing has been the major impetus beyond efforts aimed to map QTL for milk production.

### 17.3.3 Alternative Strategies: QTL Mapping in an Experimental Cross or an Outbred Population

The conventional approach towards identifying QTL affecting milk production in cattle would be to design an experimental cross (F2 or BC) from a limited number of founder parents originating from highly divergent lines or breeds supposed to have (nearly) fixed alternative alleles at the QTL to be mapped. All F1 parents would therefore be expected to segregate for the same QTL (genetic homogeneity), having relatively large substitution effects. Moreover, by crossing divergent lines, the heterozygosity of the F1 generation at the marker loci would be increased relative to the parental generations. Finally, by raising all animals in standardized conditions, the environmental contribution to the phenotypic variance could be reduced.

Performing such experimental crosses in cattle, however, is a very expensive and time-consuming effort. More importantly, the QTL that could be mapped by such an approach would not necessarily be the loci contributing to the genetic variation observed for milk production in elite dairy cattle populations. The latter loci, however, are the substrate of ongoing selection programs and those that should be identified in order to implement MAS within the existing breeding schemes. But mapping the QTL still segregating in outbred elite populations is expected to be very difficult. Because of the large number of loci likely to be involved, milk production is genetically exceedingly heterogeneous. The segregating allele substitution effects are expected to be of relatively small magnitude, as large effects would have reached near fixation under the applied intense selection. Finally, heterozygosity at the marker loci will be lower in a within-breed vs. between-breed analysis.

### 17.3.4 The Granddaughter Design: Exploiting Progeny Testing

Dairy cattle populations, however, are characterized by two features that should greatly facilitate QTL mapping efforts even in outbred pedigrees. First, because of the extensive use of artificial insemination, very large (i.e., thousands) paternal half-sib families are readily available. One could therefore attempt to map the subset of QTL for which a given founder sire is heterozygous by looking at their segregation within a single half-sib-ship. By focusing on a single informative parent one effectively reduces the genetic complexity of the trait. Second, the progeny-testing scheme that has been previously described, allows one to use the estimated breeding values of sons (based on the performances of their daughters as part of the progeny test) as phenotypes rather than milk production of daughters per se. This scheme is typically referred to as the "granddaughter design".[16] It can easily be shown that the increased heritability of the breeding value estimates when compared to the actual trait, leads to a reduction in required sample size of a factor 3.5 to 4.

### 17.3.5 QTL with Large Effects on Milk Yield and Composition Are Still Segregating in Elite Dairy Populations

In one of the first whole genome scans exploiting the granddaughter design, samples were collected from 14 paternal half-sib families with their sires representing a total of 1518 animals (33 to 208 sons per family).[17] All sons were progeny tested based on the performances of more than 150,000 daughters jointly. Breeding value estimates were available for five of the most important milk production traits: milk, fat and protein yield in kilograms as well as fat and protein percentage. One concern with the available pedigree material resulted from the selection bias characterizing it: sires with high estimated breeding values were more likely to be included in the material, as A.I. (artificial insemination) companies would often discard semen from mediocre animals as inferred from their progeny test. An evaluation of the effect of selection bias on the power to map QTLs indeed showed that selection by truncation, if ignored, could result in a considerable drop of power and underestimation of QTL effects. Paradoxically, a mixture of unselected samples and truncated samples, which is closer to what is actually observed, actually increases power in a manner analogous to selective genotyping.[17a]

The available pedigree material was genotyped with a battery of 160 microsatellite markers covering an estimated 1645 cM as bracketed segments which might amount to two thirds of the bovine genome. The average marker heterozygosity in the 14 founder sires was 56%.

The resulting genotypes were analyzed using a multipoint maximum likelihood approach. Briefly, the likelihood of the joint phenotypic and genotypic data was computed assuming that the founder sire was heterozygous for a QTL with fixed position relative to the available marker map. The half-sib breeding values were expected to be distributed as a mixture of two normals. Their mean was computed for each son as the average of the parental breeding value $\pm \frac{1}{2}$ the effect of the QTL allele substitution, $\alpha$. Their common variance was expressed as a function of $\sigma_A^2$ accounting for the reliabilities of the paternal, maternal and own breeding value estimates. Values of $\alpha$ and $\sigma_A^2$ maximizing the likelihood were obtained using an optimization routine. The analyses were performed family by family and chromosome by chromosome. The position of the hypothetical QTL was changed with respect to the marker map, and results expressed as LOD-score curves, the null hypotheses corresponding to the likelihood of the data assuming an $\alpha$ value of zero.

According to Lander and Botstein, a LOD-score threshold of 2.7 would be associated with a Type I error of 5% when screening 15 M in marker brackets of 12 cM.[18] As we were performing analyses on 14 pedigrees and for the equivalent of 3.3 independent traits, this threshold was quite arbitrarily raised to three and all LOD scores superior to this threshold were reported.

Following these rules, evidence for the presence of a QTL was obtained on five chromosomes: 1, 6, 9, 10 and 20. As expected given their known correlations, all five QTL affected more than one milk production trait. Interestingly, the different QTL were affecting these traits in very different ways. The QTL on chromosomes 6 and 20 for instance caused an increase in milk volume but without concomitant increase in protein and fat yield, but rather a decrease in protein and fat content. This QTL allele therefore acts as a "dilution" gene. As the volume of milk is primarily determined by its osmolarity and lactose is the major milk osmole, this gene could exert its effect by an increase in lactose secretion causing an influx of water to restore iso-osmolarity with blood plasma. The QTL on chromosome 9, on the contrary, was revealed by an allele that caused an increase in milk yield but without affecting its composition. This could be achieved by an increase in the number of mammary secretory cells or an increase in the mean milk yield per cell. Finally, there was some evidence that the QTL on chromosomes 1 and 10 were affecting protein and fat yield and content in a different way, which would indicate that these two correlated constituents could be manipulated independently.

The size of the identified QTL allele substitution effects ranged from 0.62 to 1.34 additive genetic standard deviations, $\sigma_A$. Given the size of the half-sib families used, this was not really unexpected as the power of the design limited us to the detection of effects of this magnitude. Therefore QTL effects that would in reality be smaller, would — if detected — be biased upward. Nevertheless, there is some evidence that the estimates of at least some of these effects are actually reasonably accurate. Indeed, for the only QTL yielding significant evidence in two independent families in this initial study (chromosome 20), the estimates of the effects were very similar: 0.08% in one family and 0.09% in the other. In addition, in an independent granddaughter design performed in the same breed, a distinct family was shown to segregate for what is thought to be the same QTL effect as the one previously identified on chromosome 6. Not only was the estimated position virtually identical, but so were the effects on the different milk production traits.[19,20]

It seems therefore that QTL alleles with unexpectedly large effects are still segregating in these populations despite the intense selection. The reasons for this are unclear at this stage, but a number of possible explanations can be envisaged. The first is that even if the present selection pressure is generally quite intense, in many countries this is a relatively recent phenomenon of the order of 10 to 15 generations. Also, while selection intensity is quite harsh at the population level, the selection pressure on individual QTL may be modest. For example, as component yields dominate the selection indices in several countries, the QTL identified on chromosomes 6 and 20, which have modest effects on component yields, may be selectively quite neutral despite their large effects on component percentages. It was pointed out by Dekkers and Dentine[22] that even under the infinitesimal model fairly large effects are expected to segregate due to constellations of favorable alleles at linked QTL. One can't exclude that some QTL with large effects are maintained at intermediate frequencies by balancing forces due to associated deleterious effects either by pleiotropy or linkage disequilibrium. This is one of the primary concerns dictating prudence when attempting to exploit mapped QTL by MAS. Finally, large effects might reflect the occurrence of recent mutations at QTL.

### 17.3.6 Towards IBD Mapping of QTL in Dairy Cattle

Several independent studies are presently being performed that will allow the testing of these first results, and undoubtedly identify additional candidate QTL.

Important progress has been made in the development of tools to increase the power of QTL mapping in outbred cattle populations. The bovine genetic map now counts more than 1000 microsatellite markers, providing very adequate coverage of the genome with approximately

2 cM intervals.[22a] Semiautomated genotyping procedures are available allowing for throughputs of the order of 2000 genotypes/person/week. Moreover, considerable improvements in analysis methods are being reported, including multiple regression methods, phenotype permutation for the determination of significance thresholds, composite interval mapping and MQM mapping to account for genetic noise due to other QTL, or Monte-Carlo importance sampling methods allowing the fitting of more complex models, particularly the inclusion of complex pedigree relationships.[22-26]

It is our contention, however, that considerable gains in QTL mapping efficiency and precision could still be achieved in dairy cattle populations by increasing the density of markers to the point required to uncover linkage disequilibrium and allow for IBD mapping. Because of the dairy population structure and concomitant reduction in effective population size, as well as the applied selection pressure, linkage disequilibrium is expected to extend over longer chromosomal regions when compared, for instance, to most human populations. The IBD mapping approach has already proven its efficiency with the available marker density for the mapping of a monogenic trait and should be explored further for quantitative traits.[27]

A simple illustration of how IBD mapping could be applied to QTL mapping in dairy populations stems from the observation that the two families shown to segregate for an apparently identical QTL effect on bovine chromosome 6 (see above), are related through a common ancestor separating both families by five meioses.[19] It is reasonable to assume that an IBD QTL allele segregates in both families which possibly traces back to the identified common ancestor. The IBD QTL allele would then be expected to be flanked by an IBD chromosome segment of size 2/n cM, n being the number of generations to coalescence (in this case possibly five). Identifying such IBD segments in both families would unambiguously define the boundaries of the chromosome region containg the QTL.

## 17.4 ANALYSIS OF AN OVINE MUSCULAR HYPERTROPHY REVEALS POLAR OVERDOMINANCE AT THE *CALLIPYGE* LOCUS

### 17.4.1 The Gene Causing the Callipyge Muscular Hypertrophy Maps to Ovine Chromosome 18

The callipyge phenotype is a generalized muscular hypertrophy appearing in sheep at 3 weeks of age, and manifesting itself primarily around the hind quarters — hence its name. Callipygous animals are characterized by a 30% increase in muscle mass when compared to controls, rendering this phenotype particularly interesting for the sheep industry.[28-30] The callipygous phenotype was first described in the early 1980s, affecting a ram called "Solid Gold". The ram was shown to transmit this unusual phenotype to some of its offspring, and subsequent controlled matings between callipygous male descendants of "Solid Gold" and unrelated normal ewes revealed a 50%:50% sex-independent segregation ratio of what was then assumed to be a dominant mutation labeled CLPG.[30,31] The same $CLPG/clpg^{+} \times clpg^{+}/clpg^{+}$ matings were then used to map the corresponding locus to the telomeric end of ovine chromosome 18 using a mini- and microsatellite-based marker map. Multipoint LOD scores >50 are now obtained from such matings positioning the *callipyge* locus at 3 cM from the closest CSSM18 microsatellite marker. These LOD scores are obtained under the assumption of full penetrance of the callipyge phenotype, and hence the *callipyge* locus accounts for all the trait variance in these crosses.[31,32]

### 17.4.2 The Callipyge Phenotype is Characterized By a Non-Mendelian Inheritance Pattern

Subsequent matings, either between callipygous ewes and normal rams, or between callipygous ewes and rams, however, did not fit the simple model of an autosomal dominant mutation. Callipygous ewes mated to normal rams would only yield normal offspring. At present more than 35 such offspring have been obtained, all of conventional phenotype. The callipyge phenotype is therefore characterized by nonequivalence of reciprocal crosses. Moreover, while matings between $CLPG/clpg^{+}$ callipygous parents is expected to result in 75% callipygous and 25% conventional offspring, 51 offspring from such crosses yielded closer to the opposite ratios: 29% callipygous and 71% conventional. The callipyge locus is therefore clearly characterized by a non-Mendelian segregation pattern.

### 17.4.3 Marker-Assisted Segregation Analysis Reveals Polar Overdominance at the Ovine Callipyge Locus

Genotyping these individuals with chromosome 18 markers flanking the callipyge locus revealed a clear pattern.[32] First, although $clpg^{+}/clpg^{+} \times CLPG/clpg^{+}$ matings generated only conventional offspring, the maternal 18 homologues were shown from marker data to segregate as expected. But offspring having inherited the *CLPG* mutation from their dam (or $clpg^{+(Pat)}/CLPG^{(Mat)}$) were unexpectedly not expressing the phenotype, contrary to individuals having inherited the same mutation from their sire (or $CLPG^{(Pat)}/clpg^{+(Mat)}$ individuals), clearly pointing towards a parent-of-origin effect.

Genotyping the offspring of the second type of matings, i.e., $CLPG/clpg^{+} \times CLPG/clpg^{+}$, revealed that $CLPG^{(Pat)}/clpg^{+(Mat)}$, $clpg^{+(Pat)}/CLPG^{(Mat)}$, and $clpg^{+(Pat)}/clpg^{+(Mat)}$ individuals generally expressed the same phenotypes as observed in previous matings, namely callipygous, normal and normal. Unexpectedly, however, $CLPG^{(Pat)}/CLPG^{(Mat)}$ offspring did not express the callipygous phenotype. The inactive $CLPG^{(Mat)}$ allele therefore seemed to dominate the active $CLPG^{(Pat)}$ allele. The resulting segregation pattern, where only heterozygous individuals having inherited the mutation from their sire express the phenotype, has been referred to as "polar overdominance".

The polar overdominance model allowed us to predict that matings between nonexpressing *CLPG/CLPG* rams and $clpg^{+}/clpg^{+}$ ewes would yield 100% callipygous offspring. This is virtually what is observed as 91% (30/33) of the offspring obtained from these matings are indeed expressing the callipygous phenotype.

### 17.4.4 Does Parental Imprinting Explain Polar Overdominance?

The parent-of-origin effect observed for the segregation of the callipyge phenotype suggests a possible role of parental imprinting. Interestingly, the homologous chromosomal regions in man (chromosome 14) and mice (chromosome 12) are known to harbor imprinted regions.[33-36] The conventional phenotype of homozygous *CLPG/CLPG* individuals, however, is difficult to reconcile with the allele-specific transcriptional silencing observed for all known imprinted genes.[37] A number of molecular models based on conventional parental imprinting can, however, be envisaged to account for the observed segregation pattern.[32] One could postulate that the callipyge mutation switches the expression pattern of the corresponding imprinted *callipyge* gene from paternal to maternal. $CLPG^{(Pat)}/clpg^{+}$ individuals would then be the only genotype with two transcriptionally silent copies of the gene, which would explain their unique phenotype. An alternative model envisages two closely linked genes, one of which being paternally expressed and coding for a transacting suppressor of the other one. The callipyge mutation would be a deletion of both genes, resulting in the expression of the otherwise suppressed gene in $CLPG^{(Pat)}/clpg^{+}$ individuals only.

To the best of our knowledge, only two other phenotypes are showing polar overdominance: P-element dependent hybrid dysgenesis in *Drosophila,* and an early embryonic lethality referred as polar lethality in DDK mice.[38,39] Hybrid dysgenesis is well understood and due to P-element transposition in the germline of offspring from P males × M females crosses. This zygotic mechanism seems difficult to reconcile with the muscle-specific expression of the callipyge phenotype. The molecular mechanisms underlying polar lethality are yet unknown, but imprinting has been invoked as a possible cause.[40,41]

Positional cloning of the callipyge gene will be required to elucidate the actual molecular mechanism underlying polar overdominance. Already, however, the polar overdominance model might help to explain complex inheritance patterns in a variety of organisms, including man.

## 17.5 CONCLUSIONS

The experiments summarized in this chapter, as well as a growing number of similar studies, demonstrate convincingly that chromosomal regions underlying the genetic variance for complex phenotypes can be identified in livestock populations using the available genomic tools. It is reasonable to predict that for loci with major phenotypic effects, such as the *mh* and callipyge loci described in this manuscript, cloning of the actual genes and identification of the culprit mutations could be achieved in the near future. Positional cloning of actual QTL, however, underlying continuously distributed quantitative traits (such as milk production in cattle), remains a major intellectual challenge. The obstacles to be expected are two-tiered. (1) Genetic fine-mapping of QTL down to the centimorgan scale, compatible with positional cloning strategies, will be very difficult to achieve with the available family material. Animal geneticists have the advantage to be able to plan matings at will. However, the costs associated with such experiments can become prohibitive when dealing with livestock species such as cattle. Some hints are given in this paper and in Georges and Andersson[19] on how intra- and inter-breed IBD mapping might improve QTL fine-mapping but the feasibility of this approach remains to be demonstrated. Moreover, trait complexities such as epistasis, gene-by-environment interactions or even non-Mendelian transmission mechanisms such as the one demonstrated for the *callipyge* locus might considerably obscure the picture. (2) Even if fine-mapping could be achieved, pinpointing the causal sequence variants among the background of DNA polymorphisms will be very difficult in outbred populations. This is especially true if one doesn't have a clear idea of the tissue(s) in which the causal mutation exerts its effect (as is often the case for complex traits) and in the case of subtle structural or even regulatory mutations. The molecular dissection of complex traits into their Mendelian components therefore promises to challenge geneticists for several decades ahead.

Fortunately for animal geneticists, the livestock industry is likely to benefit from the results of livestock genomics prior to the actual cloning of all the genes involved in the determination of production traits. As long as a sufficient proportion of the trait variance can be explained by genetic markers in linkage disequilibrium with the causal loci in the population of interest, MAS is likely to make a substantial contribution to genetic progress. Moreover, genomics will help animal breeders select for traits that were difficult to deal with using conventional breeding strategies. The virtual elimination of bovine leukocyte adhesion deficiency from the Holstein Friesian population through the systematic monitoring of a mutation in the CD18 gene nicely illustrated this assertion.[42] Moreover, mapping the genes explaining the differences observed between breeds for economically important traits will permit the exploitation of interbreed variation using marker assisted introgression schemes. These readily available advances will therefore justify the continued interest and support of the respective industries for livestock genomics, of which the impact on breeding programs is likely to increase in the future.

## REFERENCES

1. Collins, F. S., Positional cloning moves from perditional to traditional, *Nat. Genet.,* 9, 347–350, 1995.
2. Lander, E. S. and Schork, N. J., Genetic dissection of complex traits, *Science,* 265, 2037–2048, 1994.
3. Ménissier, F., Present state of knowledge about the genetic determination of muscular hypertrophy or the double muscled trait in cattle, in *Current Topics in Veterinary Medicine and Animal Science,* Vol. 16, *Muscle Hypertrophy of Genetic Origin and Its Use to Improve Beef Production,* King, J. and Ménissier, F., Eds., Martinus Nijhoff, 1982, 387–428.
4. Hanset, R. and Michaux, C., On the genetic determinism of muscular hypertrophy in the Belgian White and Blue cattle breed. II. Population data, *Génét. Sél. Evol.,* 17, 369–386, 1985.
5. Hanset, R. and Michaux, C., On the genetic determinism of muscular hypertrophy in the Belgian White and Blue cattle breed. I. Experimental data, *Génét. Sél. Evol.,* 17, 359–368, 1985.

6. Hanset, R., Michaux, C., Dessy-Doize, C., and Burtonboy G., Studies on the 7th rib in double muscled and conventional cattle, in *Current Topics in Veterinary Medicine and Animal Science,* Vol. 16, *Muscle Hypertrophy of Genetic Origin and Its Use to Improve Beef Production,* King, J. and Ménissier, F., Eds., Martinus Nijhoff, 1982, 341–349.
7. Hanset, R., The major gene of muscular hypertrophy in the Belgian Blue Cattle breed, in *Breeding for Disease Resistance in Farm Animals,* Owen and Axford, Eds., C.A.B. International, 1991, 467–478.
8. Charlier, C., Coppieters, W., Farnir, F., Grobet, L., Leroy, P., Michaux, C., Mni, M., Schwers, A., Vanmanshoven, P., Hanset, R., and Georges, M., The mh gene causing double-muscling in cattle maps to bovine chromosome. 2, *Mammalian Genome,* 6, 788–792, 1995.
9. Terwilliger, J. D., A powerful likelihood method for the analysis of linkage disequilibrium between trait loci and one or more polymorphic marker loci, *Am. J. Hum. Genet.,* 56, 777–787, 1995.
10. Dunner, S., Charlier, C., Farnir, F., Brouwers, B., Canon, J., and Georges, M., Double-muscling in the Asturiana de los Valles breed involves the same *mh* locus as in the Belgian Blue cattle breed, *Mammalian Genome,* in press.

10a. Taylor, J., personal communication, 1996.

11. Boehnke, M., Limits of resolution of linkage studies: implications for the positional cloning of disease genes, *Am. J. Hum. Genet.,* 55, 379–390, 1994.
12. Falconer, D. S. and Mackay, T., *An Introduction to Quantitative Genetics,* 4th ed., Longman, Chicago, 1996.
13. Pearson, R. E., Vinson, W. E., and Meinert, T. R., The potential for increasing productivity through selection for increased milk and component yields, in *Proc. 4th World Congress on Genetics Applied to Livestock Production,* Hill, W., Thompson, R., and Wooliams, J., Eds., Edinburgh, 14, 104–113, July 1990.
14. Van raden, P. M. and Wiggans, G. R., Derivation, calculation, and use of National Animal Model Information, *J. Dairy Sci.,* 74, 2737–2746, 1991.
15. Nicholas, F. W. and Smith, C., *Anim. Product.,* 36, 341–353, 1983.
16. Weller, J. L., Kashi, Y., and Soller, M., Power of daughter and granddaughter designs for determining linkage between marker loci and quantitatve trait loci in dairy cattle, *J. Dairy Sci.,* 73, 2525–2537, 1990.
17. Georges, M., Nielsen, D., Mackinnon, M., Mishra, A., Okimoto, R., Pasquino, A. T., Sargeant, L. S., Sorensen, A., Steele, M. R., Zhao, X., Womack, J. E., and Hoeschele, I., Mapping quantitative trait loci controlling milk production by exploiting progeny testing, *Genetics,* 139, 907–920, 1995.

17a. Coppieters, W., Mackinnon, M., and Georges, M., A note on the effects of selection on linkage analysis for quantitative traits, submitted.

18. Lander, E. S. and Botstein, D., Mapping mendelian factors underlying quantitative traits using RFLP linkage maps, *Genetics,* 121, 185–199, 1989.
19. Georges, M. and Andersson, L., Livestock genomics comes of age, *Genome Res.,* 6, 907–921, 1996.
20. Spelman, R. J., Coppieters, W., Karim, L., van Arendonk, J. A. M., and Bovenhuis, H., Quantitative trait loci analysis for five milk production traits on chromosome six in the dutch Holstein-Friesian population, *Genetics,* in press.

22. Knott, S. A., Elsen, J. M., and Haley, C. S., Multiple marker mapping of quantitative trait loci in half-sib populations, in *Proc. 5th World Congress on Genetics Applied to Livestock Production,* Guelph, 21:33–36, 1994.

22. Dekkers, J. C. M. and Dentine, M. R., Quantitatve genetic variance associated with chromosomal markers in segregating populations, *Theor. Appl. Genet.,* 81, 212–220, 1991.

22a. Beattie, C., personal communication, 1996.

23. Churchill, G. A. and Doerge, R. W., Empirical threshold values for quantitatve trait mapping, *Genetics,* 138, 963–971, 1994.
24. Zeng, Z. B., Theoretical basis for separation of multiple linked gene effects in mapping quantitative trait loci, *Proc. Natl. Acad. Sci. U.S.A.,* 90, 10,972–10,976, 1993.
25. Jansen, R., Controlling the type I and type II errors in mapping quantitative trait loci, *Genetics,* 138, 871–881, 1994.
26. Guo, S. W. and Thompson, E. A., A Monte-Carlo method for combined segregation and linkage analysis, *Am. J. Hum. Genet.,* 51, 1111–1126, 1992.

27. Charlier, C., Farnir, F., Berzi, P., Vanmanshoven, P., Brouwers, B., and Georges, M., Identity-by-descent mapping of recessive traits in livestock: application to map the bovine syndactyly locus to chromosome 15, *Genome Res.,* 6, 580–589, 1996.
28. Jackson, S. P., Miller, M. F., Green, R. D., and Brdecko, K. S., Carcass characteristics of Rambouillet ram lambs with genetic muscular hypertrophy, in *Proc. West. Sect. Am. Soc. Anim. Sci.,* 44, 167–169, 1993.
29. Jackson, S. P., Miller, M. F., and Green, R. D., The effect of a muscle hypertrophy gene on muscle weights of ram lambs, in *Proc. West. Sect. Am. Soc. Anim. Sci.,* 44, 196–169, 1993.
30. Jackson, S. P. and Green, R. D., Muscle trait inheritance, growth performance and feed efficiency of sheep exhibiting a muscle hypertropy phenotype, in *Proc. West. Sect. Am. Soc. Anim. Sci.,* 44, 364–366, 1993.
31. Cockett, N. E., Jackson, S. P., Shay, T. D., Nielsen, D., Green, R. D., and Georges, M., Chromosomal localization of the callipyge gene in sheep (*Ovis aries*) using bovine DNA markers, *Proc. Natl. Acad. Sci. U.S.A.,* 91, 3019–3023, 1994.
32. Cockett, N. E., Jackson, S. P., Shay, T. D., Farnir, F., Berghmans, S., Snowder, G., Nielsen, D., and Georges, M., Polar overdominance at the ovine callipyge locus, *Science,* 273, 236–238, 1996.
33. Temple, I. K., Cockwell, A., Hassold, T., Pettay, D., and Jacobs, P., Maternal uniparental disomy for chromosome 14, *J. Med. Genet.,* 28, 511–514, 1991.
34. Wang, J.-C. C., Passage, M. B., Yen, P. H., Shapiro, L. J., and Mohandas, T. K., Uniparental heterodisomy for chromosome 14 in a phenotypically abnormal familial balanced 13/14 robertsonian translocation carrier, *Am. J. Hum. Genet.,* 48,1069–1074, 1991.
35. Beechey, C. V. and Cattanach, B. M., Viable tertiary trisomy and monosomy within the distal chromosome 12 imprinting region, *Mouse Genome,* 92, 505–506, 1994.
36. Cattanach, B. M. and Raspberry, C., Evidence for imprinting involving the distal region of chromosome 12, *Mouse Genome,* 91, 858, 1993.
37. Efstratiadis, A., Parental imprinting of autosomal mammalian genes, *Curr. Opinion Genet. Dev.,* 4, 265–280, 1994.
38. Bregliano, J. C. and Kidwell, M. G., in *Mobile Genetic Elements,* Shapiro, J. A., Ed., Academic Press, New York, 1983, 363–410.
39. Wakasugi, N. A., Genetically determined incompatibility system between spermatozoa and eggs leading to embryonic death in mice, *J. Reprod. Fert.,* 41, 85–96, 1974.
40. Sapienza, C., Paquette, J., Pannunzio, P., Albrechtson, S., and Morgan, K., The polar-lethal ovum mutant gene maps to the distal portion of mouse chromosome 11, *Genetics,* 132, 241–246, 1992.
41. Baldacci, P. A., Richoux, V., Renard, J.-P., Guénet, J.-L., and Babinet, C., The locus Om, responsible for the DDK syndrome, maps close to Sigje on mouse chromosome 11, *Mammalian Genome,* 2, 100–105, 1992.
42. Shuster, D. E., Kehrli, M. E., Jr., Ackermann, M. R., and Gilbert, R. O., Identification and prevalence of a genetic defect that causes leukocyte adhesion deficiency in Holstein cattle, *Proc. Natl. Acad. Sci. U.S.A.,* 89, 9225-9, 1992.

# 18 Case History in Animal Improvement: Genetic Mapping of QTLs for Growth and Fatness in the Pig

*Leif Andersson, Kjell Andersson, Lena Andersson-Eklund, Inger Edfors-Lilja, Hans Ellegren, Chris S. Haley, Ingemar Hansson, Maria Johansson Moller, Sara A. Knott, Kerstin Lundström, and Lena Marklund*

## CONTENTS



## 18.1 INTRODUCTION

This study was initiated in 1988 as part of a Nordic project on animal gene mapping. At the time, only a rudimentary pig linkage map was available and no other international collaborative projects had been established. Our intention was to generate a pedigree with a high degree of heterozygosity to facilitate marker development and map construction. Moreover, we wanted a pedigree showing segregation at major loci controlling phenotypic traits. By this approach the construction of an informative linkage map could be accompanied by the mapping of trait loci. We decided to produce a three-generation pedigree by intercrossing the European wild pig with domestic Large White pigs.

There are two major strategies for quantitative trait locus (QTL) mapping in farm animals. One is to use crosses between two divergent populations which may be fixed or close to fixation for different QTL alleles due to differential selection pressures in the two populations. Alternatively, QTL mapping may be carried out using large commercial pedigree materials with extensive breeding records; for example, a particularly powerful approach is the granddaughter design utilized for

0-8493-7686-6/98/$0.00+$.50

mapping QTLs in dairy cattle populations (see Chapter 17, this volume). However, QTL mapping in a cross between two divergent populations has the following advantages:

1. Alleles with large effects may be segregating
2. Increased power in the statistical analyses due to high heterozygosity at QTLs; 100% heterozygosity in the F1 generation when there are fixed differences between breeds
3. Increased power in the statistical analyses since the linkage phase between markers and QTLs are expected to be consistent among F1 animals, in contrast to the situation for outbred populations where the linkage phase may vary between families
4. Higher heterozygosity reduces the number of marker loci needed for comprehensive genome coverage. [This was particularly important when this project started in 1988, since the use of microsatellites for gene mapping was not yet established and we planned to build our map using restriction fragment length polymorphisms (RFLPs).]

A disadvantage with QTL mapping using a divergent cross, from a practical breeding point of view, is that the QTLs detected may or may not be those controlling genetic variation within commercial populations.

Why was a wild pig-domestic pig intercross chosen for this study? The parental populations showed large phenotypic differences for almost any trait we could imagine. Analysis of this cross also had the potential to shed light on the genetics of domestication as we crossed an elite commercial pig population with its wild ancestor. The wild and domestic pigs are usually referred to as subspecies (*Sus scrofa scrofa* and *S. scrofa domesticus*, respectively), but domestic pigs have been developed from wild pigs within the last 10,000 years. Wild and domestic pigs are easy to cross and the F1 animals show normal fertility. Thus, this cross should be considered as an intraspecies cross and the F1 animals are not expected to show such extreme heterozygosity at marker loci as observed in some interspecies crosses of mice[1] or deer.[2] There are many possible alternative types of intercrosses which could be carried out in the pig. An obvious one is a cross between Chinese (or other Asiatic) domestic pigs and European domestic pigs which show marked phenotypic differences for a number of traits including growth, fatness, and fertility; a number of groups have established such intercrosses for the purpose of QTL mapping.[3]

Why was an F2 intercross produced rather than a backcross? In linkage analysis, a backcross is only more efficient than an intercross when dealing with loci showing dominance. Since we were planning to use codominant DNA markers, an intercross should give twice as many informative meioses at marker loci than a backcross. We intended to study a large number of traits and most of them showed multifactorial inheritance. Thus, segregation occurred at several or many loci per trait and we had no prior information regarding the presence of dominance or direction of dominance. The ideal design for mapping trait loci would be to develop intercrosses as well as reciprocal backcrosses. However, the limited resources available only allowed one of these options and an intercross was considered the most informative one.

## 18.2 MATERIALS AND METHODS

### 18.2.1 Development of Pedigree

The wild pig became extinct in Sweden in the 17th century but wild populations are maintained in many parts of Central and Southern Europe. Wild pigs, however, are bred in captivity under semiwild conditions in Sweden for the purpose of exotic meat production. There is a polymorphism in chromosome number among wild-pig populations in Europe.[4] The polymorphism is due to a centric fusion/fission so that the chromosome number is 2n = 38 in some populations, as in domestic pigs, whereas it is 2n = 36 in others. Two wild-pig boars from different captive herds in Sweden were selected for this study and a cytogenetic examination revealed that both animals had a 2n = 38 karyotype.[4a]

The two wild boars were mated to eight Large White (Swedish Yorkshire) sows from the experimental herd at the Swedish University of Agricultural Sciences in Uppsala. From the F1 generation, 4 males and 22 females were intercrossed to produce 200 F2 animals in two consecutive litters. The males were castrated at 2 weeks of age for practical and economical reasons. (Castrates are less aggressive than males and there is a price reduction on male meat due to the possible development of boar taint after sexual maturation.) The matings of the F1 animals were carried out on a private farm. The F2 animals were subsequently brought to the experimental pig station at about 3 months of age for fattening and trait recordings.

### 18.2.2 Phenotypic Traits

Wild and domestic pigs show striking phenotypic differences for a number of traits. Our strategy was therefore to collect data on as many traits as practically and economically possible. The traits measured on the F2 animals are listed in Table 18.1. Growth and immunological traits were recorded on the live animals whereas all other data were collected after slaughter. Blood samples, primarily intended for the immunological studies, were collected immediately before and the day after transport to the experimental farm. Serum samples were also collected 2 and 5 weeks later as well as the day before slaughter. Serum and ethylene diamine tetraacetic acid (EDTA) blood samples are currently stored at –70°C.

The project was clearly underfinanced in relation to our ambition to collect data on as many traits as possible and some interesting traits had to be excluded simply for economic reasons. An example of such a trait is behavior where we observed obvious differences in aggressiveness and alertness between our F2 animals and pure-bred domestic animals. Analysis of fertility traits would also have been most interesting to analyze as wild pigs show a pronounced seasonality in breeding in contrast to the domestic pigs. However, to collect fertility data we had to maintain a large number of the F2 animals for further breeding. Finally, this pedigree would have been excellent material for studying the genetic basis for boar taint, which is an important practical problem in pig breeding and is the major reason why male piglets are often castrated.[5] However, the F2 males were all castrated to minimize the cost for establishing the pedigree.

### 18.2.3 Development of a Marker Map

The analysis of a large number of genetic markers in our wild pig intercross was done with the aim of developing an informative linkage map for the pig as well as for mapping trait loci segregating in the pedigree. The current map includes blood group loci, protein markers, RFLPs and microsatellites. The reason why we included the classical blood group and protein markers was that we wanted to assign them to the growing DNA marker-based linkage map and it gave us around 20 markers as a start when building the map. Moreover, the classical markers represent coding sequences and at least the protein markers are useful for comparative mapping.

A considerable effort has been devoted to the development of RFLP markers using primarily pig, human, or mouse cDNA probes. The reason for this is partly that the microsatellite technology was not widely used when the current project started but, more importantly, because we wanted to assign coding sequences which are informative for comparative mapping.[6] However, a limitation with RFLP markers is their low degree of polymorphism.

Microsatellites are ideal genetic markers for linkage mapping due to the polymerase chain reaction (PCR)-based format, their abundance in the genome, and the high heterozygosity. The development of the microsatellite technology has therefore been a major breakthrough for genome analysis in vertebrate species. Soon after the first report of the presence of highly polymorphic microsatellites in humans[7,8] we decided to focus our development of genetic markers on microsatellites.[9]

We published one of the first primary linkage maps in the pig.[10] The map contained more than 100 linked markers of which a large proportion also had been physically mapped making it possible

**TABLE 18.1**
**Traits Measured in the F2 Generation of an Intercross between the European Wild Pig and Large White Domestic Pigs**

| Trait | Ref.[a] |
|---|---|
| **Growth Traits** | |
| Birth weight, kg | 11 |
| Growth rate from birth to 30 kg, g/d | 11 |
| Growth rate from 30 kg to 70 kg, g/d | 11 |
| **Body Proportions** | |
| Weight of carcass with head, kg | 34 |
| Weight of dissected carcass half, kg | 34 |
| Length of carcass, cm | 34 |
| Number of vertebra | 34 |
| Distance first thoracic-last lumbar vertebra, cm | 34 |
| Width of head, cm | 34 |
| Length of small intestine, m | 34 |
| **Weight of Internal Organs** | |
| Heart, g | — |
| Liver, g | — |
| Kidney, g | — |
| Spleen, g | — |
| **Osteochondrosis and Bone Measurements** | |
| Macroscopically osteochondrosis in medial femoral condyle, 0-6 | 35 |
| Macroscopically osteochondrosis in medial humeral condyle, 0-6 | 35 |
| Weight of femur, g | 35 |
| Length of femur, mm | 35 |
| Width of epicondyle, mm | 35 |
| Width of diaphys, mm | 35 |
| Cortex outer width, mm | 35 |
| Cortex inner width, mm | 35 |
| Cortex diff outer-inner, mm | 35 |
| Length of diaphys, mm | 35 |
| Angle of collum, degrees | 35 |
| **Carcass Composition Traits** | |
| Sidefat, mm | 34 |
| Average backfat thickness, mm | 11 |
| Abdominal fat, % of carcass | 11 |
| Weight of ham, kg | 34 |
| Lean meat in ham, % | 34 |
| Femur in lean meat of ham, % | 34 |
| Lean meat + bone in ham + back, % | 34 |
| Lean meat + bone in back, % | 34 |
| Lean meat and bone, % of back | 34 |
| Longissimus dorsi in back, % | 34 |
| Longissimus dorsi area, $cm^2$ | 34 |
| **Meat Quality** | |
| $pH_u$, longissimus dorsi | 34 |
| $pH_u$, biceps femoris | 34 |
| Drip loss, % | 34 |
| Filter paper wetness, (0–4) | 34 |

**TABLE 18.1 (continued)**
**Traits Measured in the F2 Generation of an Intercross between the European Wild Pig and Large White Domestic Pigs**

| Trait | Ref.[a] |
|---|---|
| Reflectance value, EEL | 34 |
| Pigmentation, ppm haematin in long.dorsi | 34 |
| Total protein extractability, mg/g long.dorsi | 34 |
| Sarcoplasmic protein extr., mg/g long. dorsi | 34 |
| Shear force, kg/cm$^2$ | 34 |
| **Immune Capacity Traits[b]** | |
| Total and differential white blood cell counts | 20 |
| Mitogen (ConA, PHA and PWM) induced proliferation and IL-2 production | 36 |
| ADV-induced IFN-α production | 37 |
| Phagocytosis of opsonized zymosan particles | 38 |
| Antibody (IgG) response to *E. coli* O149 and K88 antigens after immunization | 39 |

[a] For description of traits.
[b] Abbrevations used: ADV, Aujeszky's disease virus; Con A, concanavalin A; IFN-a, interferon α; IL-2, interleukin 2; PHA, phytohemagglutinin; PWM, pokeweed mitogen; IgG, Immunoglobulin G.

to conclude that the map covered a considerable portion of the pig genome. This map was used for our first QTL analysis.[11] Subsequently, we used microsatellites developed by other groups[12,13] to produce a map with almost complete coverage.[14] This map includes 236 linked loci and has a total sex-average map length of 2300 cM. It forms an excellent basis for all subsequent QTL analysis in this pedigree.

### 18.2.4 Statistical Methods

The analyses utilized a form of interval mapping, first named and popularized for crosses between inbred lines by Lander and Botstein.[15] The method requires a known map of markers covering a major proportion of the genome of the target species. The genome is then scanned for a QTL segregating in the cross of interest and linked to these markers. In the variant of interval mapping used in this study, we estimate the probabilities for every individual being each of the three possible genotypes (homozygous for alleles from the wild pig, homozygous for alleles from the Large White or inheriting one allele from each breed) for a putative QTL in a given position in the linkage group. The phenotypic score of an individual is regressed onto the additive and dominance coefficients of the QTL calculated from these genotype probabilities. We have previously shown[16] that this analysis gives almost identical results (in terms of power, estimated position, and estimated additive and dominance effect of a QTL) to the original maximum likelihood method of Lander and Botstein[15] when used for the study of inbred lines crosses. The method is simpler, faster, and easier to compute than maximum likelihood methods, hence allowing the easy evaluation of more complex and realistic models.

The QTL genotype probabilities are calculated conditional on the marker genotypes for each individual. In fact, for a particular position in the genome, if the assumption of no interference is made and the markers are completely informative (as codominant markers are in an inbred line cross), only the markers flanking the position are required to calculate these probabilities. In this situation they are relatively simple functions of the recombination fractions between a putative QTL and the two flanking markers of known genotype. In a cross between two outbred lines the situation is more complex, because individual markers are seldom fully informative. Hence it is only possible to follow the inheritance of a marker allele from a founder breed through the F1 to the F2 in a proportion of individuals. Haley et al.[17] showed how interval mapping based on regression could still be applied in this situation. The basis of the method — regression onto conditional additive and dominance coefficients for a QTL — remains the same. The calculation of the conditional QTL genotype probabilities is also relatively simple. In essence, we can now consider the two gametes of an F2 individual separately and use markers to follow their inheritance back through either the F1 sire or the F1 dam to the founder breed. Again for a particular position in the genome, only two *informative* markers are required for each gamete and the conditional probability of the gamete carrying a particular QTL allele are simple functions of recombination fractions. However, as a marker can be informative for one F1 parent but not the other, up to four markers are needed for each F2 individual (i.e., two for each F1 parent). However, the markers used may differ from individual to individual in the F2 population, so calculating the conditional QTL probabilities for some particular position in all members of the population could potentially need the use of all the markers in a linkage group. We[17] used simulated data to show how this method could increase the power to detect QTLs and reduce biases to produce reasonable estimates of their position and effect.

In the analyses applied in this study we assumed that any detectable QTL would have been fixed for alternative alleles in the two lines crossed. This seems a reasonable assumption given their long and separate selection histories. There is certainly variation for the traits analyzed segregating within each of the two breeds, but this is likely to be at QTLs which are individually of moderate or small effect and hence which would not be detectable in a study of this size.

Covariates and fixed effects were included in the analytical models along with QTL effects (see Reference 11). Including these simultaneously in the model reduces the potential for bias and should increase the power to detect QTLs by removing some of the background noise. In terms of the QTLs, three types of analyses were performed. First, we looked for a single QTL in each linkage group by fitting a putative QTL at 1-cM intervals. The position at which the test statistic (the variancc, or F, ratio in these regression analyses) is maximized (or equivalently the residual variance in minimized) is the best estimate of position of any QTL. These analyses are equivalent to those commonly used in interval mapping[15] and plotting the F-ratio against chromosomal position gives an analogous graphical display of the results. Furthermore, by transforming the F-ratio values into log-likelihood ratio statistics,[16] we can calculate the equivalent of a one or two LOD drop from the peak value, and hence plot support intervals for the estimated QTL location. Second, we looked for interactions between a single QTL and the fixed effects of feed treatment and sex. Third, we looked for evidence for two QTLs in each linkage group where one QTL had been detected. This was done by comparing the best fitting model with two QTLs anywhere on the linkage group (regressing onto the conditional probabilities for two QTLs simultaneously) with the best fitting model with only a single QTL. There was no significant evidence of interactions between QTLs and sex or feed treatment or of two QTLs in a linkage group and so only the single QTL results are reported.

The final problem in the analyses is setting the significance threshold. With tests for a single QTL being performed at 1-cM intervals through the genome, many, often highly correlated, tests are being performed. Thus using standard single test 5% significance thresholds would result in a high probability of one or more false positive results (Type I errors).[18] Hence, an appropriate significance threshold was derived by simulation. Using the real pedigree and marker data, 5000 replicates of normally distributed phenotypic data were generated in which no QTL was actually segregating. These data sets were analyzed for a single QTL and the highest test statistic from each data set was recorded. The test statistic cutting off the top 5% of the replicates provides an empirical genome-wide 5% significance threshold. For normally distributed data such simulations give a very similar significance threshold to that generated by the permutation method proposed by Churchill and Doerge.[19]

**TABLE 18.2**
**Growth and Fatness Traits in Purebred Large White Animals and in the Wild Pig F2 Intercross**

| Trait | Population | |
|---|---|---|
| | Large White n = 200 | F2 n = 191 |
| Growth rate, birth — 30 kg (g/d) | 323 (32) | 236 (38) |
| Growth rate, birth — 70 kg (g/d) | 524 (40) | 367 (49) |
| Abdominal fat (%)[a] | 1.79 (0.44) | 2.40 (0.69) |
| Average back fat thickness (mm)[b] | — | 26.4 (4.9) |
| Length of small intestine (m) | 21.2 (1.8) | 17.5 (1.9) |

[a] measured as the weight of fat tissue in the abdominal cavity as a percentage of the total carcass weight
[b] measured as the average of five measurements along the dorsal midline at shoulder, last rib, and loin; data for this trait are missing for Large White as it is not measured at a comparable weight
*Note:* Population means (and standard deviations) are given for each trait.

Reprinted from Andersson, L. et al., *Science,* 263, 1771, 1994. 

## 18.3 RESULTS OF QTL ANALYSIS

The initial QTL analysis was carried out for growth and fatness traits, using our primary linkage map comprising about 100 markers.[11] The length of the small intestine was also analyzed as this trait shows a strong correlation with growth. There has been intensive artificial selection for lean growth in domestic pigs and consequently our F2 animals were fatter, grew more slowly, and had a shorter small intestine compared with pure-bred Large White animals (Table 18.2). The trait variation was also greater in the F2 animals implying the segregation of QTLs in this population. The statistical analyses revealed evidence for QTLs on chromosome 4 with large effects on fat deposition, growth, and the length of the small intestine (Figure 18.1; Table 18.3). Wild pig alleles were associated with higher fat content, reduced growth, and shorter small intestine in line with the differences between populations (cf. Table 18.2 and 18.3). A QTL affecting early growth was also detected on chromosome 13 (Table 18.3). Gene actions at these QTLs appeared largely additive, with the dominance components being nonsignificant. The QTL for growth on chromosome 4 had an estimated additive effect of about 25 g/d, so that animals homozygous for wild pig alleles had a daily gain of almost 50 g less than the opposite homozygote. The locus thus causes a difference between the two homozygotes of about 10 kg in weight when the pig is 6 months of age. Similarly, the fatness QTL appears to control a large proportion of the genetic difference in this trait between the two founder populations. Thus, the QTLs detected on pig chromosome 4 must have been important in the response for lean growth during the evolution of the European domestic pigs.

The peaked curves for the fatness traits and intestinal length are consistent with a single QTL on chromosome 4 while the flatter curve observed for growth suggested the presence of two or more linked QTLs (Figure 18.1). However, since QTLs are mapped with relatively low precision we cannot exclude the possibility that there is a single QTL on chromosome 4 with pleiotropic effects on fatness and growth.

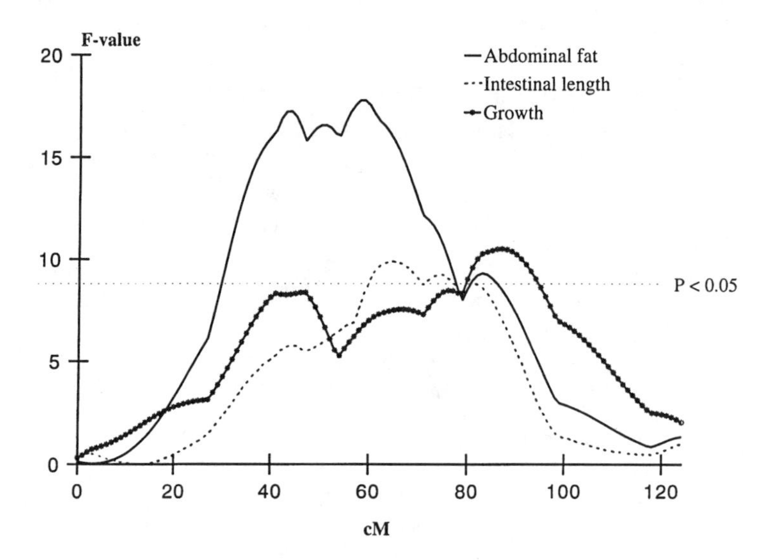


**FIGURE 18.1** Test-statistic curves (F-values) for pig chromosome 4 showing significant QTL effects on abdominal fat percentage, length of intestine, and growth from birth to 70 kg. The horizontal line marks the 5% significance threshold set by simulation.

**TABLE 18.3**
**Detected QTL Effects (Means ± SE) on Chromosomes 4 and 13 in a Wild Pig — Large White Intercross[a]**

| Trait | Additive effect[a] | Dominance effect[b] | Significance | Percent of F2 Variance |
|---|---|---|---|---|
| | **Chromosome 4** | | | |
| Growth rate, birth to 70 kg (g/d) | –23.5 ± 4.9 | 0.7 ± 7.3 | <0.01 | 11.9 |
| Abdominal fat (%) | 0.38 ± 0.06 | 0.15 ± 0.10 | <0.001 | 18.7 |
| Average back fat thickness (mm) | 2.30 ± 0.42 | 1.47 ± 0.65 | <0.001 | 17.6 |
| Length of small intestine (m) | –0.87 ± 0.18 | 0.03 ± 0.27 | <0.01 | 11.3 |
| | **Chromosome 13** | | | |
| Growth rate, birth to 30 kg (g/d) | –13.5 ± 3.6 | 6.0 ± 5.2 | <0.05 | 7.5 |

[a] Measured as half of the deviation of animals homozygous for the wild pig allele from those homozygous for the Large White allele.

[b] Measured as the deviation of heterozygous animals from the mean of the two homozygotes.



The analysis of fatness and growth traits has recently been repeated,[19a] now based on a marker map giving almost complete coverage of the pig genome[14] and a revised and more comprehensive statistical analysis. No QTL with larger effects than those described in Table 18.3 was found.

Preliminary QTL analyses of immune capacity traits have revealed significant effects for total white blood cell counts on chromosome 1, for mitogen-induced proliferation on chromosomes 4 and 7, and for immunoglobulin G response to *E. coli* K88 and O149 antigens on chromosomes 1 and 6, respectively.[20] The QTL for total white blood cell counts on chromosome 1 has been confirmed by analyzing the effect of this locus in subsequent generations of the pedigree.

A preliminary examination of the data on body proportions, carcass composition, and bone measurements revealed regions on chromosome 4 and 8 with significant effects on several traits. For instance, a QTL with an additive effect on body length of approximately 1.2 cm was found on chromosome 8 (F = 11.2).[20a]

## 18.4 STRATEGIES FOR VERIFICATION OF OBSERVED QTLs

At this initial stage of QTL analysis in farm animals it is important to verify the location and effects of observed QTLs. Given the stringent significance threshold employed in this study it is unlikely that the reported QTLs represent spurious effects (Type I error) when we consider a single trait. However, as more and more independent traits are analyzed there is an increasing risk of detecting spurious effects. In addition, as discussed by Georges et al.,[21] there is a risk that a QTL study with a low or moderate power tend to give overestimated QTL effects because only those QTLs with inflated estimates, due to chance, will reach the significance threshold. Moreover, it is an open question whether the above-described QTL effects are due to a single QTL with a large effect or several linked genes each with a smaller effect. The fact that the QTLs were observed in a cross between two divergent lines implies that we may have detected haplotype effects which may be dissociated by recombination in subsequent generations.

One way of QTL verification would be to study the effect of the locus in different pedigrees. As regards the chromosome 4 QTL(s) no other study on this subject has yet been reported. Some support for the existence of a QTL on pig chromosome 13 affecting early growth is obtained from a recent study by Yu et al.[22] based on a cross between European and Chinese pigs. They found an association between early growth and genetic polymorphism in a candidate gene *PIT1,* encoding a pituitary-specific transcription factor which is known to be important for normal growth. A recent study using our pedigree showed that *PIT1* maps close to the QTL peak on chromosome 13.[22a]

Another strategy for QTL verification would be to breed further generations of the pedigree in which the QTL was detected. We used this approach to (1) confirm the existence of a QTL for fatness on chromosome 4, (2) re-estimate its effect, and (3) refine its localization. Two F2 sows, heterozygous for the actual chromosome region, were selected for breeding. The sows were back-crossed to a Large White boar and two F3 boars were in turn backcrossed to Large White/Landrace sows. One of the F3 sires carried an intact wild-pig-derived segment for the QTL region while the other carried a recombinant haplotype including only a part of the confidence interval for the fatness QTL. A total of 88 F4 animals were produced. Phenotypic measurements included daily weight gain, ultrasound recording of fat depth at 70 and 90 kg, and several carcass traits. Genotyping was carried out with 21 chromosome 4 markers and the segregation at the QTL was deduced. Highly significant effects of this locus were observed on all fatness traits and on the length of the carcass;[22b] a small but significant effect on growth was also observed. Interestingly, the estimated effects on fatness were strikingly similar to the estimates from the F2 generation. The results suggested that the QTL does not represent a haplotype effect of several loosely linked genes. However, we cannot, of course, exclude the possibility of a cluster of closely linked genes. Further generations are being produced with the aim to establish a line of pigs segregating for the major QTL for fatness allowing further characterization of the locus (see Section 18.5). The line of pigs may even serve as a genetic model for human obesity.[23]

## 18.5 PROSPECTS FOR CLONING QTLs

Genome analysis has provided tools to genetically dissect complex multifactorial traits. The initial phase of this endeavor has involved the mapping of QTLs to specific chromosomal regions as described here. The second phase will be to clone QTLs and unravel their molecular basis. However, cloning of QTLs may often be a formidable task.

The strategy of positional cloning means that, after a trait locus has been mapped to a specific chromosomal region, various methods are employed to identify transcripts encoded in the specific region. Such transcripts are then characterized with regard to tissue distribution and the presence of mutations in order to identify the gene controlling the trait. This strategy has been very successful for cloning genes causing inherited disorders in humans and inherited defects in inbred mice. The success of the strategy has to a large extent relied on the excellent mapping tools available (i.e., well-developed marker maps and often extensive pedigree data) making it possible to map the trait locus to a narrow chromosomal region (within 1 cM). Furthermore, researchers have often been guided to the gene by the presence of chromosomal rearrangements. Finally, the identification of the causative mutation has been facilitated because it has been clear that the phenotype is due to aberrant gene expression or function. Despite these advantages, positional cloning can be a major undertaking for an advanced molecular genetics laboratory.

Positional cloning of QTLs will be even more cumbersome for two major reasons, the poor precision of QTL mapping and the fact that we do not have a clear idea of the type of mutation for which we are looking. It is obvious that segregation at QTLs rarely is due to a knock-out of gene expression, e.g., caused by a chromosomal rearrangement. As shown in Figure 18.1, the mapping of a QTL is imprecise. This is because it is not possible to score individual recombinants between a QTL and linked marker as the phenotype is influenced by the segregation at other QTLs and by environmental effects. For instance, Knott and Haley[24] showed by simulation that a QTL explaining 11% of the phenotypic variation in an F2 population of 1000 animals and mapped using completely informative markers spaced at 10-cM intervals, would be mapped to an 8-cM interval. Such a region is expected to contain more than 200 genes and is thus too large to make positional cloning feasible. It would be possible to create an isogenic line carrying a QTL allele with a major effect present in one population by backcrossing to another population in which the allele in question is not found. By selecting recombinants which are used for breeding the localization of the QTL can be improved each generation. However, this strategy is very costly for a large animal like the pig. A way out of this dilemma could be to refine the phenotypic trait. As regards the major locus for fatness on pig chromosome 4, it would be possible to make physiological characterization of the regulation of fat metabolism in these animals. Such studies may involve measuring the number and size of fat cells as well as the level of key hormones known to influence fat metabolism (e.g., growth hormone and insulin). A key to such studies is the fact that the genotype at the QTL can be predicted with high precision using a set of flanking markers. It is thus possible to define groups of animals with different QTL genotypes which can then be subjected to experimental work. Ideally, the trait should be refined so that the QTL genotype can be directly deduced from the phenotype. If so, the QTL can be mapped with the same high precision as any monogenic trait opening up the avenue of positional cloning.

The combined use of map information with the identification of candidate genes previously assigned to the actual chromosomal region is denoted positional candidate cloning. The presence of an almost complete human transcript map in the near future will shift cloning strategies in the human field from positional to positional candidate cloning.[25] The transcript maps of farm animal species are too rudimentary for this purpose. However, the conservative evolution of genome organization implies that comparative mapping can be used to exchange map information between species. The major mammalian "models" relevant for positional candidate cloning in farm animals are the human gene map containing the largest number of coding sequences and the mouse map containing the largest number of trait loci. A prominent example of the power of positional candidate cloning in farm animals based on comparative mapping is the recent identification of what appears to be the causative mutation for the dominant white coat color in pigs.[25a] The dominant white gene was previously mapped to pig chromosome 8 using this pedigree.[26] Comparative mapping revealed that the actual region shared homology with a region on mouse chromosome 5 harboring the dominant white spotting locus, and a region on human chromosome 4 harboring the locus for the dominant piebald trait. These pigmentation disorders are both due to mutations in the *KIT* gene encoding the mast/stem cell growth factor receptor. A screen for genetic polymorphism in the pig

*KIT* gene among dominant white and colored pigs revealed the presence of a duplication of *KIT*, or part of it, which were found in all white pigs tested but in no colored pigs. It was also found that the recombination rate in the actual region (close to the centromere on chromosome 8) was remarkably low which would have made a positional cloning strategy exceedingly difficult even for this monogenic trait.

Positional candidate cloning is an obvious strategy for the cloning of the major QTL for fatness on chromosome 4. Comparative linkage mapping[6] as well as Zoo-fluorescent *in situ* hybridization (Zoo-FISH) analysis[27,28] show that pig chromosome 4 is homologous to parts of human chromosomes 1 and 8. The fatness QTL is located close to the border between regions homologous to human 1 and 8 but most likely on the chromosome 1 side of the breakpoint. The exact position is uncertain because the imprecise mapping of the QTL and the fact that regions of conserved synteny often show rearranged gene orders.[6] We are currently developing a comparative map with a higher resolution to facilitate the identification of candidate genes. Several loci causing an obese phenotype in inbred mice have been cloned recently.[29-32] Comparative mapping suggests that all of these except *db* are unlikely to be homologues to the pig fatness locus. *db* encodes the leptin receptor and maps to a region of mouse chromosome 2 expected to be homologous to a region on pig chromosome 4 or 6. A chromosome 4 location would make the pig *db* homolog an important candidate gene for the fatness QTL.

As discussed above, a complicating factor for the cloning and identification of a mutation causing a QTL effect is that we do not expect the mutation to be deleterious and the mutation may be either a structural one or a regulatory one. This will make it hard to distinguish the causative mutation from a linked genetic polymorphism. It may in fact be necessary to carry out transgenic experiments in a model organism or perhaps even in pigs, to prove that an observed mutation is causing the phenotypic effect.

## 18.6 GENERAL DISCUSSION

The data reviewed here show that QTLs with large effects can be detected in crosses between divergent lines of outbred animals. The result is a strong impetus for future studies of this type which have a large potential in farm animal genetics due to the rich diversity of breeds genetically adapted to different environmental conditions and production systems. Such studies will lead to practical applications as well as advance our understanding of the genetic basis for adaptive evolution. QTLs detected in crosses between divergent populations will reveal candidate loci which may control genetic variation within commercial lines. To which extent the same loci control variation between and within selected lines is an open question and an important topic for future research. For instance, the effects of the QTLs on pig chromosome 4 influencing fatness and growth should be investigated in commercial populations. QTLs identified in crosses between divergent lines may be exploited for the development of synthetic lines carrying favorable alleles from different sources. As an example, Meishan pigs from China have superior fertility compared with major European pig breeds but clearly inferior characteristics as regards growth and fatness. Thus, it should be possible to identify QTLs for fertility, growth, and fatness, and to use this mapping information for MAS. Similarly, mapping information may be used to provide lines suitable for animal production in tropical environments which combine QTL alleles for improved production derived from elite breeds with genes important for local adaptation, including disease resistance, derived from local breeds.

A major problem in QTL mapping projects in farm animals is the rather low statistical power due to the limited sample sizes used. The sizes of the experiments are limited by the costs of producing and/or collecting pedigree material as well as the cost of genotyping. This situation leads to inflated estimates of the effect of those QTLs which are detected, although this is often not a major problem as the estimates can be revised in confirmatory studies. More seriously it leads to a high rate of Type II errors (i.e., we miss real QTLs). For instance, simulation studies indicate that only QTL effects explaining at least 5% of the total variation have a reasonable chance of being detected in a study like the one described in this paper[33] (see also Chapter 10, in this volume). Therefore, it is also important to collect information on less significant QTL effects (for example those significant at the standard 5% threshold or at the 'suggestive' level proposed by Lander and Kruglyak[18]) perhaps in the form of a database in order to allow future global meta-analyses summarizing data across several similar studies.

The infinitesimal model of quantitative genetic variation assumes that quantitative traits are controlled by innumerable number of loci each with a very small effect. This is of course an oversimplification as illustrated by the results in this study. However, the results are consistent with a large number of loci each with a small effect since we cannot exclude the possibility that the large QTL effects are due to the combined effect of many linked loci. To resolve this issue it would probably be necessary to clone the QTLs. Cloning of QTLs will be a major future challenge in this field of research. Such studies are important because of our current ignorance regarding the molecular basis for quantitative genetic variation. The extent to which QTLs reflect polymorphism in coding sequences vis-à-vis polymorphism in regulatory sequences remains unknown. Moreover, it is not known whether QTLs with large effects, such as the fatness QTL detected in this study, are caused by a single mutation, or the accumulation of several consecutive mutations at the same locus, or mutations at several closely linked loci. Molecular characterization of QTLs will advance quantitative genetics theory and may lead to the development of more realistic models for the evolution of phenotypic traits in domestic and natural populations. It will also be of major importance for practical applications of genome research in agriculture because it may be possible to select directly at the trait locus rather than to rely on marker assisted selection. Moreover, it may open up possibilities for the use of transgenic technologies in future animal improvement programs.

## REFERENCES


1. Copeland, N. G. and Jenkins, N. A., Development and applications of a molecular genetic linkage map of the mouse genome, *Trends Genet.*, 7, 113, 1991.
2. Tate, M. L., Mathias, H. C., Fennessy, P. F., Dodds, K. G., Penty, J. M., and Hill, D. F., A new gene mapping resource: interspecies hybrids between Père David's deer (*Elaphurus davidianus*) and red deer (*Cervus elaphus*), *Genetics*, 139, 1383, 1995.
3. Archibald, A. L., Fat pigs can blame their genes, *Curr. Biol.*, 4, 728, 1994.
4. Bosma, A. A., Chromosomal polymorphism and G-banding patterns in the wild boar (*Sus scrofa* L.) from the Netherlands, *Genetica*, 46, 391, 1976.

4a. Gustavsson, I., Unpublished data, 1989.

5. Lundström, K., Malmfors, B., Stern, S., Rydhmer, L., Eliasson-Selling, L., Mortensen, A. B., and Mortensen, H. P., Skatole levels in pigs selected for high lean tissue growth rate on different protein levels, *Livestock Prod. Sci.,* 38, 125, 1994.
6. Johansson, M., Ellegren, H., and Andersson, L., Comparative mapping reveals extensive linkage conservation — but with gene order rearrangements — between the pig and human genomes, *Genomics*, 25, 682, 1995.
7. Weber, J. L. and May, P. E., Abundant class of human DNA polymorphism which can be typed using the polymerase chain reaction, *Am. J. Hum. Genet.*, 44, 388, 1989.
8. Litt, M. and Luty, J. A., A hypervariable microsatellite revealed by *in vitro* amplification of a dinucleotide repeat within the cardiac muscle actin gene, *Am. J. Hum. Genet.*, 44, 397, 1989.
9. Johansson, M., Ellegren, H., and Andersson, L., Cloning and characterization of highly polymorphic porcine microsatellites, *J. Heredity*, 83, 196, 1992.
10. Ellegren, H., Chowdhary, B. P., Johansson, M., Marklund, L., Fredholm, M., Gustavsson, I., and Andersson, L., A primary linkage map of the porcine genome reveals a low rate of genetic recombination in the pig, *Genetics*, 137, 1089, 1994.
11. Andersson, L., Haley, C. S., Ellegren, H., Knott, S. A., Johansson, M., Andersson, K., Andersson-Eklund, L., Edfors-Lilja, I., Fredholm, M., Hansson, I., Håkansson, J., and Lundström, K., Genetic mapping of quantitative trait loci for growth and fatness in pigs, *Science,* 263, 1771, 1994.

12. Rohrer, G. A., Leeson, J. A., Keele, J. W., Smith, T. P., and Beattie, C. W., A microsatellite linkage map of the porcine genome, *Genetics*, 136, 231, 1994.
13. Archibald, A. L., Haley, C. S., Brown, J. F., Couperwhite, S., McQueen, H. A., Nicholson, D., Coppieters, W., Van de Weghe, A., Winterö, A. K., Fredholm, M., Larsen, N. J., Nielsen, V. H., Milan, D., Woloszyn, N., Robic, A., Dalens, M., Riquet, J., Gellin, J., Caritez, J.-C., Hue, D., Burgaud, G., Ollivier, L., Bidanel, J.-P., Vaiman, M., Renard, C., Gelderman, H., Davoli, R., Ruyter, D., Vestege, E. J. M., Groenen, M. A. M., Davies, W., Höyheim, B., Keiserud, A., Andersson, L., Ellegren, H., Johansson, M., Marklund, L., Miller, R. J., Anderson-Dear, D. V., Signer, E., and Jeffreys, A. J., The PiGMaP consortium linkage map of the pig (*Sus scrofa*), *Mammalian Genome*, 6, 157, 1995.
14. Marklund, L., Davies, W., Ellegren, H., Fredholm, M., Höyheim, B., Johansson Moller, M., Juneja, K., Mariani, P., Coppetiers, W., and Andersson, L., A comprehensive pig linkage map based on a wild pig — Large White intercross, *Anim. Genet.*, 27, 255, 1996.
15. Lander, E. S. and Botstein, D. B., Mapping Mendelian factors underlying quantitative traits using RFLP linkage maps, *Genetics*, 185, 1989.
16. Haley, C. S. and Knott, S. A., A simple regression method for mapping quantitative trait loci in line crosses using flanking markers, *Heredity*, 69, 315, 1992.
17. Haley, C. S., Knott, S. A., and Elsen, J.-M., Mapping quantitative trait loci in crosses between outbred lines using least squares, *Genetics*, 136, 1195, 1994.
18. Lander, E. S. and Kruglyak, L., Genetic dissection of complex traits: guidelines for interpreting and reporting linkage results, *Nat. Genet.*, 11, 241, 1995.
19. Churchill, G. A. and Doerge, R. W., Empirical threshold values for quantitative trait mapping, *Genetics*, 138, 963, 1994.
19a. Knott, S. A., et al., in preparation, 1997.
20. Edfors-Lilja, I., Fossum, C., Wattrang, E., Gustafsson, U., Ellegren, H., Johansson, M., Marklund, L., and Andersson, L., Genetic influence on total and differential white blood cell counts in pigs, *Proc. 5th World Congress on Genetics to Livestock Production, Guelph, Canada, August 7–12, 1994*, 19, 360, 1994.
20a. Andersson-Eklund, L., in preparation, 1997.
21. Georges, M., Nielsen, D., Mackinnon, M., Mishra, A., Okimoto, R., Pasquino, A. T., Sargeant, L. S., Sorensen, A., Steele, M. R., Zhao, X., Womack, J. E., and Hoeschele, I., Mapping quantitative trait loci controlling milk production in dairy cattle by exploiting progeny testing, *Genetics*, 139, 907, 1995.
22. Yu, T.-P., Tuggle, C. K., Schmitz, C. B., and Rothschild, M. F., Association of PIT1 polymorphisms with growth and carcass traits in pigs, *J. Anim. Sci.*, 73, 1282, 1995.
22a. Marklund, L., Unpublished data, 1996.
22b. Marklund, L., Nyström, P. E., Stern, S., and Andersson, L., submitted, 1997.
23. Andersson, L., Genes and obesity, *Ann. Med.*, 28, 5, 1996.
24. Knott, S. A. and Haley, C. S., Aspects of maximum likelihood methods for the mapping of quantitative trait loci in line crosses, *Genet. Res.*, 60, 139, 1992.
25. Collins, F. S., Positional cloning moves from perditional to traditional, *Nat. Genet.*, 9, 347, 1995.
25a. Johansson Moller, M., Chaudhary, R., Hellmén, E., Höyheim, B., Chowdhary, B., and Andersson, L., Pigs with the dominant white coat color phenotype carry a duplication of the *KIT* gene encoding the mast/stem cell growth factor receptor, *Mammalian Genome* 7, 822, 1996.
26. Johansson, M., Ellegren, H., Marklund, L., Gustavsson, U., Ringmar-Cederberg, E., Andersson, K., Edfors-Lilja, I., and Andersson L., The gene for dominant white color in the pig is closely linked to *ALB* and *PDGFRA* on chromosome 8, *Genomics*, 14, 965, 1992.
27. Rettenberger, G., Klett, C., Zechner, U., Kunz, J., Vogel, W., and Hameister, H., Visualization of the conservation of the synteny between pigs and humans by heterologous chromosomal painting, *Genomics*, 26, 372, 1995.
28. Frönicke, L., Chowdhary, B. P., Scherthan, H., and Gustavsson, I., A comparative map of the porcine and human genomes demonstrates ZOO-FISH and gene-mapping based chromosomal homologies, *Mammalian Genome*, 7, 285, 1996.
29. Zhang, Y., Proenca, R., Maffei, M., Barone, M., Leopold, L., and Friedman, J. M., Positional cloning of the mouse *obese* gene and its human homologue, *Nature*, 372, 425, 1995.
30. Naggert, J. K., Fricker, L. D., Varlamov, O., Nishina, P. M., Rouille, Y., Steiner, D. F., Carrol, R. J., Paigen, B. J., and Leiter, E. H., Hyperproinsulinaemia in obese *fat/fat* mice associated with a carboxypeptidase E mutation which reduces enzyme activity, *Nat. Genet.*, 10, 135, 1995.

31. Tartaglia, L. A., Dembski, M., Weng, X., Deng, N., Culpepper, J., Devos, R., Richards, G. J., Campfield, L. A., Clark, F. T., Deeds, J., Muir, C., Sanker, S., Moriarty, A., Moore, K. J., Smutko, J. S., Mays, G. G., Woolf, E. A., Monroe, C. A., and Tepper, R. I., Identification and expression cloning of a leptin receptor, OB-R, *Cell*, 83, 1263, 1995.
32. Kleyn, P. W., Fan, W., Kovats, S. G., Lee, J. J., Pulido, J. C., Wu, Y., Berkemeier, L. R., Misumi, D. J., Holmgren, L., Charlat, O., Woolf, E. A., Tayber, O., Brody, T., Shu, P., Hawkins, F., Kennedy, B., Baldini, L., Ebeling, C., Alperin, G. D., Deeds, J., Lakey, N. D., Culpepper, J., Chen, H., Glücksmann-Kuis, M. A., Carlson, G. A., Duyk, G. M., and Moore, K. J., Identification and characterization of the mouse obesity gene *tubby*: a member of a novel gene family, *Cell,* 85, 281, 1996.
33. van Ooijen, J. W., Accuracy of mapping quantitative trait loci in autogamous species, *Theor. Appl. Genet.*, 84, 803, 1992.
34. Lundström, K., Karlsson, A., Håkansson, J., Hansson, I., Johansson, M., Andersson, L., and Andersson, K., Production, carcass and meat quality traits of $F_2$-crosses between European Wild Pigs and domestic pigs including halothane gene carriers, *Anim. Sci.,* 61, 325, 1995.
35. Uhlhorn, H., Dalin, G., Lundeheim, N., and Ekman, S., Osteochondrosis in Wild Boar — Swedish Yorkshire Crossbred Pigs (F2 Generation), *Acta Vet. Scand.*, 36, 41–53, 1995.
36. Edfors-Lilja, I., Bergström, M., Gustafsson, U., Magnusson, U., and Fossum, C. Genetic variation in Con A induced production of interleukin 2 by porcine peripheral blood mononuclear cells, *Vet. Immunol. Immunopathol.*, 27, 351, 1991.
37. Artursson, K., Lindersson, M., Varela, N., Scheynius, A., and Alm, G., Interferon-a production and tissue localization of interferon-$\alpha/\beta$ producing cells after intradermal administration of Aujeszky's disease virus-infected cells in pigs, *Scand. J. Immunol.*, 41, 121, 1995.
38. Edfors-Lilja, I., Wattrang, E., Magnusson, U., and Fossum, C., Genetic variation in parameters reflecting immune competence of swine, *Vet. Immunol. Immunopathol.,* 40, 1, 1994.
39. Edfors-Lilja, I., Gustafsson, U., Duval-Iflah, Y., Ellegren, H., Johansson, M., Juneja, R. K., Marklund, L., and Andersson, L., The porcine intestinal receptor for *Escherichia coli* K88*ab*, K88*ac*: regional localization on chromosome 13 and influence on IgG response to the K88 antigen, *Anim. Genet.,* 26, 237, 1995.

# 19 Case History in Humans: Mapping QTLs for Complex Traits in Humans

*Hakan Sakul and Lon R. Cardon*

## CONTENTS



## 19.1 INTRODUCTION

Genetic studies of complex traits in humans are currently being undertaken at an unprecedented scale of funding, competition, and collaboration. Human genetics laboratories worldwide are expanding, combining efforts, and/or collaborating with industry to conduct heretofore untenable studies to search for and characterize susceptibility genes in breast, lung, and prostate cancer, insulin- and non-insulin-dependent diabetes, obesity, cardiovascular disease, asthma, psychiatric disorders (including schizophrenia and bipolar disorder), and other diseases.

Unfortunately, the number of successes in common complex traits is very small relative to the effort expended. Two breast cancer genes have been identified,[1,2] at least one gene involved in hypertension has been characterized,[3] and obesity genes initially localized in rodents are being evaluated for function in humans.[4,5] However, the lack of success for other complex traits, and indeed for other genes within these disease areas, does not necessarily imply a failure of the ongoing studies. Rather, complex trait mapping in humans is quite young compared to genetic research using experimental animals. Much of the human research underway at present is in early stages of genotyping, linkage analysis and fine-mapping and, therefore, has not yet reached the levels of specific gene identification, mutation characterization, and functional analysis of proteins.

The rapid growth of complex trait studies in humans is attributable in large part to advances in methods, hardware, and automation of microsatellite genotyping and DNA sequencing.[6,7] There

0-8493-7686-6/98/$0.00+$.50

have been corresponding advances in statistical genetics, cloning procedures, analysis of DNA sequence data, and functional genomics, but these developments have not had as broad an impact as large-scale genotyping and sequencing. The impact of these genotyping and sequencing advances lie primarily in the magnitude of throughput. Highly automated genotyping and sequencing machines are now available (e.g., Applied Biosystems 373/377, Pharmacia automated laser fluorescence sequencer) which allow rapid processing of large amounts of genetic material in relatively short amounts of time. With respect to genotyping, these innovations have led to a new model for disease gene studies in complex traits: the genome-screen. Genome screens of 200 or more markers yielding a map density of ≤20 cM form the cornerstone of many of the large studies of human complex diseases.[8-11] Following such scans, DNA sequencing is used to locate new microsatellite repeats or to identify single nucleotide polymorphisms for use as biallelic markers in fine-mapping experiments and for mutation detection in gene localization studies. Sequencing capacities are approaching megabase levels of base-perfect DNA sequence for these types of studies, which should facilitate characterization of the role of subtle mutations likely to be involved in human complex diseases.

The ability to conduct high throughput genotyping and sequencing has led to a shift away from assessment of a small number of candidate genes and more toward evaluation of the entire genome, thereby eliminating potentially erroneous assumptions concerning which genes are likely candidates for conferring susceptibility to complex traits. This approach defines a full positional cloning design; a model which begins with family collection, proceeds through genotyping, linkage analysis and fine-scale genetic mapping, and finally proceeds into localized gene cloning, sequencing, and mutation detection.

Here we describe progress in the field of Type II diabetes using this positional cloning approach. As noted above, much of the current research in complex diseases, and in non-insulin-dependent diabetes mellitus (NIDDM) in particular, is focused on the first stages of the full positional cloning model. Therefore, our focus here is necessarily on these areas.

## 19.2 PHENOTYPIC CLASSIFICATION OF DIABETES

Diabetes mellitus is a genetic disorder that is highly prevalent in Western societies. Its prevalence in the Third World countries is increasing as their lifestyle evolves towards the western lifestyle. The disorder involves metabolisms of carbohydrates, protein and fat, is associated with insufficiency of insulin secretion, and is accompanied by various degrees of insulin resistance.

Diabetes mellitus has different forms, and the clinical courses also include other categories of glucose intolerance such as impaired glucose tolerance and gestational diabetes. The present classification of diabetes mellitus includes type I (insulin-dependent) and type II (non-insulin-dependent) diabetes. Both forms of diabetes are the current focus of several intensive genetic research programs. We will briefly define Type I diabetes here, but our primary emphasis will be on non-insulin-dependent diabetes mellitus, the most common form of diabetes.

### 19.2.1 Type I Diabetes (IDDM)

Type I or insulin-dependent diabetes mellitus (IDDM) is a T-cell-dependent autoimmune disease characterized by infiltration and destruction of the pancreatic islets, leading to absolute dependence on exogenous insulin.[12,13] It occurs in approximately 10% of all diabetics in the Western world, with the highest prevalence rates being in northern European countries.[14] Classically, this type occurs most commonly in childhood and adolescence (type IA), but it can also become symptomatic for the first time at any age. The affected individuals display, among other things, fatigue and weight loss, and are dependent on administration of insulin to prevent ketosis, coma, and death. It has been suggested that environmental factors, such as certain viral infections, nutritional or chemical agents may lead to cell-mediated autoimmune destruction of β-cells.[15]

Despite the strong evidence for inherited susceptibility, studies with monozygotic ("identical") twins suggest that only 30 to 40% of the disease susceptibility is caused by genetics, the remaining 60 to 70% is attributed to environmental factors.[16]

### 19.2.2 Type II Diabetes (NIDDM)

Type II or NIDDM affects approximately 90% of diabetics in the Western world. Its genetic basis is commonly expressed by a more frequent familial pattern of occurrence than in IDDM. The complications of NIDDM primarily affect the vascular system and lead to excessive rates of coronary heart disease, renal failure, retinopathy and blindness, neuropathy, and amputation. These complications give rise to most of the morbidity and mortality associated with diabetes.

#### 19.2.2.1 Maturity-Onset Diabetes of the Young (MODY)

A subtype of NIDDM that has an early age of onset and autosomal dominant inheritance is known as maturity-onset diabetes of the young (MODY).[17] The autosomal dominant mode of inheritance is the hallmark of MODY, and clearly distinguishes it from other types of NIDDM. To overcome the difficulties of mapping genes in the polygenic forms of NIDDM, MODY has been used as a model to identify NIDDM susceptibility genes. In MODY patients, diagnosis of diabetes is usually made by the age of 25, and frequently between 9 and 13 years of age. This makes it possible to collect multigenerational families which contrasts with NIDDM in which the late age-of-onset makes familial ascertainment very difficult. An example of the power of MODY pedigrees is the "RW" pedigree that has been followed since 1958.[18] This family comprises more than 360 family members from over five generations.

## 19.3 DIAGNOSIS

Diabetes mellitus is defined as fasting plasma hyperglycemia (≥140 mg/100 dL, 7.8 mM) and/or hyperglycemia (200 mg/100 dL, 11.1 mM) 2 h after orally ingesting 75 g of glucose (OGTT, oral glucose tolerance test).[19] Other tests that measure specific components include the IV glucose tolerance test (IVGTT)[20] where 25 g of glucose are infused into the bloodstream over a 3-min period. This test measures acute insulin response (AIR) in 1-min intervals from 0 to 10 min, and further clarifies the relationships between glycemia and insulin secretory function. Because people with NIDDM are generally insulin-resistant,[21,22] the "clamp" test[23] was developed to measure insulin resistance. In this test, whole body insulin action is quantitated as the rate of a variable glucose infusion required to maintain a constant plasma glucose concentration during a constant insulin infusion.

Because these tests measure different variables related to NIDDM status, a question of which phenotype(s) to analyze becomes important. In most studies, OGTT data are the only available test results collected from diabetics as well as non-diabetics. The remaining two tests are usually performed only on non-diabetic individuals. In fact, most studies do not collect "clamp" data as it is an invasive and expensive procedure. On the other hand, OGTT results may have a large variation even for the same individual, depending on the time of the test. This phenotypic unreliability is one of the significant challenges for genetic studies of NIDDM, and for most common complex diseases in humans.

### 19.3.1 Genetic Basis of NIDDM

There is a large between-population variation in prevalence of NIDDM throughout the world. The Pima Indians of Phoenix, Arizona, have the highest recorded prevalence of NIDDM.[24] In this population, approximately 50% of males and 40% of females become diabetic between the ages of 35 and 44.[25] Extraordinarily high rates are also found in the Micronesian population living in Nauru in the Central Pacific. Although these high frequencies are useful for epidemiology studies, the reduced phenotypic variability contributes to a decreased heritability, making genetic studies much more difficult.

The most convincing evidence for a genetic basis of NIDDM comes from studies of twins. Concordance rates for NIDDM in older monozygotic twins are 50 to 90%, which is much higher than dizygotic twins, siblings or other first-degree relatives.[26] However, this does not provide any information on the number of NIDDM-causing genes or their mode of inheritance. There is also a large variation in occurrence of NIDDM between diverse ethnic groups living in the same geographical locations, such as Asian Indians, Chinese and Malays living in Singapore, Asian Indians and Europeans living in Great Britain, and different ethnic groups living in the U.S.[27] More direct evidence of genetic susceptibility comes from populations within which there is a genetic admixture, when they reside in the same environment. An example of this is the Gila River Indian Community in Arizona. The prevalence of NIDDM is twice as high in full-blooded Pima Indians as in non-Indians while those with half-Pima and half non-Indian ancestry have an intermediate population prevalence for NIDDM.[28]

In a segregation analysis[29] of NIDDM, carried out in 2697 subjects from 653 nuclear families of Pima Indian heritage, a major gene was found to influence the risk for NIDDM in Pima Indians by affecting age of onset. The authors allowed for the effects of age, birth cohort and obesity as important covariates, and concluded that the expression of this gene may depend on environmental factors that have become more prevalent in recent-birth cohorts. It is important to note that these analyses did not indicate the location of such a major gene, but, instead, demonstrated that the pattern of familial transmission was consistent with that of a single gene of large effect. The location of any major gene for NIDDM remains elusive.

Our own data[30] on heritability ($h^2$) of several prediabetic phenotypes also suggests a strong additive genetic component in the total phenotypic variation ($h^2$ of AIR at 3, 4, 5, 6, 8 and 10 min ranging from 0.43 to 0.80).

## 19.4 QTL MAPPING AND LINKAGE ANALYSIS OF DIABETES

As previously described, the advent of high-throughput genotyping, coupled with algorithmic developments in linkage analysis and dramatic increases in computing power, has made it possible to carry out genome-wide scans for common complex disease loci. In diabetes, this approach has led to some recent publications reporting significant linkage findings which are now being actively pursued by fine-scale genetic mapping and positional cloning. Following is a summary of the results from a few such studies.

### 19.4.1 IDDM

Type I diabetes has been shown to be a polygenic trait in mice, with the major locus encoded by the major histocompatibility complex (MHC) and at least nine other loci contributing to disease development. In a U.K. sib-pair study,[31] a semi-automated fluorescence-based technology and linkage analysis were used in an effort to identify the susceptibility genes for Type I diabetes (a description of the sib-pair design is given below). In addition to MHC region on chromosome 6p21 (IDDM1) and the insulin receptor region (IDDM2), they identified 18 different chromosomal regions with positive evidence of linkage to the disease. The authors concluded that the average marker spacing of 11 cM in this study was not dense enough to identify all of the major genes for type I diabetes by linkage analysis, and that a 3-cM marker spacing along with several hundred sib pairs would be required to properly locate all the major genes involved. This study illustrates a ubiquitous difficulty in complex trait genetics: lack of statistical power due to very large sampling and genotyping requirements.

### 19.4.2 MODY

The large RW pedigree mentioned above has been extensively studied in the search for MODY-causing genes. In one such study, 185 individuals from this pedigree were tested for diabetes. A DNA polymorphism in the adenosine deaminase gene (ADA) on the long arm of chromosome 20 (MODY1) was found to cosegregate with MODY.[32] While mutations in this gene have not yet been identified, the gene results in impaired insulin secretion that leads to severe hyperglycemia.

Although families with MODY are diverse with respect to clinical and metabolic profiles, most MODY patients have a decreased insulin response to glucose, suggesting a primary pancreatic β-cell defect. Therefore, genes whose products seem to be involved in insulin secretion are candidate genes for MODY. Such genes include the liver or pancreatic β-cell glucose transporter (GLUT2) and glucokinase (GCK), the key enzyme for glucose metabolism in liver and β-cells. Linkages between the glucokinase locus on chromosome 7 and MODY (MODY2) in 16 French MODY families were found[33] using the candidate gene approach and assuming an autosomal dominant mode of inheritance. In a majority of these families, there was evidence for linkage to GCK, suggesting that this gene or another closely linked locus is principally predisposing for the disease in this population.

In another study of French families, a MODY gene (MODY3) was localized to a 10-cM interval between D12S86 and D12S342.[34] Further work with families from Denmark, U.S., Germany, and Japan was carried out[35] using two-point and multipoint linkage analyses with an autosomal dominant mode of inheritance. The authors confirmed this finding and localized the MODY3 gene in a 5-cM interval between D12S86 and D12S807/D12S820. These pedigrees are being used to facilitate cloning of MODY3 and identification of mutations that impair its function.

### 19.4.3 NIDDM

Although the last few years have witnessed considerable success in identifying the linkages putatively responsible for early onset, monogenic forms of NIDDM, there has been little progress in identifying the genes responsible for the more common late-onset form of NIDDM. Besides the polygenic nature of NIDDM, a further complication comes from the interaction of genotype and environmental factors. Most genetic studies of NIDDM to date have focused on candidate genes in pathways that are impaired in subjects with NIDDM, such as glucose-stimulated insulin secretion in the pancreatic β-cell and nonoxidative metabolism of glucose in skeletal muscle.[36]

Linkage studies of common diseases with a late-age of onset, such as NIDDM, pose many challenges. The mean age at diagnosis is usually such that it is often not possible to collect data from all members of a nuclear family. Often one or both parents are not available due to early mortality associated with these diseases. Further compounding these issues is the apparent lack of an appropriate genetic model that would adequately represent the familial aggregation. These issues make the analysis of such data complicated at best. The traditional LOD-score methods[37] used in successfully identifying monogenic disease genes are not readily applicable to these polygenic diseases. Consequently, the use of affected sib-pairs in a model-free approach has received much attention.[38]

Sib-pair methods are considered model-free because, unlike the traditional LOD-score approach, they require no assumptions of disease-gene frequency, penetrance and mode of inheritance, and are (reasonably) robust to misspecification of marker allele frequencies. For discrete traits the basic premise of the approach is that, on average, siblings share 50% of their autosomal genes identical by descent, and therefore allele sharing significantly greater than 50% by affected siblings should reflect disease gene-marker cosegregation. For such traits, investigators typically describe familiality in terms of a relative risk ratio ($\lambda_S$)[39] as the risk ratio for siblings of affected probands divided by the population prevalence for the trait. For quantitative traits, the premise of the sib-pair design is equally simple: differences between siblings which are negatively correlated with allele sharing are considered to be indicative of linkage.[40]

Recently, a sib-pair based genome scan[10] was performed using 174 autosomal and 16 X-linked markers (approximately 8.6 cM map density) in a group of Mexican American families. In addition to two-point linkage analyses, multipoint exclusion maps were generated at three different values of $\lambda_s$. An upper limit was set at 2.8, a lower limit at 1.2, and a moderate value of 1.6. Linkage analyses, using affected sib pairs only, showed suggestive evidence for linkage (maximum LOD score >2.6) between NIDDM and D2S125, a marker located in 2q-ter (NIDDM1). Additionally, two regions (on chromosomes 3 and 15) were reported to show compelling evidence of linkage to NIDDM. The authors have not been able to replicate this finding in non-Hispanic whites or Japanese Americans. Whether this may be a linkage specific to Mexican Americans still needs to be evaluated in other collections of Mexican American families.

A similar genome scan was designed recently to search for NIDDM-causing genes in 26 multiplex Finnish families.[41] The ascertainment criteria were three or more living affected individuals per family, with at least one patient having the disease before age 65, no parents or offspring in primary sibship having IDDM, at least one parent and one offspring not diagnosed with NIDDM, and no individuals with MODY. Preliminary analyses showed no evidence of linkage. The authors then ranked the families based on 30-min insulin levels from OGTT, and concentrated on the lowest quartile, as the Finnish population do not exhibit extreme insulin resistance but instead they fit the model of defective insulin secretion. Results from these follow-up analyses suggested linkage (NIDDM2) to chromosome 1, near D12S1349 where MODY3 has previously been mapped. The authors concluded that NIDDM2 and MODY3 may represent different alleles of the same gene where severe mutations cause MODY3 and milder mutations cause NIDDM. This finding also awaits replication in other populations.

Several other researchers have looked at NIDDM and related quantitative phenotypes in diverse populations using the genome scan approach or by saturating regions of known or suspected linkage.[42,43] These and related efforts in finding QTLs predisposing humans to common diseases have as a common denominator vastly improved genotyping capacity that makes it possible to perform genome scans with a 10 to 15 cM coverage in a reasonable amount of time, as well as the improved linkage algorithms and publically available linkage software that can perform data analysis efficiently.

## 19.5 FUTURE DIRECTIONS AND CHALLENGES

A common theme throughout the many genetic studies of diabetes is that a large number of linkages have been reported, yet no complex disease genes have been cloned. Thus, as genetic research in diabetes is still in its infancy, many studies are still required in order to clone disease genes and characterize the function of mutations. In general, this is true of most common complex diseases in humans. Although the genome-scan has led to important advancements in genetic research, there are several significant obstacles that must be overcome for successful cloning.

Perhaps the most notable obstacle yet remaining is the inability to conclusively define and reduce linked regions. The literature is replete with reports of linkage which are unreplicated in follow-up studies. This situation, which is clearly apparent in diabetes, is the same in nearly every complex disease under investigation. Part of this may be due to genetic heterogeneity, as different loci may be involved in different populations. Part may also be due to imperfectly measured phenotypes, a characteristic of nearly all complex human traits. In addition, many reports are likely to be false positives, as complex diseases rarely yield unambiguous evidence for linkage. The linkages described for NIDDM illustrate this point well, as most were obtained using marginal thresholds for statistical significance. Thus, the study of human complex traits has a serious problem with statistical power.

As we have described in the context of diabetes, many researchers have turned to the model-free sib-pair approach to avoid incorrect assumptions about the frequencies, penetrances, and mode of transmission of complex disease loci. While this approach has some demonstrated utility, it carries a significant disadvantage with respect to power. Most ongoing studies of complex traits in sib-pairs are directed toward identifying genes responsible for at least 30% of the phenotypic variance. These studies typically involve 100 to 400 siblings. With larger samples (approximately 500 to 1000), genes of 15 to 30% heritability may be identifiable.[44] Unfortunately, in the absence of other information, sib-pair studies of complex traits have almost no chance of identifying genes accounting for less than 10% of the phenotypic variance. In disease areas such as diabetes, such "minor" genes may well be the rule, rather than the exception.

Some current solutions to this problem include: (1) sampling from isolated or otherwise unique populations, such as the Pima Indian study and the MODY RW pedigree described above, in expectation that the genetic heterogeneity may be reduced; (2) adopting a brute force approach of collecting thousands of sibling pairs, at great expense in time and resources; (3) conducting more extensive subphenotyping in order to identify disease-related traits of higher heritability and to reduce phenotypic measurement error; (4) increasing the density of markers in genome-scans; and (5) attempting to replicate marginally significant findings in other populations via collaborations with other investigators.

In diabetes, these alternatives are exemplified by two recent studies. The first is what is known as the GENNID (genetics of NIDDM) study. GENNID is a multicenter effort[45] with the objective to develop a resource consisting of comprehensive data and lymphoblastoid cell lines of well-characterized NIDDM families that will be available to the scientific community for genetic studies of NIDDM. The ethnic groups included in the study are of non-Hispanic white, Hispanic, African-American, and Japanese-American origin. These multiplex NIDDM families include a minimum of one affected sib-pair. Detailed family and medical histories are obtained from all participants. Family members with diabetes have fasting blood samples drawn while non-diabetic family members have OGTT and, when possible, insulin sensitivity and insulin secretion measurements by frequently sampled IVGTT or the clamp test. Over 1400 individuals from approximately 220 families have been studied since the start of the program in July 1993. The goal is to collect data from 300 non-Hispanic white families, more than 100 Hispanic families, more than 100 African-American families, and 15 Japanese-American families by July 1997.

Another group effort aimed at identifying NIDDM-causing genes is the Pima Diabetes Study. Data from over 700 microsatellite markers typed on approximately 1300 individuals of Pima Indian origin are being analyzed in this study. Extensive phenotypic measurements, collected over several years, include OGTT for all subjects, and IVGTT and clamp measurements on non-diabetics. Most patients in this study have data from multiple visits, as they participated in annual follow-up exams.[46,47] Results from these analyses will be invaluable towards understanding the physiology of NIDDM, and as such, will be useful in the design and interpretation of other studies.

Another means by which investigators are addressing the statistical power problem with complex disease designs is a change in strategy from a positional cloning/linkage approach to a candidate gene/association approach. A shift toward association studies of candidate genes has recently been advocated[44] and the notion was expanded to propose genome-wide association studies. Such studies would not be burdened by requirements of extensive family sampling and they hold the promise of greater statistical power when markers are very near or embedded within trait loci. The underlying assumption of this design is that most predisposing mutations were likely to have occurred more than about 50 generations ago, and therefore, there have been a sufficient number of meioses to recombine away alleles which are not physically close to the mutated allele(s). Historically, there have been many problems with false positives in this design due to admixture or sampling stratification.[48] Recently, however, methods have been developed which employ family-based disequilibrium structures to minimize these problems.[49-51] These approaches make use of specific transmission patterns of parental alleles in order to construct "controls" which are optimally matched to the cases. As the capacity for genome-wide genotyping continues, methods such as these will become increasingly important for the delineation of complex traits in humans.

As recent improvements in techniques of molecular biology and in lab automation systems continue to provide vast amounts of genotypic data for the study of common and complex diseases, mapping, cloning and sequencing of disease-causing QTLs seem a more reachable target now than in the recent past. Studies summarized here exemplify the thought processes that laid the ground for the advances made in the analysis of such genotypic data. Although further improvements in all components involved in the hunt for QTLs is far from over, the advances made in the study of complex diseases thus far is a good indication that it is conceivable to find the QTLs causing complex diseases in humans.

## REFERENCES


1. Miki, Y., Swensen, J., Shattuck-Eidens, D., Futreal, P. A., Harshman, K., Tavtigian, S., Liu, Q., Cochran, C., Bennett, L. M., Ding, W. et al., A strong candidate for the breast and ovarian cancer susceptibility gene BRCA1, *Science*, 266, 66, 1994.
2. Neuhausen, S. L., Mazoyer, S., Friedman, L., Stratton, M., Offit, K., Caligo, A., Tomlinson, G., Cannon-Albright, L., Bishop, T., and Kelsell D., Haplotype and phenotype analysis of six recurrent BRCA1 mutations in 61 families: results of an international study, *Am. J. Hum. Genet.*, 58, 271,1996.
3. Lifton, R. P. and Jeunemaitre, X., Finding genes that cause human hypertension. *J. Hypertens.*, 3, 231, 1993.
4. Zhang, Y., Proenca, R., Maffei, M., Barone, M., Leopold, L., and Friedman, J. M., Positional cloning of the mouse obese gene and its human homologue, *Nature*, 372, 425, 1994.
5. Noben-Trauth, K., Naggert, J. K., North, M. A., and Nishina, P. M., A candidate gene for the mouse mutation tubby, *Nature*, 380, 524, 1996.
6. Adams, M. D., Fields, C., and Venter, J. C., Eds., *Automated DNA Sequencing and Analysis*, Academic Press, San Diego, 1994.
7. Hall, J. M., LeDuc, C. A., Watson, A. R., and Roter, A. H., An approach to high-throughput genotyping, *Genome Res.*, 6, 781,1996.
8. Davies, J. L., Kawaguchi, Y., Bennett, S. T., Copeman, J. B., Cordell, H. J., Pritchard, L. E., Reed, P. W., Gough, S. C., Jenkins, S. C., Palmer, S. M. et. al., A genome-wide search for human type 1 diabetes susceptibility genes, *Nature*, 371, 130, 1994.
9. Daniels, S. E., Bhattacharrya, S., James, A., Leaves, N. I., Young, A., Hill, M. R., Faux, J. A., Ryan, G. F., le Souef, P. N., Lathrop, G. M., Musk, A. W., and Cookson, W. O. C. M., A genome-wide search for quantitative trait loci underlying asthma, *Nature*, 383, 247, 1996.
10. Hanis, C. L., Boerwinkle, E., Chakraborty, R., Ellsworth, D. L., Concannon, P., Stirling, B., Morrison, V. A., Wapelhorst, B., Spielman, R. S., Gogolin-Ewens, K. J., Shephard, J. M., Williams, S. R., Risch, N., Hinds, D., Iwasaki, N., Ogata, M., Omori, Y., Petzold, C., Rietzch, H., Schroder, H. E., Schulze, J., Cox, N. J., Menzel, S., Boriraj, C. V. V., Chen, X., Lim, L. R., Lindner, T., Mereu, L. E., Wang, Y. Q., Xiang, K., Yamagata, K., Yang, Y., and Bell, G. I., A genome-wide search for human non-insulin-dependent (type 2) diabetes genes reveals a major susceptibility locus on chromosome 2, *Nat. Genet.*, 13, 161, 1996.
11. Satsangi, J., Parkes, M., Louis, E., Hashimoto, L., Kato, N., Welsh, K., Terwilliger, J. D., Lathrop, G. M., Bell, J. I., and Jewell, D. P., Two stage genome-wide search in inflammatory bowel disease provides evidence for susceptibility loci on chromosomes 3, 7, and 12, *Nat. Genet.*, 14, 199, 1996.
12. Todd, J. A., Genetic control of autoimmunity in type I diabetes, *Immunol. Today*, 11, 122, 1990.
13. Todd, J. A., Genetic analysis of type 1 diabetes using whole genome approaches, *Proc. Natl. Acad. Sci. U.S.A.*, 92, 8560, 1995.
14. Odugbesan, O. and Barnett, A. H., Racial differences, in *Immunogenetics of Insulin Dependent Diabetes*, Barnett, A. H., Ed., MTP Press, Lancester, 1987, 91.
15. Fajans, S. S., Definition and classification of diabetes including maturity-onset diabetes of the young, in *Diabetes Mellitus*, LeRoith, D., Taylor, S. I., and Olefsky, J. M., Eds., Lippincott-Raven, New York, 1996, chap. 27.
16. Barnett, A. H., Eff, C., Leslie, R. D. G., and Pyke, D. A., Diabetes in identical twins: a study of 200 pairs, *Diabetologia*, 20, 87, 1981.
17. Fajans, S. S., Scope and heterogeneous nature of maturity-onset diabetes of the young (MODY), *Diabetes Care*, 13, 49, 1990.
18. Fajans, S. S., Bell, G. I., and Bowden, D. W., MODY: a model for the study of the molecular genetics of NIDDM, *J. Lab. Clin. Med.*, 119, 206, 1992.
19. WHO Study Group: Diabetes Mellitus. World Health Organization Technical Report Series 727, Geneva: WHO, 1985.

20. DeFronzo, R. A. and Ferrannini, E., The pathogenesis of non-insulin-dependent diabetes. An update, *Medicine*, 62, 125, 1982.
21. DeFronzo, R. A., Lilly Lecture 1987, The triumvirate: B-cell, muscle liver. A collusion responsible for NIDDM, *Diabetes*, 37, 667, 1988.
22. Reaven, G. M., Role of insulin resistance in human disease, *Diabetes*, 37, 1595, 1988.
23. DeFronzo, R. A., Tobin, J. D., and Andres, R., Glucose clamp technique: a method for quantifying insulin secretion and resistance, *Am. J. Physiol.*, 6, E214, 1979.
24. Knowler, W. C., Bennett, P. H., Hamman, R. F., and Miller, M., Diabetes incidence and prevalence in Pima Indians: a 19-fold greater incidence than in Rochester, Minn., *Am. J. Epidemiol.*, 108, 497, 1978.
25. Knowler, W. C., Pettitt, D. J., Saad, M. F., and Bennett, P. H., Diabetes mellitus in the Pima Indians: incidence, risk factors and pathogenesis, *Diabetes/Metab. Rev.*, 6, 1, 1990.
26. Zimmet, P. Z., Kelly West Lecture 1991, Challenges in diabetes epidemiology — from West to the rest, *Diabetes Care*, 15, 232, 1992.
27. King, H., Rewers, M., and WHO Ad Hoc Diabetes Reporting Group, Global estimates for prevalence of diabetes mellitus and impaired glucose tolerance in adults, *Diabetes Care*, 16, 157, 1993.
28. Knowler, W. C., Williams, R. C., Pettitt, D. J., and Steinberg, A. G., Gm3,5,13,14 and type 2 diabetes mellitus: an association in American Indians with genetic admixture, *Am. J. Hum. Genet.*, 43, 520, 1988.
29. Hanson, R. L., Elston, R. C., Pettitt, D. J., Bennett, P. H., and Knowler, W. C., Segregation analysis of non-insulin-dependent diabetes mellitus in Pima Indians: evidence for a major-gene effect, *Am. J. Hum. Genet.*, 57, 160, 1995.
30. Sakul, H., Pratley, R., Cardon, L., Ravussin, E., Mott, D., and Bogardus, C., Familiality of physical and metabolic characteristics that predict the development of non-insulin-dependent diabetes mellitus in Pima Indians, *Am. J. Hum. Genet.,* 60, 651, 1997.
31. Davies, J. L., Kawaguchi, Y., Bennett, S. T., Copeman, J. B., Cordell, H. J., Pritchard, L. E., Reed, P. W., Gough, S. C. L., Jenkins, S. C., Palmer, S. M., Balfour, K. M., Rowe, B. R., Farrall, M., Barnett, A. H., Bain, S. C., and Todd, J. A., A genome-wide search for human type 1 diabetes susceptibility genes, *Nature*, 371, 130, 1994.
32. Bell, G. I., Xiang, K. S., Newman, M. V., Wu, S. H., Wright, L. G., Fajans, S. S., Spielman, R. S., and Cox, N. J., Gene for non-insulin-dependent diabetes mellitus (maturity onset diabetes of the young subtype) is linked to DNA polymorphism on human chromosome 20q, *Proc. Natl. Acad. Sci. U.S.A.*, 88, 1484, 1991.
33. Froguel, P., Vaxillaire, M., Sun, F., Velho, G., Zouali, H., Butel, M. O., Lesage, S., Vionnet, N., Clement, K., Fougerousse, F., Tanizawa, Y., Weissenbach, J., Beckmann, J. S., Lathrop, G. M., Passa, P., Permutt, M. A., and Cohen, D., Close linkage of glucokinase locus on chromosome 7p to early-onset non-insulin-dependent diabetes mellitus, *Nature*, 356, 162, 1992.
34. Vaxillaire, M., Boccio, V., Philippi, A., Vigouroux, C., Terwilliger, J., Passa, P., Beckmann, J. S., Velho, G., Lathrop, G. M., and Froguel, P., A gene for maturity onset diabetes of the young (MODY) maps to chromosome 12q, *Nat. Genet.,* 9, 418, 1995.
35. Menzel, S., Yamagata, K., Trabb, J. B., Nerup, J., Permutt, M. A., Fajans, S. S., Menzel, R., Iwasaki, N., Omori, Y., Cox, N. J., and Bell, G. I., Localization of MODY3 to a 5-cM region of human chromosome 12, *Diabetes*, 44, 1408, 1995.
36. Hamman, R. F., Genetic and environmental determinants of non-insulin-dependent diabetes mellitus (NIDDM), *Diabetes Metab. Rev.*, 8, 287, 1992.
37. Morton, N. E., Sequential tests for the detection of linkage, *Am. J. Hum. Genet.*, 7, 277, 1955.
38. Lander, E. and Kruglyak, L., Genetic dissection of complex traits: guidelines for interpreting and reporting linkage results, *Nat. Genet.,* 11, 241, 1995.
39. Risch, N., Linkage strategies for genetically complex traits. II. The power of affected relative pairs, *Am. J. Hum. Genet.*, 46, 229, 1990.
40. Haseman, J. K. and Elston, R. C., The investigation of linkage between a quantitative trait and a marker locus, *Behav. Genet.*, 2, 3, 1972.
41. Mahtani, M. M., Widen, E., Lehto, M., Thomas, J., McCarthy, M., Brayer, J., Bryant, B., Chan, G., Daly, M., Forsblom, C., Kanninen, T., Kirby, A., Kruglyak, L., Munnelly, K., Parkkonen, M., Reeve-Daly, M. P., Weaver, A., Brettin, T., Duyk, G., Lander, E. S., and Groop, L. C., Mapping of a gene for type 2 diabetes associated with an insulin secretion defect by a genome scan in Finnish families, *Nat. Genet.,* 14, 90, 1996.

42. Stern, M. P., Duggirala, R., Mitchell, B. D., Reinhart, L. J., Shivakumar, S., Shipman, P. A., Uresandi, O. C., Benavides, E., Blangero, J., and O'Connell, P., Evidence for linkage of regions on chromosomes 6 and 11 to plasma glucose concentrations in Mexican Americans, *Genome Res.*, 6, 724, 1996.
43. Prochazka, M., Thompson, D. B., Knowler, W. C., Bennett, P. H., Bogardus, C., Scherer, S. W., and Tsui, L. C., Linkage and association of markers at 7121.3–922.1 with non-insulin dependent diabetes mellitus (NIDDM) in Pima Indians, *Cytogenet. Cell Genet.*, 71, 30, 1995.
44. Risch, N. and Merikangas, K., The future of genetic studies of complex human diseases, *Science*, 273, 1516, 1996.
45. Raffel, L. J., Robbins, D. C., Norris, L. M., Boerwinkle, E., DeFronzo, R. A., Elbein, S. C., Fujimoto, W., Hanis, C. L., Kahn, S. E., Permutt, M. A., Chiu, K. C., Cruz, J., Ehrmann, D. A., Robertson, R. P., Rotter, J. I., and Buse, J., The GENNID study, *Diabetes Care*, 19, 864, 1996.
46. Bennett, P. H., Burch, T. A., and Miller, M., Diabetes mellitus in American (Pima) Indians, *Lancet*, 2, 125, 1971.
47. Bogardus, C. and Lillioja, S., Pima Indians as a model to study the genetics of NIDDM, *J. Cell Biochem.*, 48, 337, 1992.
48. Lander, E. S. and Schork, N. J., Genetic dissection of complex traits, *Science*, 265, 2037, 1994.
49. Spielman, R. S., McGinnis, R. E., and Ewens, W. J., Transmission test for linkage disequilibrium: the insulin gene region and insulin-dependent diabetes mellitus (IDDM), *Am. J. Hum. Genet.*, 52, 506, 1993.
50. Knapp, M., Seuchter, S. A., and Baur, M. P., The haplotype-relative-risk (HRR) method for analysis of association in nuclear families, *Am. J. Hum. Genet.*, 52, 1085, 1993.
51. Terwilliger, J. D., A powerful likelihood method for the analysis of linkage disequilibrium between trait loci and one or more polymorphic marker loci, *Am J. Hum. Genet.*, 56, 777, 1995.

# PART III

## SOCIAL IMPACT OF QTL MAPPING

# 20 From Malthus to Molecular Mapping: Prospects for the Utilization of Genome Analysis to Enhance the World Food Supply

*Susan R. McCouch and Jinhua Xiao*

**CONTENTS**



## 20.1 INTRODUCTION

We are frequently reminded that there is no such thing as a free lunch. But for many people, there is no such thing as lunch. Exponential increases in human populations are straining the resource base of the planet and seriously threatening our ability to keep up with the demand for food. There is disagreement as to whether the continuous increase in the number of human beings on the planet is a cause or an effect of hunger and poverty, and why hunger persists in a world where global food supplies are sufficient to provide an adequate diet for all. Malthusian analysts will claim that the inevitable is near, that we are reaching the point where population simply overwhelms the carrying capacity of our finite planet.[1,2] Within this world view, mass starvation is the natural consequence of continuing population growth. On the other side of the coin, optimists suggest that solutions are bound to emerge once the inescapability of the food and hunger crisis is fully comprehended and reported. The basis for this argument is that populations tend to be self-limiting as people gain consciousness about the status of resource limitations,[3] that demand catalyzes the development of appropriate technology, and that human ingenuity is capable of devising both technical and political solutions that will result in a healthy and abundant food supply for all.

At present, the world's population hovers around 5.7 billion.[4] It is expected to reach 8 billion by the year 2025, and 10 billion by the middle of the 21st century (Figure 20.1). To keep pace with

0-8493-7686-6/98/$0.00+$.50

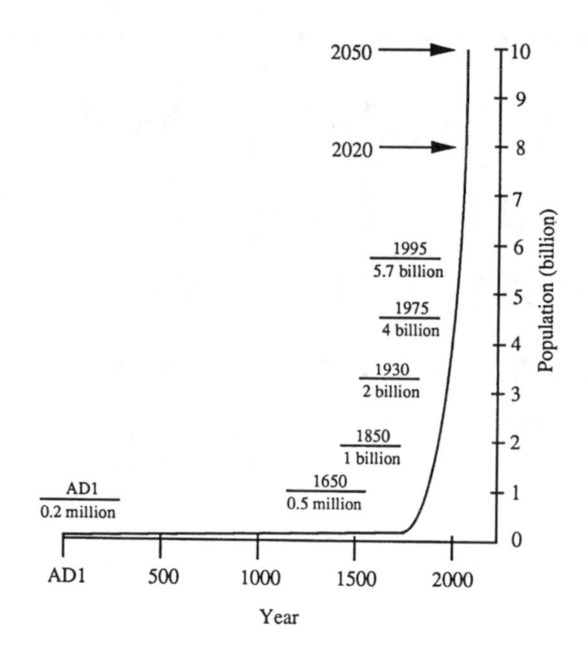


**FIGURE 20.1** The world's population growth and projections from now to 2050.

this increase in population, global food production will have to expand by at least 50% in the next 25 years, and will have to double before 2050. One point upon which most people agree is that achieving these goals depends largely on the ingenuity with which technical innovations can be coupled with political solutions to help meet the demand. The kinds of technical innovations that are needed include the ability to increase agricultural productivity per se both globally and locally, to develop genetic and management strategies that minimize fluctuations in food production patterns, and to develop regenerable systems that maintain or enhance the natural resource base upon which food production depends. Simultaneously, political forces must be focused on restricting the birth rate, on targeting basic foodstuffs to areas of the world and segments of the population where they are most needed, and on raising consciousness about consumption patterns in relation to health of individuals and health of the planet.

A debate rages over whether increases in agricultural productivity can be achieved without severe ecological damage. In reality, it is likely that without increases in agricultural productivity, ecological damage will only be accelerated because of the links between population density, poverty, hunger, and resource degradation.[5] Many of the world's poorest people live in the most fragile ecosystems, and population pressure magnifies the desperation of their situation. What is urgently needed is a way to couple increased food production with the alleviation of hunger and a limit to population growth so that protection of the natural resource base can be seriously addressed.

In the past, global food production has increased because cropped area has expanded, the number of crops per year has increased, and productivity per unit area has increased. Few of these options for increasing production are available today. The resource based upon which agriculture depends is itself at risk. Fertile land, fresh water, clean air, and biodiversity are no longer expandable units. In the future, most of the anticipated increases in crop production must come from increases in the efficiency of production with emphasis on enhancing genetic potential.

In this chapter, we will address the question of how molecular genetic-assisted manipulation of food crops can be expected to increase food production per se, how molecular breeding can help stabilize crop production by targeting resistance to pests and diseases and to environmental stress, how nutritional and quality traits can be addressed genetically, and what kinds of crop improvement strategies can contribute to the sustainable utilization of biological resources on the planet. We will also comment on the networks of people that are needed if these efforts are to have a positive impact on global food supply.

## 20.2 GENES, GENOMES, AND AGRICULTURAL PRODUCTION

The essential nature of the gene in all biological systems means that this unit of heredity has become one of the major focal points for attempts to enhance the productivity and sustainability of agricultural systems. Molecular markers contribute to our understanding of where genes are located along the chromosomes, what they contribute to phenotype, and the ways in which they interact with each other and with the environment. Molecular markers also provide a strategy for isolating genes of interest, offering a toolbox of creative opportunities for optimizing productivity and environmental adaptation of agriculturally important species.

There is no other area of biotechnology today that parallels molecular mapping, or genome analysis, in the ability to link structural analysis of genes and chromosomes with process-oriented sciences such as biochemistry and physiology, and ultimately with phenotypic performance of whole plants or animals in production environments. In conjunction with traditional plant and animal breeding, genome analysis can be used to develop new crop and livestock strains that outperform the "best" existing varieties and breeds. The definition of "best" is not restrictive; performance can be evaluated in either intensively managed, high-input environments or in heterogeneous, resource-limited environments. Locally adapted materials can be improved for specific traits while retaining their unique genetic identity. The location of genes that contribute to target phenotypes may be inferred from both species-specific studies and comparative mapping studies that link the evolutionary history of major food crops. By establishing specific research objectives and bringing together the resources and talent of people in different domains of molecular biology, variety development, and food production, molecular marker-based breeding can make an important contribution to the future of our food production systems.

## 20.3 MOLECULAR MARKER-ASSISTED VARIETY DEVELOPMENT

In agriculture, most traits of interest, such as yield of harvestable produce in plants, growth of domestic animal species, milk production of cattle, and oil or protein content of seed crops, forages, or legumes are quantitative and complex. These characters differ from traits that are simply inherited and measured qualitatively. The advent of DNA markers greatly facilitated the study of quantitative traits and made it possible to dissect such traits into discrete quantitative trait loci (QTL).[6] QTL resemble individual Mendelian loci, except that the direction and magnitude of their effect is often obscured because they act in concordance with other QTL in producing an observed phenotypic effect. With the help of molecular markers, the number of genetic loci contributing to an observable phenotype can be estimated, the magnitude of the effect of each locus and the specific map positions of those loci can be determined, and their gene action, pleitropic effects and epistatic interactions can be characterized.

Almost any trait that can be reliably measured is amenable to QTL analysis, and once the map position of a QTL has been determined, linked markers can be used to transfer these loci to new genetic backgrounds for evaluation in a breeding program, or as the basis for in-depth genetic characterization and map-based gene isolation. So the question arises, what traits are worth the investment in QTL analysis and how will the investment pay off in terms of increasing global food production?

### 20.3.1 INCREASING YIELD POTENTIAL

Yield, or harvestable production per unit of input, provides an essential indicator of agricultural productivity and is central to most breeding programs. However, because of the polygenic inheritance and the large influence of environment on this trait, it is difficult for breeders to manipulate efficiently. In traditional variety improvement, decisions about plant selection are based on the phenotypic performance of individuals or families, and replication is necessary to determine what part of the phenotypic variation is the result of genetics and what part is due to environment. QTL mapping provides a way to link the field performance in a given environment with the presence or absence of specific alleles at discrete genetic loci, and identifies molecular markers that can be used to monitor the inheritance of these loci in a segregating population. The use of markers linked to target QTL can accelerate genetic progress by facilitating the selection of individuals containing the optimum combination of positive alleles at critical loci.

The ability to enhance the yield of crop plants by selecting for positive QTL would be of tremendous value in plant improvement. The first question is whether yield potential, as a vitally important trait in agriculture, is amenable to genetic analysis. There is convincing evidence in the literature that it is. QTL underlying yield have been reported in maize,[7-11] rice,[12,13] wheat,[14,15] barley,[16-19] and tomato.[20-22] QTL for growth rate have been reported in cattle[23] and pigs,[24,25] for milk production in dairy cattle[26] and sheep,[27] and for egg production in chickens.[28] These studies suggest that yield QTL can be identified, and as a consequence, that markers could be used to assist in the accumulation of yield-enhancing QTL alleles in a breeding program.

This leads directly to the second question under consideration, which is whether the investment in QTL analysis of a trait such as yield will contribute to increasing global food production. To date there is no convincing example of a commercially available high-yielding plant variety or high-producing animal breed that has been developed based on marker assisted selection (MAS) of QTL for yield or production capacity. However, there are several projects underway that are likely to achieve this objective within the next 5 years in both plants[29,30] and animals (see Chapters 7, 17, and 18, this volume). If marker-based selection for yield and production QTL shortens the breeding process, makes it more efficient, and allows breeders to identify novel yield-enhancing alleles from new sources, and/or to combine yield-enhancing genes in productive new ways, then the investment will certainly pay for itself and the approach can be expected to be rapidly incorporated into public and private breeding programs.

While marker-based schemes for genetic selection empower the breeder, they do not alter the natural qualities of the product for farmers and consumers. Therefore the public is not likely to be aware when genome analysis is used in variety development, any more than the public is aware of the *in vitro* methods now commonly employed in plant and animal breeding. Because molecular marker assisted breeding is essentially a selection tool and does not interfere with the natural process of meiosis during genetic recombination, it introduces no new biosafety issues and the product should require no special labeling or registration. This means that there should be no lag time between the development of new varieties based on molecular-MAS strategies and the marketing and distribution of those varieties. It also means that the cost of implementing a breeding MAS will be largely defined by the cost of the program itself, with few of the external costs associated with genetically engineered forms of plant improvement.

Crop performance is determined by the interaction between the genes in the plant and the environment in which the plant is grown. In cases where optimal yield is important, plants have been selected to prosper with high inputs of nitrogen fertilizer, continuous water supply, chemical control of pests and diseases, high solar radiation, and intensive management. It is critical that food production in the most productive farmland be encouraged to help ensure the availability of basic foodstuffs and to take some of the pressure off of production in fragile ecosystems. However, to fulfill the dietary requirements of many people in the world, it is also essential that local food production continue to be viable, that it provide stable yields for subsistence communities, and that a variety of foodstuffs be encouraged in local food production systems. Plant selections that are better able to withstand stresses such as pest and disease attack, drought or flooding, low nutrient deficiencies or toxicities, cold, heat and low-light intensities are needed. In traditional agricultural settings where the availability of water, fertilizer and chemical pest control is limited by biological and economic constraints, genotypes have been developed that embody fine-tuned adaptations to these suboptimal environments. However, yield of traditional varieties is typically quite low and it is of interest to identify genes and combinations of genes that provide both adaptation and enhanced yield potential in resource-limiting environments.

QTL analysis provides a way to identify genes responsible for superior performance in just about any environment of interest and to quantify the magnitude of the effect of those loci. The results of QTL mapping to date indicate that a QTL important in one environment does not always contribute significantly to phenotypic expression in another environment, a phenomenon referred to as genotype by environment interaction (G × E). By using information about which genes are optimal for use in specific environments, breeders can target individual QTL in their efforts to develop new varieties of crops and livestock that are optimal for production in a range of different environments.

### 20.3.2 Stabilizing Crop Performance

There is considerable loss of potential crop production due to diseases and insects in every environment. Breeders, pathologists, and entomologists have devoted a tremendous amount of time and resources to the development of resistant varieties because these varieties offer the most effective and economical way to control crop pests. This involves combining multiple resistance genes into highly productive crop varieties based on phenotypic selection. Stable resistance for many of the most devastating diseases and insects involves genes that contribute partial as well as complete resistance to a given pest or disease.[31,32] Increasingly, there is interest in combining quantitatively inherited resistance characters and this poses formidable logistical problems for a traditional breeding program.

Molecular mapping technology makes it possible to study the inheritance of quantitative resistance and to characterize individual genetic factors underlying the resistance. Using QTL mapping, Wang et al.[33] and Li et al.[34] located and characterized QTL for partial resistance to blast and sheath blight, respectively, both of which represent major disease problems in rice. Schön et al.[35] and Bubeck et al.[36] mapped QTL for resistance against the European corn borer, and against gray leaf spot, respectively, both major limitations to maize production worldwide. Chen et al.[37] identified QTL for resistance to stripe rust in barley, Dion et al.[38] located QTL for resistance to the blackleg disease in canola, and Webb et al.[39] mapped quantitative resistance to the cyst nematode in soybean. In-depth characterization of individual QTL will help breeders determine which alleles to introgress into promising elite breeding lines in their efforts to develop durably resistant varieties. Genetic resistance to pests and diseases is a core requirement for stable production of food crops in both high and low input environments. Breeding efforts targeting crop improvement for these kinds of production environments can be expected to benefit directly from the identification of QTL for resistance by using the linked markers to introgress specific polygenes of interest into locally adapted varieties for further testing *in situ*.

Yield loss due to environmental stress is also considerable. This includes extremes of water, nutrient availability, and temperature regime. Fresh water is one of the most severely limiting natural resources on the planet today. For agricultural purposes, water is limited or fluctuates dramatically in many areas, due to seasonal variation and to the geographical distribution of water relative to arable land. Access to water is a critical feature distinguishing high production systems from marginally productive cropping systems. Various types of water deficit result in drought and consequent loss of yield potential. Breeding for drought tolerance is complex and difficult. Using QTL mapping, Champoux et al.[40] located genes associated with root parameters and vegetative stage drought avoidance in rice. Lebreton et al.[41] and Quarrie et al.[42] reported the identification of QTL for drought responses in maize and wheat, respectively. QTL for winter hardiness in barley have also been reported.[43] These results represent primary genetic analysis that must be followed up with more in-depth characterization of the putative QTL over years and environments to determine whether the polygenes that have been identified will be broadly useful in variety development. If so, the use of linked markers for these traits will greatly facilitate the development of drought and cold resistant crop varieties.

Continued work in this area promises to help unravel the genetics of many environmental stress-resistance or avoidance strategies and lays the foundation for comparative QTL analysis among related crop genera.[44,45] There are opportunities for utilizing the markers linked to stress-resistance QTL directly for selection of useful genes both within a species and across genera. The fact that breeders and geneticists can now exchange information about how defined genetic loci affect crop performance points to new possibilities for designing genotypes that will help stabilize crop production under conditions of environmental stress.

Another abiotic stress problem that tends to affect the poorest farmers is loss of productivity due to inadequate or inappropriate plant nutrition. The use of chemical fertilizer has made significant contributions to increased crop productivity, but for poor farmers, fertilizer is expensive and is not always available. On the opposite extreme, overuse of chemical fertilizers leads to runoff and subsequent contamination of water supplies and to the degradation of cropland due to "salting out". Genetic variation for nitrogen use efficiency has been documented in sorghum,[46] maize,[47] and bean,[48] and marker based approaches to genetic improvement of this trait are likely to be productive. Biological nitrogen fixation by soil-inhabiting bacteria provides a naturally occurring N source for plant nutrition. The system of $N_2$ fixation by nodule forming *Rhizobium* bacteria in legumes has been successfully exploited in agriculture,[49,50] but many other crop species have also been found to stimulate biological $N_2$ fixation. Rice has the ability to stimulate $N_2$ fixation in the rhizosphere[51,52] and a recent study reported QTL for this quantitatively inherited trait.[53] Molecular marker-assisted QTL mapping of the stimulation of $N_2$ fixation in the rhizosphere of rice provides a new approach to the development of rice varieties better able to survive under conditions of low soil nitrogen availability. Such an approach is suggestive of the possibilities of using genetic improvement to target traits of importance in stabilizing the food supply in many of the poorest regions of the world.

### 20.3.3 Improving Quality and Safety of the Food Supply

The widespread use of pesticides for control of crop pests causes serious environmental and public health problems. These problems include food contamination and water pollution with pesticide and fertilizer residues, nontarget effects on fish and beneficial natural enemies, and human and livestock poisonings. The use of integrated pest management strategies aims to minimize the use of pesticides while assuring the availability and profitability of the harvest. The availability of resistant crop varieties is one of the cornerstones of integrated pest management and the implementation of molecular marker facilitated selection for durable resistance, as discussed above, can be seen as part of a broader effort to develop agricultural products that improve the health and well being of both the human population and the environment. Thus, developing strategies to recombine complete and partial resistance genes, as proposed by Wang et al.,[33] or to obtain stable resistance through the development and deployment of multilines,[31] are both likely to involve the use of molecular maps and markers. For the consumer, harvesting or purchasing food from genetically resistant crop varieties that do not require extensive use of chemical pesticides represents an important contribution to reducing the contamination of food with pesticide residues.

The food we eat not only provides the energy (calories), but also the essential nutrients, such as iron, iodine, and vitamins required by the human body. In cases where these micronutrients are deficient in the human body, diseases such as diarrhea, anemia, measles, malaria, parasites, and other nutritional impairments develop more easily and to a greater extent.[54] Therefore, the nutritional quality of food is very important to the health of human beings. Most traits related to nutritional quality, such as vitamin and protein content of cereal grains are quantitatively inherited and complex. Cereal grains are the major source of protein in the diets of a large proportion of people in developing countries. Breeding for higher protein and micronutrient content of cereal crops has been proposed as a strategy for improving the nutrition of these people,[55] especially that of children and pregnant women, whose nutritional demands are most intense.[56] In addition to diversifying the diet of people who lack vital micronutrients, it may be necessary to improve the inherent levels of available proteins and vitamins in a breeding program if the prerequisite genetic variation can be found in sexually compatible germplasm. QTL mapping offers breeders an obvious opportunity to try to manipulate the genetic component such as flavor and appearance of these nutritional characters while holding on to other characteristics that have traditionally made the food attractive to consumers.

Oil extracted from oil seeds and corn kernels is one of the most important food constituents and provides a source of fatty acids. The composition of fatty acids in the oil has important effects on the health of humans. For example, oleic acid (18:1) is believed to reduce the probability of heart disease. The concentration of fatty acids in the oil is conditioned by polygenes, like other quantitative traits. QTL mapping makes it possible to characterize discrete genetic factors underlying the concentration of fatty acids in specific oils. Alrefai et al.[57] mapped QTL for the concentration of five important fatty acids, including oleic acid, in maize. Molecular markers can now be used for selection in breeding programs in developing varieties with high oleic acid concentration.

Most traits of aesthetic value, such as eating quality, cooking quality, and appearance are also quantitatively inherited. It is likely that many of them can be adjusted to suit the preferences of specific cultures or regions through QTL mapping and molecular MAS. An example that is relevant to world food production is the "stay soft" character in rice. This aspect of grain quality means that cooked rice stays soft and does not dry out and harden as the rice cools after cooking. This trait allows people to eat rice that was cooked several hours before. Such a character is very important in a country such as the Philippines where rice provides the staple food three meals a day for many people, who do not have fuel enough to cook fresh rice for each meal. Previous research has shown that the "stay soft" trait is related to the amylase content in the grain,[58] a quantitatively inherited character. Thus, QTL mapping of amylase content in rice would enable breeders to determine how many loci contributed significantly to this trait, where they were along the rice chromosomes, and what proportion of the phenotypic variance was explained by each QTL. The availability of markers linked to the "stay soft" QTL would facilitate the transfer of this valuable trait into new rice varieties for staple food consumption among the world's poorest people. In addition to this character, QTL analysis of many other quality characteristics highly valued by consumers is likely to provide valuable selection tools for improving both the nutritional and eating qualities of the foods that people consume.

### 20.3.4 Detecting Novel QTL Alleles

For many of the characters mentioned above, germplasm containing the genes, or QTL, that condition a target trait has already been identified. In such cases, the role of genome analysis is to localize the target genes along the chromosomes and to provide markers to facilitate selection in breeding, as well as to provide the foundation for map-based isolation of key loci. An exciting alternative to this approach involves the use of molecular maps and markers to discover new genes. Where germplasm resources have not been well characterized, genome analysis provides a set of tools and map coordinates that facilitate the discovery of useful genes within large pools of otherwise inferior alleles. By analyzing segregants from wide crosses, many unusual alleles can be discovered and rapidly brought into cultivated gene pools.

Transgressive variation is often observed in a segregating population derived from genetically divergent parents. This phenomenon refers to the appearance of segregants that fall outside of the range of phenotypes observed in the parents.[59] The occurrence of transgressive segregants is well known to breeders, but it has been difficult to identify and capture the genetic factors responsible for the positive transgressive phenotypes and fix them through traditional breeding. QTL mapping in populations showing transgressive variation has made it possible to determine which QTL contribute to the superior performance, to identify the specific parental alleles that make a positive contribution at each QTL, and to transfer the better alleles from one genetic background to another using molecular MAS.[60] The approach is more analytical and systematic than phenotypic selection and makes it possible for a breeder to efficiently accumulate favorable alleles in a preferred genetic background.

If transgressive variation is observed in the offspring of a cross between two parents, it follows that parental phenotypes are not always good predictors of the performance of their progeny. Two individuals that have the same phenotype for a particular trait may produce offspring that perform either much better or much worse than either parent.[13] This is because, though the parents may resemble each other in specific traits, they may not have the same compositions of genes or alleles conditioning the trait. The probability that they contain different QTL is largely a function of their genetic distance, with more distantly related individuals being more likely to contain different QTL. The question then arises, at what genetic distance is transgressive variation most likely to be positive for agriculture? By phenotypically evaluating intra- and interspecific crosses, traditional breeders generally agree that there is too much deleterious linkage drag associated with the use of wild ancestors to improve all but a few qualitatively inherited traits.[61] Reliance on phenotypic analysis alone requires that many positive alleles be present together in a line in order to observe a phenotypic advantage. This requirement has made the process of combining superior alleles less efficient than it would be if the breeder had a way to identify individual alleles at the genetic level and then use that information to sequentially combine the superior alleles into a new variety. The use of genetic maps and markers makes it possible to efficiently identify the positive QTL alleles that reside in wild ancestors and to selectively transfer them into elite cultivars.

Using advanced backcross analysis, Tanksley et al.[20] discovered a positive QTL allele from *Lycopersicon pimpinellifolium*, a low-yielding wild relative of cultivated tomato, that increased yield by 17% in an elite commercial variety. The implication of this work is that advanced backcross QTL analysis may be applicable to other crop species such as rice or maize for which molecular maps are also available. If so, it represents a model for effectively utilizing the vast collections of wild and exotic germplasm that have been painstakingly collected and maintained over the last 30 to 50 years. These collections represent a treasure trove of useful genetic variation that has been largely inaccessible to traditional breeding efforts involving quantitative traits. An important outcome of the use of such a marker-based breeding strategy would increase the overall level of genetic diversity in cultivated crop varieties. This would contribute to the sustainability of our production systems by making available new sources of naturally occurring genes for optimizing plant performance in a wild range of biological and cultural environments.

## 20.4 CONCLUSION AND FUTURE PROSPECTS

QTL mapping brings power and resolution to the process of genetic analysis. It finds immediate application as a tool for selection in plant and animal improvement, and also provides the foundation for map-based gene isolation. In an international setting where independent research groups work on locally adapted breeds or varieties, the use of a common set of maps and markers for a specific species provides a common language and a systematic way of storing, retrieving, and interpreting information about the genes that underlie valuable phenotypes. Databases that facilitate access to the vast amounts of information generated by scientists dispersed in different programs around the world are indispensable to the interpretation of QTL analysis and to efficient implementation of marker-assisted variety improvement. Within the U.S., the National Agricultural Library (NAL) maintains plant and animal genome databases on many agriculturally important species that are publicly accessible via the Gopher or the World Wide Web (http://probe.nalusda.gov:8300). Similar databases are being developed in Japan (http://www.staff.or.jp), Europe (http://www.sanger.ac.uk/embnet/embnet.html), Korea (http://168.126.188.2/Nasti and http://168.126.188.2/Nasti/OtherService/RiceGenome.html), and at the International Crop Improvement Centers (http://www.cimmyt.mx). These databases are continually updated and serve as references for both national and international researchers.

As increasing numbers of agronomically valuable genes and QTLs are discovered and characterized at the molecular level, the potential for cross referencing information in a variety of species and across levels of biological complexity (i.e., nucleotide sequence, biochemical function, phenotypic effect, agricultural relevance) is leading to new understandings of gene structure, gene function, and gene evolution. Genome analysis currently plays a vital role in this process because it serves both as a method for gene identification and provides tools for directly manipulating traits of importance in plant and animal agriculture.

The impact that molecular marker analysis is to have on increasing global food supplies is largely dependent on the establishment of effective networks of people around the globe who recognize the potential of the approach and have the knowledge and resources to manage both molecular genetics and traditional breeding technology. Projects aimed at providing opportunities for young plant and animal breeders to use genome analysis as part of their variety development programs are in their infancy. Success of these efforts will depend on designing experiments that target carefully prioritized and realistic goals, and involving the appropriate combination of researchers who have the incentive and determination to follow through on variety development while integrating information obtained from initial QTL analyses.

Primary QTL analysis provides a set of markers putatively linked to genes controlling quantitative traits, but further generations of crossing and selection are required to evaluate the stability of those QTLs and to develop improved varieties. Many well designed QTL studies stagnate following publication of the first set of results because the follow-up work is often perceived as less glamorous than the original genetic studies, and because personal contacts and continuity of funding between molecular genetics laboratories and applied breeding programs are often lacking. One way around this problem would be to establish a public competitive grants program where grant renewal was conditional upon development of improved breeding lines rather than merely publication of gene/QTL mapping or cloning results. With a few well-organized and well-funded networks in place, the contributions of genome analysis to increasing food production around the world could be convincingly demonstrated.

## REFERENCES

1. Brown, L. R., *Full House: Reassessing the Earth's Population Carrying Capacity,* W. W. Norton & Co., New York, 1994.
2. Malthus, T. R., *An Essay on the Principle of Population; or, a View of Its Past and Present Effects on Human Happiness: With an Inquiry into our Prospects Respecting the Future Removal or Mitigation of the Evils Which It Occasions,* J. Murray, London, 1817.
3. Abernethy, V., *Population Politics: The Choices that Shape Our Future,* Insight Books, New York, 1993.
4. FAO, *Conducting Agricultural Censuses and Surveys,* Food and Agriculture Organization of the United Nations, Rome, 1996.
5. Pinstrup-Andersen, Per., *World Food Trends and Future Food Security,* International Food Policy Research Institute, Washington D. C., 1994.
6. Tanksley. S. D., Mapping polygenes, *Annu. Rev. Genet.,* 27, 205, 1993.
7. Stuber, C. W., Lincoln, S. E., Wolff, D. W., Helentjaris, T., and Lander, E. S., Identification of genetic factors contributing to heterosis in a hybrid from two elite maize inbred lines using molecular markers, *Genetics,* 132, 823, 1992.
8. Rogot, M., Sisco, P. H., Hoisington, D. A., and Stuber, C. W., Molecular-marker-mediated characterization of favorable exotic alleles at quantitative trait loci in maize, *Crop Sci.,* 35, 1306, 1995.
9. Ajmone-Marsan, P., Monfredini, G., Ludwig, W. F., Melchinger, A. E., Franceschini, P., Pagnotto, G., and Motto, M., In an elite cross of maize a major quantitative trait locus controls one-fourth of the genetic variation for grain yield, *Theor. Appl. Genet.,* 90, 415, 1995.
10. Veldboom, L. R. and Lee, M., Molecular-marker-facilitated studies of morphological traits in maize. II: Determination of QTL for grain yield and yield components, *Theor. Appl. Genet.,* 89, 451, 1994.
11. Beavis, W. D., Smith, O. S., Grant, D., and Fincher, R., Identification of quantitative trait loci using a small sample of topcrossed and F4 progeny from maize, *Crop Sci.,* 34, 882, 1994.
12. Xiao, J., Li, J., Yuan, L., and Tanksley, S. D., Dominance is the major genetic basis of heterosis in rice as revealed by QTL analysis using molecular markers, *Genetics,* 140, 745, 1995.

13. Xiao, J., Li, J., Yuan, L., and Tanksley, S. D., Identification of QTL affecting traits of agronomic importance in a recombinant inbred population derived from a subspecific rice cross, *Theor. Appl. Genet.*, 92, 230, 1996.
14. Hyne, V. M., Kearsey, J., Martinez, O., Gang, W., and Snape, J. W., A partial genome assay for quantitative trait loci in wheat (*Triticum aestivum*) using different analytical techniques, *Theor. Appl. Genet.*, 89, 735, 1994.
15. Schlegel, R. and Meinel, A., A quantitative trait locus (QTL) on chromosome arm 1RS of rye and its effects on yield performance of hexaploid wheat, *Cereal Res. Commun.*, 22, 7, 1994.
16. Thomas, W. T. B., Powell, W., Waugh, R., Chalmers, K. J., Barua, U. M., Jack, P., Lea, V., Forster, B. P., Swanston, J. S., Ellis, R. P., Hanson, P. R., and Lance, R. C. M., Detection of quantitative trait loci for agronomic, yield, grain and disease characters in spring barley (*Hordeum vulgare* L.), *Theor. Appl. Genet.*, 91, 1037, 1995.
17. Iyamabo, O. E. and Hayes, P. M., Effects of plot type on detection of quantitative-trait-locus effects in barley (*Hordeum vulgare* L.), *Plant Breed.*, 114, 55, 1995.
18. Backes, G., Graner, A., Foroughi-Wehr, B., Fischbeck, G., Wenzel, and Jahoor, G. A., Localization of quantitative trait loci (QTL) for agronomic important characters by the use of a RFLP map in barley (*Hordeum vulgare* L.), *Theor. Appl. Genet.*, 90, 294, 1995.
19. Hayes, P. M., Liu, B. H., Knapp, S. J., Chen, F., Jones, B., Blake, T., Franckowiak, J., Rasmusson, D., Sorrells, M., Ullrich, S. E., Wesenberg, D., and Kleinhofs, A., Quantitative trait locus effects and environmental interaction in a sample of North American barley germplasm, *Theor. Appl. Genet.*, 87, 392, 1993.
20. Tanksley, S. D., Grandillo, S., Fulton, T. M., Zamir, D., Eshed, Y., Petiard, V., Lopez, J., and Beck-Bunn, T., Advanced backcross QTL analysis in a cross between an elite processing line of tomato and its wild relative *L. pimpinellifolium*, *Theor. Appl. Genet.*, 92, 213, 1996.
21. Eshed, Y. and Zamir, D., An introgression line population of *Lycopersicon pennellii* in the cultivated tomato enables the identification and fine mapping of yield-associated QTL, *Genetics*, 141, 1147, 1995.
22. Paterson, A. H., Deverna, J. W., Lanini, B., and Tanksley, S. D., Fine mapping of quantitative trait loci using selected overlapping recombinant chromosomes in an interspecies cross of tomato, *Genetics*, 124, 735, 1990.
23. Grignola, F., Mao, I. L., Mejdell, C., Lie, O., and Solbu, H., Association of Alleles for the Class I Bovine Lymphocyte Antigens with Conformation, Semen Traits, and Growth Rate of Young Bulls, *J. Dairy Sci.*, 78, 908, 1995.
24. Rothschild, M. F., Liu, H. C., Tuggle, C. K., Yu, T. P., and Wang, L., Analysis of pig chromosome 7 genetic markers for growth and carcass performance traits, *J. Anim. Breed. Genet.*, 112, 341, 1995.
25. Andersson, L., Haley, C. S., Ellegren, H., Knott, S. A., Johansson, M., Andersson, K., Andersson-Eklund, L., Edfors-Lilja, I., Fredholm, M., Hansson, I., Hakansson, J., and Lundstrom, K., Genetic mapping of quantitative trait loci for growth and fatness in pigs, *Science*, 263, 1771, 1994.
26. Georges, M., Nielsen, D., Mackinnon, M., Mishra, A., Okimoto, R., Pasuino, A. T., Sargeant, L. S., Sorensen, A., Steele, M. R., Zhao, X., Womack, J. E., and Hoeschele, I., Mapping quantitative trait loci controlling milk production in dairy cattle by exploiting progeny testing, *Genetics*, 139, 907, 1995.
27. Moioli, B., Pilla, F., Rando, A., and Tripaldi, C., Possible exploitation of DNA polymorphisms as marker genes for quantitative trait loci to improve milk production traits in sheep and goats, in *Book of Abstracts of the Annual Meeting of the European Association for Animal Production*, Vol. 1. Book of Abstracts of the 46th Annual Meeting of the European Association for Animal Production; Meeting, Prague, Czech Republic, Wageningen Pers, Wageningen, Netherlands, 1995.
28. Wei, M. and Van-Der-Werf, J. H. J., Genetic correlation and heritabilities for purebred and crossbred performance in poultry egg production traits, *J. Anim. Sci.*, 73, 2220, 1995.
29. Stuber, C. W., Mapping and manipulating quantitative traits in maize, *Trends Genet.*, 11, 477, 1995.
30. Xiao, J., Li, J., Grandillo, S., Ahn, S. N., McCouch, S. R., Tanksley, S. D., and Yuan, L., A wild species contains genes that may significantly increase the yield of rice, *Nature*, in press.
31. Browning, J.A. and Frey, K.J., Multiline cultivars as a means of disease control, *Annu. Rev. Phytopathol.*, 7, 355, 1969.
32. Leonard, K. J. and Fry, W. E., *Plant Disease, Epidemiology, Genetics, Resistance, and Management*, Macmillan, New York, 1989.

33. Wang, G., Mackill, D. J., Bonman, J. M., McCouch, S. R., Champoux, M. C., and Nelson, R. J., RFLP mapping of genes conferring complete and partial resistance to blast in a durably resistant rice cultivar, *Genetics,* 136, 1421, 1994.
34. Li, Z., Pinson, S. R. M., Marchetti, M. A., Stansel, J. W., and Park, W. D., Characterization of quantitative trait loci (QTLs) in cultivated rice contributing to field resistance to sheath blight *(Rhizoctonia solani), Theor. Appl. Genet.,* 91, 382, 1995.
35. Schön, C. C., Lee, M., Melchinger, A. E., Guthrie, W. D., and Woodman, W. L., Mapping and characterization of quantitative trait loci affecting resistance against second-generation European corn borer in maize with the aid of RFLPs, *Heredity,* 70, 648, 1993.
36. Bubeck, D. M., Goodman, M. M., Beavis, W. D., and Grant, D., Quantitative trait loci controlling resistance to gray leaf spot in maize, *Crop Sci.,* 33, 838, 1993.
37. Chen, F. Q., Prehn, D., Hayes, P. M., Mulrooney, D., Corey, A., and Vivar, H., Mapping genes for resistance to barley stripe rust (*Puccinia striformis f.* sp. *hordei*), *Theor. Appl. Genet.,* 88, 215, 1994.
38. Dion, Y., Gugel, R. K., Rakow, G. F. W., Seguin-Swartz, G., and Landry, B. S., RFLP mapping of resistance to the blackleg disease (causal agent, *Leptosphaeria maculans* (Desm.) Ces. et de Not.) in canola (*Brassica napus L.*), *Theor. Appl. Genet.,* 91, 1190, 1995.
39. Webb, D. M., Baltazar, B. M., Rao-Arelli, A. P., Schupp, J., Clayton, K., Keim, P., and Beavis, W. D., Genetic mapping of soybean cyst nematode race-3 resistance loci in the soybean PI 437.654, *Theor. Appl. Genet.,* 91, 574, 1995.
40. Champoux, M. C., Wang, G., Sarkarung, S., Mackill, D. J., O'Toole, J. C., Huang, N., and McCouch, S. R., Locating genes associated with root morphology and drought avoidance in rice via linkage to molecular markers, *Theor. Appl. Genet.,* 90, 968, 1995.
41. Lebreton, C., Lazic-Jancic, V., Steed, A., Pekic, S., and Quarrie, S. A., Identification of QTL for drought responses in maize and their use in testing causal relationships between traits, *J. Exp. Bot.,* 46, 853, 1995.
42. Quarrie, S. A., Gulli, M., Calestani, C., Steed, A., and Marmiroli, N., Location of a gene regulating drought-induced abscisic acid production on the long arm of chromosome 5A of wheat, *Theor. Appl. Genet.,* 89, 794, 1994.
43. Hayes, P. M., Chen, F., Pan, A., Oziel, O., Blake, T. K., and Chen, T. H., Quantitative trait locus analyses of malting quality and winter hardiness traits in barley, *Cereal Foods World,* 38, 604, 1993.
44. McCouch, S. R., and Doerge, R. W., QTL mapping in rice, *Trends Genet.,* 11, 482, 1995.
45. Yu, G. X., Bush, A. L., and Wise, R. P., Comparative mapping of homoeologous group 1 regions and genes for resistance to obligate biotrophs in *Avena, Hordeum,* and *Zea mays, Genome,* 39, 155, 1996.
46. Gardner, J., C., Maranville, J., W., and Paparozzi, E., T., Nitrogen use efficiency among diverse sorghum cultivars, *Crop Sci.,* 34, 728, 1994.
47. Teyker, R., H., Seedling responses to band applied ammonium hydroxide rates and to nitrogen form in two maize hybrids, *Plant Soil,* 144, 289, 1992.
48. Lynch, J., Gonzales, A., Tohme, J. M., and Garcia, J. A., Variation in characters related to leaf photosynthesis in wild bean populations, *Crop Sci.,* 32, 633, 1992.
49. Hardarson, G., Methods for enhancing symbiotic nitrogen fixation, *Plant Soil,* 152, 1, 1993.
50. Senaratne, R. and Ratnasinghe, D. S., Ontogenic variation in nitrogen fixation and accumulation of nitrogen in mungbean blackgram cowpea and groundnut, *Biol. Fertil. Soils,* 16, 125, 1993.
51. Santiago-Ventura, T., Bravo, M., Dae, C., Ventura, V., Watanabe, I., and App, A. A., Effects of nitrogen fertilizers straw and dry fallow on the nitrogen balance of a flooded soil planted with rice, *Oryza sativa, Plant Soil,* 93, 405, 1986.
52. Tirol-Padre, A., Ladha, J. K., Punzalan, G. C., and Watanabe, I., A plant sampling procedure for acetylene reduction assay to detect rice varietal differences in ability to stimulate nitrogen fixation, *Soil Biol. Biochem.,* 20, 175, 1988.
53. Wu, P., Zhang, G., Ladha, J. K., McCouch, S. R., and Huang, N., Molecular-marker-facilitated investigation on the ability to stimulate N-2 fixation on the rhizosphere by irrigated rice plants, *Theor. Appl. Genet.,* 91, 1177, 1995.
54. Chen, R. S. and Kates, R. W., World food security: prospects and trends, *Food Policy,* 19, 192, 1994.
55. Corke, H., Protein content and composition in crosses between wild and cultivated barley, *Cereal Res. Commun.,* 23, 411, 1995.
56. Morales, E. and Graham, G. G., Effect of amounts consumed on the digestion of cassava by young children, *J. Nutr.,* 117, 2116, 1987.

57. Alrefai, R., Berke, T. G., and Rocheford, T. R., Quantitative trait locus analysis of fatty acid concentrations in maize, *Genome,* 38, 894, 1995.
58. IRRI, *Proc. Workshop on Chemical Aspects of Rice Grain Quality,* International Rice Research Institute, Los Banos, Philippines, 1979.
59. Darlington, C. D. and Mather, K., *The Elements of Genetics,* Allen & Uniwin, London, UK, 1949.
60. Grandillo, S., Ku, H. M., and Tanksley, S. D., Characterization of *fs8.1*, a major QTL influencing fruit shape in tomato, *Mol. Breed.,* in press.
61. Tanksley, S. D. and Nelson, J. C., Advanced backcross QTL analysis: a method for the simultaneous discovery and transfer of valuable QTLs from unadapted germplasm into elite breeding lines, *Theor. Appl. Genet.,* 92, 191, 1996.

# 21 Ethical Consequences of Mapping QTLs for Complex Human Traits

*John C. Crabbe and John K. Belknap*

**CONTENTS**



## 21.1 INTRODUCTION

Ethical concerns surrounding human biology are certainly not new.[1] Debate over possible threats to the dignity and rights of the individual have been with us for some time and has been an important driving force in the development of modern ethical practices in medical research and treatment.[2] Two of the more important of these developments are informed consent and patient (or subject) confidentiality. It is difficult for the younger generation of researchers to remember a time when concern for these issues did not constitute a normal part of research practices. Nonetheless, the urgency for and complexity of dealing with many of these ethical concerns will increase with advancements in technology, such as those foreshadowed in this volume. This has led to much increased interest in bioethics in recent years, particularly with regard to human genomics. The genetic issues surrounding human disease are critical and highly complex. Genetic contributions to disease risk typically reflect the actions of multiple genes, making positive diagnosis of increased (or decreased) genetic risk extraordinarily difficult.

**TABLE 21.1**
**World Wide Web (WWW) sites on the Internet where data and activities of the Ethical, Legal and Social Implications (ELSI) programs can be found, including comprehensive bibliographies of bioethics publications relevant to human genomics**

| | |
|---|---|
| The National Center for Genome Resources Home Page | http://www.ncgr.org |
| Department of Energy Program Report | http://www.ornl.gov/TechResources/Human_Genome/.html |
| Lawrence Berkeley National Laboratory ELSI Project | http://www.lbl.gov/Education/ELSI/ELSI.html |

In the case of genetically influenced diseases, some new ethical dimensions have emerged not often seen in other research areas. Two key examples should be noted. First, the capability to diagnose increased (or decreased) risk for many diseases is occurring at an increasingly earlier age, sometimes even preceding birth. This poses many vexing questions concerning the ethical use of such information: for example, when is termination of pregnancy justified? Should genetic risk become a factor in decisions about employability, insurability, or reproduction? Second, the family (pedigree) is usually the unit of analysis rather than the individual, vastly complicating concerns about confidentiality and informed consent.[2-4] For example, pedigree studies may gather information about individuals who did not give informed consent, and some subjects may be asked to provide information about relatives who strongly oppose disclosure. At a practical level, some resolution of these problems in pedigree research is important to maintain the basic trust between researchers and subjects essential for sound research.

Largely for these and other reasons, a component of the Human Genome Project (HGP), known as the Ethical, Legal, and Social Implications (ELSI) program, has been funded to study questions posed by advances in human genomics, to educate the public concerning these, and to propose policies to mitigate potentially adverse ethical, legal, and social consequences. ELSI represents the first, and presently the largest, federally funded bioethics program.[2] The most current reviews of ELSI activities can be found at three World Wide Web (WWW) sites on the Internet. These are listed in Table 21.1. Over 50 other WWW sites providing related information and bibliographies have been summarized and published by Kaltenbach and McCain.[5]

Until recently, the role of genetic influences was clear and demonstrable only for traits with relatively simple genetic determination. As molecular analyses have increasingly been applied to the analysis of complex traits, we are beginning to sort through the web of genetic and environmental influences on behaviors, personality, and complex disease states, which has led to an intensified interest in the application of bioethical principles in these areas. As quantitative trait locus (QTL) mapping strategies have begun to identify the location of genes influencing such traits, it is necessary to consider how such knowledge might be applied, and with what consequences.

Excellent reviews have appeared recently, where detailed consideration of the many related issues may be found (see References 2 through 8 and the sources listed in Table 21.1). In order to limit the range of our discussion to those points most salient for immediate consideration by scientists working in the area of genetic mapping, we will exclude several topics from consideration, for example those relating to intellectual/industrial property pursuant to discovery.[6,9] We will raise questions in several areas of ethical concern, and then turn our attention to how different approaches to answering scientific questions might have ethical impact.

## 21.2 PRIVACY AND CONFIDENTIALITY ISSUES

### 21.2.1 Consequences for Individuals

Genetic testing has or will soon have the power to be able to identify individuals who possess genetic risk markers predictive of their individual expression of complex traits. DNA can be extracted from vanishingly small samples of tissue, saliva, blood, hair, and potentially skin cells shed during incidental contact. The polymerase chain reaction (PCR) allows identification both of the individual and of the possession of risk markers, or disease genes. The availability of this information raises several questions.

The APOE-E4 allele is an excellent example of the difficulties associated with genetic testing for complex traits. The APOE-E4 allele at the APOE locus is known to be associated with Alzheimer's disease.[10] Since most Alzheimer's patients do not possess this allele, a negative finding for a particular individual has little prognostic value. On the other hand, a positive finding does not accurately predict Alzheimer's, since most positive individuals do not develop the disease. The relationship between the APOE-E4 allele and Alzheimer's is a statistical association apparent in a large population sample, but translating this to the individual patient, wanting to know his or her prognosis, is fraught with difficulties. Even if accurate prediction were possible, no preventive treatment currently exists for those at high risk. Genetic counselors deal with these problems routinely, but the risk of needless alarm or false expectations is high outside the counseling relationship. With these concerns in mind, a committee of The American College of Medical Genetics/American Society of Human Genetics has published a position paper recommending against genotyping for APOE-E4 as a predictive test for Alzheimer's in the majority of cases.[11]

#### 21.2.1.1 Informed Consent

Samples of DNA may be extremely easy to gather incidentally. Furthermore, most people receiving regular medical care donate blood at some point for many routine diagnostic procedures. Whether tissue collected for one purpose may be used to generate genetic data for another is open to question. In some instances (e.g., mandatory testing of newborns for phenylketonuria) testing results are given whether requested or not. In many cases, however, the link between a DNA sample and the identity of the donor must be severed before such a sample can be used for other purposes.

#### 21.2.1.2 Impact on Relationships and Families

Even assuming that such testing is sought and performed, and risk markers then identified, the consequences of that knowledge need to be considered. Where privacy is currently a consideration, test results are currently reported only to the individual seeking test results. What are their rights (or the obligations) to inform (or not to inform) spouses/partners, or other close family members? In most circumstances, the answers to these questions are not codified.

#### 21.2.1.3 Status of Children

Does a newborn have rights regarding genetic testing? Minor children? If these are limited, when does a child achieve control over her own genetic data? Should parents have the right to seek selective termination of pregnancy based on prenatal diagnosis? What about a case where putative risk markers for disease or "undesirable behavior" were identified? What about complex traits, where any individual marker might add only a small percentage to risk?

### 21.2.2 Employment, Insurance, and Medical Care

Much attention has been paid of late to the financial consequences of information derived from genetic testing. Insurers are literally in the business of risk management, and argue that their entire financial strategy is based on accurate estimation of their actual risk of having to pay an insured. Is it acceptable to allow life or medical insurance policies to offer lower rates to nonsmokers, but not to allow charging higher rates to individuals who bear genetic risk markers for atherosclerosis? Will the transition to managed care, which places a great premium on minimizing costs, lead to medical decisions of not to treat certain individuals for whom genetic risk is deemed unacceptably high?

In our home state, the Oregon Legislature last session passed a Genetic Privacy Act. This declares genetic information to be personal property, and prohibits employment and health insurance discrimination based on such information. It further requires informed consent before such information can be obtained, stored, or shared. Several similar bills have been introduced in the U.S. Congress. Doubtless, genetic information will be legally regulated before the consequences of that regulation are fully understood.

## 21.3 SCIENTIFIC ISSUES

Scientists engaged in studies seeking to map complex traits face several practical issues every day. Those working directly with human populations, and those employing animal models to identify complex trait genes, have some different challenges, but a number of complications are similar. We will first discuss issues of a more general nature, and then treat the special problems of genetic animal models.

### 21.3.1 Understanding Gene-Environment Interaction

#### 21.3.1.1 One QTL, Two QTLs, How Many are Enough QTLs?

Most complex traits are multigenic or polygenic (many genes have a demonstrable influence on the trait, each with a relatively small influence). When QTL analyses are performed, each analysis typically identifies several potential QTLs, which may have individually small influences, but in the aggregate contribute substantially to individual differences in the trait. How many QTLs are necessary to identify in order to establish "risk?" Put another way, what proportion of trait (phenotypic) variability should be predictable from a QTL analysis for it to be of practical value? For a disease caused by a fully penetrant single locus, the question is not relevant, but for complex traits, it is often crucial. It will be increasingly important to apply multivariate strategies to the analysis of the contribution of multiple QTLs to complex phenotypes.

#### 21.3.1.2 False Positives and False Negatives: What are the Relevant Risks?

In addition to the issue of the adequacy of prediction in a "complete" QTL analysis, we must consider the accuracy (or validity) of the mapping enterprise. There is much current discussion regarding the levels of statistical certainty appropriate to establish linkage of an individual QTL marker with a trait. The "conservative" opinion voiced by Eric Lander and associates has tended to carry the day.[12,13] The strategy proposes extremely high confidence levels for accepting any given linkage, which essentially provide reasonable certainty (95%) that not even one QTL association in a genome-wide search will be a false positive finding (Type I error). However, the cost of this stringency is paid in the consequent increase in false negatives (Type II error) whenever sample size is limited. Thus, many QTL associations that may indeed indicate the presence of a gene affecting the trait, but one which controls a relatively small proportion of the trait variance, may be missed.

We and others have argued for a balanced approach to QTL mapping that depends upon replication in separate studies.[14-16] The use of such an approach allows a more modest level of stringency for associations in any single study, but uses a stringent joint probability of association considered in the aggregate across studies to protect against Type I error. With this approach, fewer Type II errors are often made because of a more efficient use of multiple limited samples. Therefore, it can be expected that a more complete picture of the QTL architecture underlying the trait can be gained when very large scale studies are not the norm, as is often the case in human research.

If a QTL analysis is seen as a precursor for identifying an individual as being "high-risk," with the potential consequences mentioned in previous sections, the importance of considering Type I and Type II errors becomes more focused. One would not want to classify individuals based on risk markers that were not rigorously linked with the disease phenotype, for example (supporting

Lander's argument[12,13] for stringent alpha levels). On the other hand, analysis of all genetic markers even putatively associated with risk might yield a better picture of an individual's true genetic risk. Such an analysis favors an approach that explicitly retains both Type I and Type II error protection.

#### 21.3.1.3 Difficulties in Defining Complex Human Traits

As with any genetic enterprise, the ultimate validity of the project depends upon the accuracy of phenotype definition. One difficulty of complex trait analyses derives from their complexity. After more than 100 years, debate still surrounds the nosology of schizophrenia-related disorders, and their genetic basis. It is conceivable that the various schizophrenia diagnoses in DSM-IV (the standard diagnostic guide of the American Psychiatric Association) present different points on a continuum ranging from eccentricity to severe mental disorder. If this is the case, then QTL analyses may prove up to the task of mapping this complex trait. What if the various diagnoses and subdiagnoses in this spectrum represent different disorders with different etiologies? The genetic heterogeneity that threatens to underly each complex trait may make it extremely difficult to map any genes with certainty.

#### 21.3.1.4 Risk or Protective Markers Vs. Functional Genes

Analysis of the ethical concerns raised by genetic mapping may take on a different flavor depending upon whether the QTLs are represented by risk markers or by functional genes underlying a QTL. While any reliably mapped QTL presumably represents the effects of a linked gene or genes that is associated with the trait mapped, in some instances QTL mapping may lead directly to a candidate gene of apparent relevance for the trait. For example, QTL analyses of morphine-induced analgesia and drinking of morphine in saccharin solutions suggested the influence of a QTL on mouse chromosome 10.[17,18] The close proximity of the gene coding for the μ-opioid receptor to this region of chromosome 10 suggests that the QTL mapped may have been the μ-opioid receptor gene in each case. However, this region and other regions of the genome mapped by QTL analysis are likely to contain hundreds of genes. Use of this information in the first instance might plausibly draw an identified individual at risk into a successful therapy, whereas in the latter case, the individual might have been better off not knowing at all.

### 21.3.2 Animal Models of Complex Human Traits

#### 21.3.2.1 Partial Vs. Complete Models

It is safe to say that there is no entirely adequate, complete animal model for any complex human trait. Despite the great degree of co-linearity between the chromosomes of humans and other mammalian species (for example, Reference 19), there are nonetheless pronounced species differences in neurobiology and behavior. However, the more limited goal of modeling some discrete parts or aspects of a complex human trait have been achieved with animal models in many instances. In our area of expertise, for example, there are several lines of rats that have been genetically selected based on their willingness to drink 10% ethanol solution in water even when pure water is available as an alternative (for review, see Reference 20). The high preference lines may drink to intoxication as a result of the self-administration of relatively large amounts of alcohol. Studies with such genetic animal models have been quite useful in advancing our understanding of the neurobiological and behavioral effects of chronic ethanol self-administration.[21] However, they cannot reasonably be expected to model the powerful social and environmental influences on human alcohol consumption.

When the identification of risk markers or genes in animals is considered, the reason for concern about the adequacy of the animal model becomes clear. If we identify a gene or genes associated with a preference for ethanol solutions in rodents (see Reference 22), is the homologous gene likely to be of importance for humans? At this stage, this is an open question, and empirical observations underway will allow us to answer the question. In the meantime, how should we treat the finding of any such gene? To announce that a "gene for alcohol drinking" has been discovered gives a rather different message to the lay public than to the scientific community — the latter presumably understands the caveats and limits of generalizability across species.

#### 21.3.2.2 Strengths and Weaknesses of Animal-Based Approaches

Species differences notwithstanding, there are some clear advantages of pursuing genetic mapping studies in animals. First, our ability to control the genetic populations under investigation is markedly enhanced. Second, it is possible to analyze the biological substrates of the behavior, even if it means removing brain tissue for anatomical or neurochemical investigations, for example. Finally, genetic manipulations can currently be performed on animal populations that are infeasible in humans. On the other hand, rats and mice are not people, so the isomorphism of the complex behaviors studied in rodents with the human traits modeled is often limited. One of the interesting findings emerging from animal models is the restricted range of genetic covariance across different drug response traits.[23] For example, the debilitating effect of alcohol on a mouse's ability to maintain balance on a rotating rod and on the tendency to stagger when walking appear to the human experimenter to be highly related behaviors that might generally be called "ataxia." However, QTL analyses revealed that genetic control over these traits was largely independent.[24] This discrete genetic control illustrates the importance of extrapolating with caution from animal models to the complex human traits modeled.

## 21.4 SUMMARY

The field of bioethics surrounding human genomics is clearly in a phase where far more questions are asked than are answered. This will continue to be the case until a consensus can be reached concerning many of the above-posed questions, and a practical framework is established paralleling that which is now in place for informed consent and patient (subject) confidentiality in many nongenetic areas of research. At a minimum, the consensus required is that policy makers and both the general public and the scientific community must be involved if new policies are to engender respect and compliance. This can only occur if frank questions are asked and openly debated. One thing that is clear is that scientists engaged in the study of the genetic basis for complex traits, whether using animal models or human populations, will not be able to escape the necessity of examining their work under the searchlight of its ethical implications. It is to be hoped that the interaction of science, politics, and culture will be productive in achieving ethical consensus as we move into the future of genetic understanding.

## ACKNOWLEDGMENTS

Preparation of this article was supported by grants from the Department of Veterans Affairs, and by PHS Grants AA10760, AA06243, and DA05228.

## REFERENCES

1. Mappes, T. A. and Zembaty, J. S., *Biomedical Ethics*, McGraw-Hill, NY, 1991.
2. Parker, L. S., Bioethics for human geneticists: models for reasoning and methods for teaching, *Am. J. Hum. Genet.*, 54, 137, 1994.
3. Parker, L. S., Ethical concerns in the research and treatment of complex disease, *Trends Genet.*, 11, 520, 1995.
4. Frankel, M. S. and Teich, A. H., *Ethical and Legal Issues in Pedigree Research*, A. A. A. S., Washington, D.C., 1993.

5. Kaltenbach, E. A. and McCain, L., Ethical, legal and social implications of genome research, *Molec. Med. Today*, 1, 400, 1995.
6. Frankel, M. S. and Teich, A. H., *The Genetic Frontier: Ethics, Law and Policy*, A. A. A. S., Washington, D.C., 1994.
7. Kevles, D. J. and Hood, L., *The Code of Codes: Scientific and Social Issues in the Human Genome Project*, Harvard Univ. Press, Cambridge, MA, 1992.
8. Parker, L. S. and Gettig, E., Ethical issues: genetic screening, gene therapy and scientific conduct, in *Handbook of Psychiatric Genetics,* Blum, K. and Noble, E. P., Eds., CRC Press, Boca Raton, FL, 1997, 469.
9. Pompidou, A., Research on the human genome and patentability — the ethical consequences (editorial), *J. Med. Ethics*, 21, 69, 1995.
10. Corder, E. H., Saunders, A. M., Strittmatter, W. J., Schmechel, D. E., Gaskell, P. C., Small, G. W., Roses, A. D., Haines, J. L., and Pericak-Vance, M. A., Gene dose of Apolipoprotein E Type 4 allele and the risk of Alzheimer's Disease in late onset families, *Science,* 261, 921, 1993.
11. American College of Medical Genetics/American Society of Human Genetics Working Group, *JAMA*, 274, 1627, 1995.
12. Lander, E. S. and Kruglyak, L., Genetic dissection of complex traits: guidelines for interpreting and reporting linkage results, *Nat. Genet.,* 11, 241, 1995.
13. Lander, E. S. and Schork, N. J., Genetic dissection of complex traits, *Science,* 265, 2037, 1994.
14. Belknap, J. K., Mitchell, S. R., O'Toole, L. A., Helms, M. L. and Crabbe, J. C., Type I and Type II error rates for quantitative trait loci (QTL) mapping studies using recombinant inbred mouse strains, *Behav. Genet.*, 26, 149, 1996.
15. Thomson, G., Identifying complex disease genes: Progress and paradigms, *Nat. Genet.,* 8, 108, 1994.
16. Weeks, D. E. and Lathrop, G. M., Polygenic disease: methods for mapping complex disease traits, *Trends Genet.,* 11, 513, 1995.
17. Belknap, J. K., Mogil, J. S., Helms, M. L., Richards, S. P., O'Toole, L. A., Bergeson, S. E., and Buck, K. J., Localization to chromosome 10 of a locus influencing morphine-induced analgesia in crosses derived from C57BL/6 and DBA/2 mice, *Life Sci. (Pharmacol. Lett.)*, 57, PL117, 1995.
18. Berrettini, W. H., Ferraro, T. N., Alexander, R. C., Buchberg, A. M., and Vogel, W. H., Quantitative trait loci mapping of three loci controlling morphine preference using inbred mouse strains, *Nat. Genet.,* 7, 54, 1994.
19. Copeland, N. G., Jenkins, N. A., Gilbert, D. J., Eppig, J. T., Maltais, L. J., Miller, J. C., Dietrich, W. F., Weaver, A., Lincoln, S. E., Steen, R. G., Stein, L. D., Nadeau, J. H., and Lander, E. S., A genetic linkage map of the mouse: current applications and future prospects, *Science,* 262, 57, 1993.
20. Phillips, T. J. and Crabbe, J. C., Behavioral studies of genetic differences in alcohol action, in *The Genetic Basis of Alcohol and Drug Actions*, Crabbe, J. C. and Harris, R. A., Eds., Plenum Press, New York, 1991, 25.
21. Crabbe, J. C. and Li, T.-K., Genetic strategies in preclinical substance abuse research, in *Psychopharmacology: The Fourth Generation of Progress*, Bloom, F. E. and Kupfer, D. J., Eds., Raven Press, New York, 1995, 799.
22. Phillips, T. J., Crabbe, J. C., Metten, P., and Belknap, J. K., Localization of genes affecting alcohol drinking in mice, *Alcohol. Clin. Exp. Res.*, 18, 931, 1994.
23. Crabbe, J. C., Belknap, J. K., and Buck, K. J., Genetic animal models of alcohol and drug abuse, *Science,* 264, 1715, 1994.
24. Crabbe, J. C., Phillips, T. J., Gallaher, E. J., Crawshaw, L. I., and Mitchell, S. R., Common genetic determinants of the ataxic and hypothermic effects of ethanol in BXD/Ty recombinant inbred mice: genetic correlations and quantitative trait loci, *J. Pharmacol. Exper. Therap.*, 277, 624, 1996.

# EPILOGUE

# 22 Prospects for Cloning the Genetic Determinants of QTLs

*Andrew H. Paterson*

## CONTENTS



## 22.1 WHY CLONE QUANTITATIVE TRAIT LOCI (QTLs)?

The molecular basis of QTLs remains a missing link in our understanding of the relationship between genotype and phenotype. Many ideas exist regarding what we will find at a QTL, ranging from the relatively conservative notion that QTLs simply represent subtle allelic variants in the same genes that segregate for discrete variants, to more exotic ideas such as the influence of nearby heterochromatin or differential methylation on levels and/or timing of gene expression.

The past decade has seen tremendous progress in our ability to study and manipulate individual determinants of complex traits — with an understanding of the molecular basis of QTLs, new avenues for investigation and manipulation would surely be created. As illustrated in the preceding chapters, QTLs are a class of genetic factor that has tremendous importance across a wide range of life sciences. Cloning of a few examples, even using the cumbersome technology available today (see below), is a worthwhile investment of scarce resources.

There is every reason to anticipate that genes imparting quantitative variation will soon be cloned, using methods similar to those that have succeeded in cloning of numerous mutations with discrete phenotypic effects. Once we clone enough QTLs to develop a paradigm for their molecular basis, then justification for cloning of additional QTLs will rely upon (1) the commercial or intellectual value of their phenotype, and (2) more efficient methods that reduce the cost of cloning QTLs.

## 22.2 WHAT WILL REPRESENT A CLONED QTL?

Throughout this volume, QTLs have been considered to be "chromosomal locations of individual genes or groups of genes that influence complex traits," a definition that heretofore adequately served our purposes. Essentially, we agreed to define QTLs as a place, without speculating about just what it was that sat at that place.

0-8493-7686-6/98/$0.00+$.50

What we seek to clone is not a place, but a Polygenic Trait Determinant (PTD), a genetic factor that resides at that place. While it is often tactily assumed that a QTL represents a single genetic determinant, already there are several examples of individual QTLs that have been resolved into multiple PTDs by genetic recombination. For example, several putatively overdominant QTLs in maize and rice have been shown to reflect two (or more) loci with dominant alleles in repulsion-phase linkage (see Chapter 14, this volume).

Moreover, the distinction between discrete mutations and QTLs is often unclear. In this volume, QTLs have been very broadly defined as "genes that influence complex traits," noting also that such genes can account for either large or small portions of phenotypic variance. By using advanced-backcross breeding approaches (see Chapter 15, this volume), we can engineer populations that segregate only for a single QTL, and render that QTL virtually discrete. Consequently, the context in which a genetically-variable locus is discovered may influence whether or not we define alleles at that locus as QTLs. Had we discovered the locus in a population that did not segregate at any other loci affecting the same trait, we might have called it a discrete mutant!

For this discussion, I will address the molecular cloning of individual genetic determinants for which allelic variation does **not** result in discrete phenotypic classes in the primary mapping population where the variants are discovered. Such genes are the bread and butter of plant and animal breeding programs, the basis of variation in fitness among individuals in natural populations, and the foundation of predispositions to a host of medical disorders. Such genes have sometimes been a stumbling block, obscuring the study of unlinked discrete mutations affecting specific attributes. Given modern molecular tools, it is no longer necessary to treat QTLs as an incongruity — instead, we can incorporate them into more precise and robust genetic models.

## 22.3 THE CURRENT PARADIGM FOR CLONING QTLs

The first molecular cloning of a QTL is likely to follow a paradigm similar to that outlined in several preceding chapters (see Fig. 11.3). Briefly, QTLs are initially discovered in a primary mapping population comprised of individual backcross or F2 genotypes, or recombinant inbred lines. Multiple QTL models, and breeding experiments to create secondary populations segregating for only one or a few QTLs, may each improve the power, precision and accuracy of QTL analysis by eliminating many factors that previously contributed to error variance. Fine-mapping of QTLs to small chromosomal regions involves two steps: (1) the identification of unique recombinants that differ in genome composition near the QTL, and (2) phenotypic evaluation of numerous progeny from these recombinants to obtain a reliable measure of the true QTL genotype for each recombinant. Once the QTL is mapped to a suitably small chromosomal segment, chromosome walking/landing is employed to make a contig including the determinant(s), providing the basis to isolate transcripts or genomic subclones for use in mutant complementation experiments.

Sounds great ... but it reads better than it lives. In several well-studied plants, we are well into our second decade of genome mapping and are just now reaching the point at which QTL cloning may soon become a reality. Cloning of the first QTL may not be the hardest one — rather, it may be the most fortuitous one, and the technological difficulty of this goal (as most) may be reflected by the overall frequency of failures rather than by the achievement of a few successes.

### 22.3.1 Streamlining QTL Cloning

Several emerging sources of information virtually assure the obsolescence of the current paradigm for positional cloning, not only of QTLs but of most genes.

(1) *High-density, unified molecular maps accelerate fine-mapping, and broaden the scope of our search for discrete mutants to target.* Closely-related (confamilial) genera often retain large chromosomal tracts in which gene order is co-linear, punctuated by structural mutations such as inversions and translocations. This information enables us to simultaneously advance the state of

genome analysis in many different taxa. By alignment of the chromosomes of different taxa by cross-mapping of selected DNA probes, we can achieve order-of-magnitude improvements in the density of DNA markers available for specific chromosomal regions. A few years ago, it remained an ordeal to find closely-spaced DNA markers for "fine mapping" of specific chromosomal regions. Today, every new marker added to the sorghum map (for example), also enriches the maps of maize, rice, wheat, and other grasses; the collective efforts of the genetics community will soon (if not already) obviate the need for tedious marker enrichment based on techniques such as arbitrarily primed PCR methods (see Chapter 2, this volume). In less well-characterized genomes where marker enrichment experiments remain necessary, more efficient techniques that can simultaneously screen hundreds (AFLPs) or thousands (RLGS: see Chapter 2, this volume) of DNA loci are being implemented.

Chromosome alignment across taxa also enables us to evaluate correspondence of QTLs in one taxon, to either QTLs or discrete mutants in other taxa. It was suggested long ago that such correspondence, if it existed, might provide a mechanism by which one could use discrete mutants as a means to obtain the genic DNA that accounted for QTLs.[1]

Unified maps expand the scope of our search for discrete mutants, to serve as facilitators of QTL cloning. Rather than have to search for a discrete mutant of (for example) maize that appears to be allelic to a QTL, we can also evaluate correspondence in genetic map location to genes in sorghum or rice. For example, one such experiment has revealed a sorghum mutant explaining nearly 100% of the phenotypic variation in "disarticulation" (shattering) of the mature inflorescence, that corresponds (at least in location) to two putatively homoeologous maize QTLs on chromosomes 1 and 5, and a rice QTL on chromosome 9.[2] A "chromosome walk" to the sorghum gene is in progress, and should afford a test of the Robertson[1] proposal. The recent demonstration that small chromosomal regions may remain co-linear (or at least similar) in taxa as divergent as monocots and dicots[3] suggests that we may be able to further extend the scope of our search for discrete mutants.

(2) *High-throughput DNA sequencing reduces the need for chromosome walking and isolation of transcripts.* Complete sequences for several model plant genomes would eliminate many of the most cumbersome steps in the current paradigm for positional cloning. Today, to isolate candidate transcripts that might account for a QTL, we perform a series of mapping experiments, followed by chromosome walking, then by one or more of a series of complicated techniques to try to isolate all transcripts near a locus of interest.

Consider the consequences of having a complete DNA sequence for the target organism (or even a closely-related organism that has been aligned with the target by comparative mapping). Genetic mapping may remain the most convenient "first step" in scanning the genome for loci that affect a target phenotype. However, the information provided by a genetic mapping experiment would be profoundly different from what is available today; data about the approximate location of a target gene(s) would equate to having a list of open reading frames within the target region, and putative functions for many of them. Further inferences about likely candidate genes might be based on the presence of tissue-specific promoters, or other clues regarding when and where specific transcripts might be expressed, even if the gene product is not yet known. The sequence itself provides the basis for developing PCR primers that might be used to quickly sequence alleles from wild-type and mutant phenotypes, to seek any at which loss-of-function mutations are found. While such a search may sometimes be fruitless, many mutants isolated to date could, in principle, have been identified based in part on such mutations. At the least, many transcripts that are **non**-mutant might have been deemed unlikely candidates, quickly and efficiently.

## 22.4 WHAT WILL BE THE SCOPE OF QTL CLONING?

Initially, the cloning of a select subset of QTLs may be justified by our lack of knowledge of their molecular basis. High-priority candidate QTLs might be chosen based on economic or medical importance, evolutionary significance as reflected by apparent correspondence across diverse taxa, or simply based on falling at well-characterized locations in genomes that are facile models.

As we begin to develop an understanding of the molecular basis of QTLs, justification for arduous and expensive "walks" to QTL may become increasingly difficult. For example, manipulation of the vast majority of QTLs in plant breeding programs does not require cloning — so why bother?

QTL cloning offers a host of prospective applications. Rapid progress in isolation of the complete repertoire of human genes and sequencing of the human genome provides a tremendous resource for identifying the molecular basis of predispositions to a host of medical disorders and behaviors, such as are exemplified in earlier chapters.

Alignment of the human genome with that of other mammals will enable the sequence to be applied directly to isolation of genes important to agriculture. *Arabidopsis* and rice offer like opportunities for plants. The ability to assemble groups of desirable QTL alleles into cassettes that might be integrated into a plant or animal genome as an artificial chromosome would enable one to avoid the necessity of dealing with many independently-segregating loci — rendering complex traits discrete.

The extent to which QTL cloning becomes routine will depend on technological improvements that (hopefully) render the current paradigm obsolete. Complete sequences for model genomes will be a starting point, as addressed above. A second key step may be the ability to stably integrate intact large DNA clones into plants and animals, obviating the need to dissect (for example) YACs or BACs into manageable chunks; development of such technology is being actively pursued both in plants and in animals. A third important step might be the ability to quickly make complete libraries of megabase DNA that do not require handling as individual isolates, but might be screened in bulk with the efficacy of phage. This would be especially important in applications to breeding and genetics, enabling one to quickly isolate desirable alleles from specific elite genotypes. P1 phage libraries are perhaps the closest fit to this need that is presently available, but their relatively small insert sizes remain an encumbrance; the ~90 kb limit of P1 corresponds roughly to 0.1 cM in many higher plants and animals, a level of genetic map resolution that is not easily attainable for QTLs.

By merging improved capabilities for integration of exogenous DNA into genomes, together with routine development of large DNA libraries, one can envision a "QTL shotgun screening" approach that eliminates sexual hybridization altogether. Rather than spend many generations developing AB-QTL stocks, one might simply make plant populations that have been transformed with bits of genomic DNA from an exotic relative (or even another taxon!), then use classical plant breeding procedures to decide what is worth keeping. The discovery that many valuable genes appear to reside in some unlikely places (see Chapters 15 and 20, this volume), suggests that we have not been casting the net wide enough in our search for variation, and impels efforts to reach further. The number of transformants that must be screened is a limiting factor; however, artificial chromosomes capable of carrying large inserts (say 1 cM) could obviate this.

## 22.5 IN CLOSING ...

We will soon enter our second century of quantitative genetics, commemorating the development of particulate models for blending inheritance, with the implementation of tools and techniques for cloning its individual determinants. The progress of the past decade alone has revolutionized how I personally view my chosen field of crop improvement, and my colleagues in a wide range of fields echo the same viewpoint throughout this volume. Speaking for the contributors to this volume, we hope you now share our view of QTLs as examples of "little things that mean a lot." Moreover, we hope that this volume helps to stimulate your interest, and/or accelerate your progress, in applying these principles to gaining a better understanding of the biotic world, how it came to arrive at its present state, and what its future might hold.

## REFERENCES

1. Robertson, D. S., A possible technique for isolating genetic DNA for quantitative traits in plants, *J. Theor. Biol.,* 117, 1, 1985.
2. Paterson, A. H., Lin, Y. R., Li, Z., Schertz, K. F., Doebley, J. F., Pinson, S. R. M., Liu, S. C., Stansel, J. W., Irvine, J. E., Convergent domestication of cereal crops by independent mutations at corresponding genetic loci, *Science,* 269, 1714, 1995.
3. Paterson, A. H., Lan, T. H., Reischmann, K. P., Chang, C., Lin, Y. R., Liu, S. C., Burow, M. D., Kowalski, S. P., Katsar, C. S., DelMonte, T. A., Feldmann, K. A., Schertz, K. F., Wendel, J. F., Toward a unified map of higher plant chromosomes, transcending the monocot-dicot divergence, *Nat. Genet.,* 14, 380, 1996.

# Index

# Index

## A



## B

## C



## D

## H



## I



## J



## L



## M

## N



## O



## P

## Q

## R



## S